Morphology of flowers and inflorescences

Morphology of flowers and inflorescences

Professor Dr F. Weberling

Department of Biology, University of Ulm

Translated by R. J. Pankhurst

Botany Department, British Museum (Natural History), London

The right of the
University of Cambridge
to print and sell
all manner of books
was granted by
Henry VIII in 1534.
The University has printed
and published continuously
since 1584.

CAMBRIDGE UNIVERSITY PRESS
Cambridge
New York   Port Chester
Melbourne   Sydney

Published by the Press Syndicate of the University of Cambridge
The Pitt Building, Trumpington Street, Cambridge CB2 1RP
40 West 20th Street, New York, NY 110011, USA
10 Stamford Road, Oakleigh, Melbourne 3166, Australia

Originally published in German as *Morphologie der Blüten und der Blütenstände* by Prof. Dr. Focko Weberling, Ulm, 1981 and © Eugen Ulmer 1981

First published in English by Cambridge University Press 1989 as *Morphology of flowers and inflorescences*

English translation © Cambridge University Press 1989

Printed in Great Britain at the University Press, Cambridge

British Library cataloguing in publication data

Weberling, F.
Morphology of flowers and inflorescences.
1. Flowering plants. Anatomy
I. Title   II. Morphologie der Blüten
und der Blütenstände *English*
582.13′044

Library of Congress cataloguing in publication data

Weberling, Focko.
[Morphologie der Blüten und der Blütenstände. English]
Morphology of flowers and inflorescences / F. Weberling;
translated by R. J. Pankhurst.
   p.   cm.
Translation of: Morphologie der Blüten und der Blütenstände.
ISBN 0–521–25134–6
1. Flowers—Morphology.   2. Inflorescences—Morphology.
I. Title.
QK653.W3913 1989
582′.0463—dc19   88–23792 CIP

ISBN 0 521 25134 6

HC

Dedicated to my beloved and venerated wife

Contents

Preface

It may very well be thought that the most interesting part of the subject of plant morphology is represented by the study of the morphology of the flower. At the same time its results are of fundamental significance for the systematics of flowering plants. It might therefore be expected that the fundamentals of flower morphology, as well as the numerous and varied results of researches into it, would be described in comprehensive texts. If we discount the works of A. W. Eichler ("Floral diagrams", 1875/78), which although extensive and unsurpassed in many respects are now more than a hundred years old, we can nevertheless only cite the chapter devoted to flower morphology in K. von Goebel's "Organographie der Pflanzen" (Plant organography) with the inclusion of the two supplementary volumes, along with the work "Organisation und Gestalt im Bereich der Blüte" (literally, "Organisation and Gestalt in the region of the flower") by W. Troll, which deal with some fundamental concepts of flower morphology. Of course the most important details relating to new results are given in general textbooks such as the "Lehrbuch der Botanik für Hochschulen" (Textbook of Botany) founded by E. Strasburger, where they are presented briefly and with characteristic precision, but nevertheless there is no space left for a thorough discussion of the fundamentals and the more recent results and problems of flower morphology.

This rather critical gap in academic teaching was at least partially closed by means of the "Praktische Einführung in die Pflanzenmorphologie" (Practical Introduction to Plant Morphology) by W. Troll, the second part of which is devoted to the structure of the flowering plant. This work was intended as an introduction to an elementary course in morphological systematics but it contains much else besides. We have taken over quite a number of the individual illustrations from it, and surely as one might expect a student of Troll to do. Explicit reference may be made to the detailed illustrations of numerous individual examples from a great variety of morphological contexts which are contained in this work of Troll's.

In the present work we have attempted to combine the results of numerous earlier and more recent individual studies on the morphology and anatomy of flowers and floral organs into a single synthesis. Necessarily, therefore, only some of the many pieces of research which have been concerned with the clarification of questions of floral morphology during the past hundred years could be considered. However, this is perhaps justified if selection from this work leads to a clearly defined and basic understanding. We have nevertheless tried to give a key to the different directions in which the further literature leads by means of particular citations and references. Above all we hope that we have correctly interpreted all the research which we have made use of despite the limits imposed. In the selection of material and by concentrating on the elucidation of particular problem areas we have not pursued perfection but tried to allow access to the different areas of plant morphology. The experience which we have been able to gain through practicals in morphology and systematics and special lectures in flower and inflorescence morphology was therefore relevant.

In the second part of this book we would like to fulfil a promise, made long ago and often repeated, to give a short and comprehensive description of the characterisation of the inflorescence according to the investigations of W. Troll and certain of his students in the past decades. With the exception of some research results we could also make reference to Troll's notes of unpublished work for the discussion of certain questions, and moreover make use of his splendid photograph collection which he bequeathed to us as part of his comprehensive scientific legacy for the continuation of his work on the inflorescence.

In an introduction to the morphology of the flower and the inflorescence one cannot avoid referring to current knowledge and problems of the biology of the flower and of plant dispersal, since without consideration of these functional aspects no satisfactory "explanation" of morphological phenomena can be given. In the third part of our book we present chapters on self- and cross-pollination, about the varied interrelationships between flowers and pollinators, the systematics of the fruit and aspects of the biology of dispersal which is meant to open the door, although only with a short summary, to this fascinating area of biology, and to give references to the relevant literature.

"Perhaps it may be considered rather bold of me to have undertaken such work as this, when others – whom I hardly need name – might be thought to be better qualified for it. In fact this publication does not appear without some trepidation on my part". If A. W. Eichler already thinks it proper to make such a remark in the foreword to the first volume of his "Floral Diagrams", then it is even more true for us. We believe we have to go along further with Eichler "But I am encouraged by the awareness that what I say here, in spite of all its shortcomings, may not be quite useless."

In the compilation of this book we received help of all kinds, for which we are sincerely grateful. In the first place our thanks go to Fräulein Ursula Schultheis, our collaborator of long standing, who as illustrator took care of the production of the figures. A series of copies of illustrations were obligingly volunteered to us by certain colleagues. These were: Herr Prof. Dr. H. A. Froebe, Aachen, Dr. W. van Heel, Rijksherbarium Leiden, Prof. Dr. P. Leins, Bonn, Pater Dr. A. Trapp, Gymnasium Mariengarden, Borken, Prof. Dr. St. Vogel, Vienna, Frau Dr. Katharina Urmi-König and Herr Dr. E. Urmi, Ulm and Zürich, respectively, to whom we offer our thanks for their kind support. In this we also include Herr. Dr. U. Hecker, Mainz, who lent us some of the original illustrations from the books of W. Troll, as well as Frau Ursula Sperling, Mainz, who kindly undertook the production of suitable prints from negatives from the photographic collection of W. Troll. For all kinds of assistance, especially for the provision of literature and plant material, we would like to thank Herr Prof. Dr. H. Merxmüller, Frau Dr. Anneliese Schreiber and Herr J. Bogner, Botanische Staatssammlung und Botanischer Garten, Munich. For helpful information, and friendly criticism, we also thank Herr Prof. Dr. D. Hartl, Mainz, Herr Prof. D. P. Leins, Bonn, Herr Dr. N. Magin, Aachen, Herr Prof. Dr. D. Müller-Doblies, Berlin, and Herr Prof. Dr. O. Rohweder, Zurich. We are grateful to Herr J. R. Hoppe and Herr Th. Stützel, for proof-reading the corrections, and Frau Christel Necker and other helpers for the preparation of the index.

We would like to acknowledge the help given by Dr. Roland Ulmer and the staff of the publisher of the German edition, in particular Herr Dr. Steffen Volk, who has assisted us with expert advice and much understanding.

F. Weberling.
Ulm, January 1981.

Translator's notes

There are two kinds of difficulty which the translation of this book has posed which are worth comment, since they go beyond the normal problems of translation from German to English. The first point is that, inevitably, there are places where the expressiveness of the original language cannot be conveyed. This is particularly noticeable in the terminology for the parts of the flower. In German, the words for sepal, petal, stamen, carpel etc. all use the word 'Blatt' for 'leaf' in compound form, e.g. Fruchtblatt = carpel, thereby neatly expressing the underlying theory that all these organs are fundamentally modified leaves. This is lost in English. Another example is in the use of several different words in German for what in English is loosely termed a 'bract' or 'bracteole'. Botanical terminology in English lacks any way of distinguishing the bracts which relate to the structure of the inflorescence as a whole from those which relate to the individual flowers. A glossary has been added in which new definitions of terms are marked so that the differences in usage are quite clear. As an example, notice the unfamiliar definition of 'prophyll'.

Secondly, German-speaking botanists have investigated the morphology of plants in a characteristically thorough fashion over several centuries and the vocabulary of morphological description is larger in German than it is in English. Consequently, I have often been faced with the temptation to coin new words, which I have resisted on principle, and have made considerable efforts to find suitable equivalents instead. There was one section however where the case for new words was particularly compelling, and I acknowledge the help of Dr. W. T. Stearn in inventing the new words "dentonection" and "capillinection" which appear in section 1.4.2.4. Finally, my thanks are due to Dr. E. Launert for much assistance with those numerous words and phrases which were 'not in the dictionary'. I also wish to record that Prof. Weberling has spent an unprecedented amount of time in careful checking of the proofs.

R. J. Pankhurst
London, 26th January 1986

Glossary

The meaning of botanical terms in German differs in some cases from English, and these differences are at times important to a correct scientific understanding. For the convenience of the English-speaking reader, Prof. Weberling and Prof. Müller-Doblies have provided a glossary for the English edition of his book. Terms whose definition may be unfamiliar are marked with an asterisk. The original reference to the term is given, if known.

abaxial positioned away from the axis (Goebel 1913: 273)

accessory bud, Ger. "Beiknospe", bud occurring in the same leaf axil in addition to the regular bud, see section 2.1.7.

accessory flower see above.

accessory shoot, Ger. "Beisproß", bud occurring in the same leaf axil in addition to the regular bud, see section 2.2.8.

acrotony*, adj. acrotonic, giving preference to the apical region, ; opp. basitony (Rickett 1944: 225 refers to Goebel 1931: 80; 86–88).

adaxial positioned towards the axis.

addorsed prophyll, adaxial prophyll, see sections 1.2.1, 2.1.4.

ananthic paraclades paraclades which develop leaves, but which do not reach flowering (Troll 1964: 203).

anthela a panicle in which the lower branches overtop the upper ones (Meyer 1819: 19).

anthoclade, cymose branch of a usually acrotonic inflorescence with foliage prophylls retaining the mode of cymose branching and thus forming "alternately terminal flowers and foliose leaves" (Goebel), in extreme cases with adventitious roots and continuing to grow (Goebel 1931: 2 emend. Troll 1964: 157).

anthotelic*, applied to inflorescences, parts of inflorescences, or axes; ending in a flower or in an aborted but distinctly floral bud; same as definite, determinate (Briggs and Johnson 1979: 241).

axonoscopic*, referring to serial accessory buds which develop acropetally above the regular axillary shoot, opp. phylloscopic.

basal internode*, the internode which precedes the main florescence in a polytelic system (Troll 1951: 379).

basal shoot, see section 2.2.4 (Troll 1964: 251).

basitony*, adj. basitonic, giving preference to the basal region; opp. acrotonic (Goebel 1931: 88).

blastotelic*, not ending in a flower, i.e. ending in a non-floral bud; applied to inflorescences, parts of inflorescences or axes; indeterminate (Briggs and Johnson 1979: 241).

boragoid, the cincinnus of Boraginaceae and other families, imitating a raceme, because of a straightening of the axis taking place simultaneously with the initiation of the individual flowers.

bostryx, a monochasium in which the continuation takes place from the axil of a more or less lateral prophyll and in a helicoid manner i.e. in a cyme in which the successive pedicels describe a spiral around the sympodial axis; a helicoid cyme.

botryoid, adj. botryoidal, like a botrys (qv) but with a terminal flower, see section 2.1.4 (Troll 1959: 117, 1964: 52 sensu Troll 1969: 454; adj. due to Briggs and Johnson, 1979).

botrys, simple blastotelic inflorescence with a variable number of lateral flowers with well-developed pedicels along an elongated inflorescence axis, a raceme.

brachyblast, a short shoot (Hartwig 1852: 76).

bract, a reduced leaf, commonly a subtending leaf in inflorescences (see Rickett 1954: 194). See also pherophyll.

bracteose, or bracteate, 1) of a shoot; bearing bracts, 2) of a leaf; having the reduced nature of a bract. (Troll 1964: 6 and Briggs and Johnson 1979.)

catalepsis, adj. cataleptic, development of a bud taking place later after at least one resting period (Müller-Doblies 1976: 177–178; Müller-Doblies and Weberling 1984: 133).

cataphyll, Ger. "Niederblatt", a more or less scale-like morphologically reduced leaf, preceding the foliage leaves and commonly functioning as a protective organ of a bud.

cauliflory, adj. cauliflorous, the production of flowers from the old wood.

cincinnus, a monochasium in which the continuation takes place from the axil of a more or less lateral prophyll in a zigzag fashion, i.e. in a cyme in which all the pedicels are on one side (unilateral) of the sympodial axis.

coflorescence, see section 2.2.1

conjunct-heterocladic, see section 2.1.4

cyme*, a cymose partial inflorescence; which develops no further leaves apart from the prophylls and thus has only one or two possibilities for lateral branches.

cymoid, an acrotonic thyrsoid, reduced to the distal pair of paraclades, giving the impression of a cyme (Troll 1964, p. 33).

cymose*, describes a branching pattern of inflorescences, where the branching only takes place from the axil of the prophylls, i.e. branching is confined to 1 or 2 lateral branches.

definite, same as determinate, anthotelic.

determinate, of an axis which is terminated by a flower.

dichasial cyme, a cymose partial inflorescence with two branches developing from the (usually opposite) axils of the prophylls and continuing the same manner of branching.

dichasium, adj. dichasial, a branching pattern such that only the two uppermost (often opposite) branches develop below the apex. If these (decussate or alternate) lateral axes only develop the two prophylls and their axillary shoots below their terminal flower, the dichasium appears to be similar to cymose branching.

diplobotryum, or dibotryum, a twice racemose inflorescence (Čelakovský 1892: 8).

diplothyrse, a twice thyrsic inflorescence.

disjunct–heterocladic, see section 2.1.4

dolichoblast, a long shoot, see section 2.2.4

drepanium, a monochasial branching pattern in which the continuation takes place from the axil of a more or less abaxial leaf and in which all consecutive branches occur in one plane and fall to the same side (Eichler 1875: 35). (Since the branching does not take place in the axil of a prophyll in the strict sense, the drepanium does not belong to the cymose branching pattern, strictly speaking.)

dyad, partial inflorescence with a terminal and a lateral flower.

effloration*, the order of flower opening.

enrichment branch, a flower-bearing paraclade whose flowers reach full development (A. Braun 1851: 41 emend. Troll 1964: 231).

enrichment zone, part of a monotelic or polytelic inflorescence which bears the enrichment branches (Troll 1951: 368, 1964: 148); see section 2.2.1

epipodium, (K. F. Schimper, fide Troll 1937: 203), see section 1.2.1

epitony, giving preference to the upper side of a dorsiventral organ, opp. of hypotony (Troll 1937: 17).

final internode, internode preceding the terminal flower in a monotelic system.

florescence, the basic unit of a polytelic inflorescence, either a blastotelic botrys or a blastotelic thyrse (Agardh 1858:lxii emend. Troll 1964: 145).

foliation*, collective word for the position and character of leaves of various kinds.

frondose* see section 2.1.1

frondulose* see section 2.1.1, 2.2.6

helicoid cyme, cyme, bostryx, see section 2.1.4

heterocladic thyrse, (Troll 1969: 3), see section 2.1.4

homaxone, congenital cincinnus formation, see section 2.1.8

homocladic thyrse, (Troll 1969: 3), see section 2.1.4

hypopodium, first internode of a lateral axis (below the first or single prophyll), see section 1.2.1

hypotagma, Ger. "Unterbau", that part of an inflorescence which includes all the nodes of the enrichment zone, the inhibition zone and the innovation zone (Goebel 1931: 3, Troll 1951: 383).

hypsophyll, Ger. "Hochblatt", a bracteose (or sometimes a more or less metamorphosed frondose) leaf within the inflorescence, opp. of cataphyll (Troll 1950: 395, 1964: 6).

indefinite, indeterminate, blastotelic.

indeterminate, an inflorescence axis not terminated by a flower is said to be indefinite (Roeper 1826: 439), same as indefinite, blastotelic.

inflorescence, any aggregation of flowers (L. 1751: 112). Troll's approach(1950, 1964) includes also flowering shoots under the term inflorescence, considering the foliation, whether bracteose, frondulose or frondose, as a secondary character. There is no definition of "an inflorescence", but the term is deliberately kept rather imprecise in order to have a convenient word for each inductive attempt at the analysis of an unknown inflorescential system. On the other hand, the use of the term **synflorescence** requires the analysis of the structure of a given inflorescence within the context of the whole branching system of the taxon and often also its allies.

inhibition zone, Ger. "Hemmungszone", see section 2.2.3

innovation bud, see section 2.2.3 (Troll 1951: 383, 1964: 282).

innovation shoot, in perennials only, see section 2.2.3 (Troll 1951: 383, 1964: 282).

innovation zone, in perennials, the basal zone of a flowering shoot comprising the innovation buds, see section 2.2.3 (Troll 1951: 383, 1964: 283).

intercalated sterile bracts, metaxyphyll, Ger. "Zwischenblätter", sterile bracts intercalated next to the terminal flower, see section 2.2.8 (Buchenau 1865: 392, Nordhagen 1937: 12).

main florescence, see section 2.2.1, 2.2.6 (Troll 1953: 40).

mesopodium, the internode between two prophylls, c.f. hypopodium (Troll 1937: 203).

metatopy, dislocation of organs by unequal growth.

metaxyphyll, (Briggs and Johnson 1979: 24), see intercalated sterile bracts.

monad, of an inflorescence: a solitary flower together with its axis and the prophylls (if any) of that axis.

monobotryum, or haplobotryum, a simple inflorescence as opposed to a compound one (pleiobotryum). (Troll 1964: 42).

monochasial cyme, a cymose partial inflorescence with a single lateral branch in each order of ramification, opp. of dichasial cyme.

monochasium, adj. monochasial, a branching pattern with a single lateral continuation shoot. (Eichler 1875: 34, 36 from Bravais and Bravais 1837: 196 sensu Troll 1937: 104, 1954: 207, 1959: 83).

monopodium, adj. monopodial, 1) in the strict sense, an axis of a cormophyte which is built up by the same growing point. 2) in the wider sense, the whole branching system with such a monopodial main axis may be said to be monopodial.

monotelic, see section 2.2, 2.2.8 (Troll 1961).

paraclade, in monotelic systems, flower-bearing (anthotelic) branches below the terminal flower, or in polytelic systems, branches ending in a coflorescence. These are therefore not strictly speaking homologous structures, but play a similar role in the symmetry of each of the two synflorescences. See section 2.2.2 (Schulz 1847: 4 emend. Troll 1964: 146).

partial florescence, cymosely branched element of a florescence (Troll 1960).

partial inflorescence, any more or less separated lateral part of an inflorescence; see the remarks under the definition of inflorescence.

pedicel, the stalk of a single flower in an inflorescence of several flowers. (L. 1751: 40).

peduncle, stalk of an inflorescence of several flowers; a more or less extended internode preceding the inflorescence. (L. 1751: 40).

pherophyll, foliage leaf or bract, subtending an axillary shoot (or flower). (Briggs and Johnson 1979: 246).

phylloscopic*, referring to serial accessory buds which develop basipetally below the axillary shoot, cf. axonoscopic.

pleiobotryum, polytelic synflorescence bearing paraclades of first, second, and higher orders with racemose florescences, see section 2.1.4 (Troll 1964: 42).

pleiochasium, an acrotonic branching pattern in which the relative main axis stops its growth (anthotelic or blastotelic), being overtopped by three or more lateral branches, forming a more or less loose whorl e.g. in species of *Euphorbia, Sedum* **and** *Damasonium.* (Eichler 1875: 34, 36 from Bravais and Bravais 1837).

pleiothyrse, pleiothyrsoid, a compound thyrse, or thyrsoid, respectively, including diplothyrse or diplothyrsoid and those of higher order (Troll 1964: 87).

polytelic type, see section 2.2 (Troll 1961, 1962, 1964).

primary flower, the first flower of a cymosely branched partial inflorescence.

prolepsis, adj. proleptic, development taking place earlier, by at least one resting period. (Müller-Doblies and Weberling 1984: 121).

proliferation, the return of an inflorescence apex to vegetative growth. Same as prolification.

proliferation of the florescence, Ger. "Spätprolifikation", retarded proliferation, occurs only in polytelic systems, see proliferation (Troll 1959: 116).

proliferation of the enrichment zone, Ger. "Frühprolifikation", precocious or premature proliferation, occurs in monotelic as well as in polytelic systems, see proliferation (Troll 1959: 116).

prophyll*, Ger. "Vorblatt", see section 1.2.1.

rhipidium, a monochasium in which the continuation takes place from the axil of an addorsed prophyll, i.e. the successive pedicels all lie in one plane and follow a zigzag pattern (Buchenau 1865: 392; Eichler 1875: 35).

scape*, Ger. "Schaft", a prolonged or elongated internode in or below an inflorescence.

sciadioid, or pseudumbel; thyrsoid which is contracted into an umbel-like inflorescence (e.g. a contracted cymoid as in *Holosteum umbellatum*). (Troll 1964).

scorpioid cyme, same as cincinnus. (DC. 1827, 1: 415).

short shoot, see brachyblast.

spadix, a spike with a fleshy axis. (L. 1751: 55, 77).

spike, simple inflorescence with sessile flowers along an elongated main axis without a terminal flower (L. 1751: 41).

stachyoid, similar to a spike but with a terminal flower, see section 2.1.4

starved forms, Ger. "Hungerformen", see section 2.2.7

stunted forms, same as starved forms.

syllepsis, adj. sylleptic, applies only to lateral shoots; development without an intervening resting period and together with the main shoot (mostly without preceding bud scales). (Müller-Doblies and Weberling 1984: 128–133, 143).

sympodium, adj. sympodial, in the strict sense; a seemingly uniform axis of a cormophyte that is not built up by the same growing point, but by a

sequence of axillary buds (i.e. several shoot generations) is called a sympodium and said to be sympodial.

synflorescence, a system of florescences (in the polytelic type) or a system of a
terminal flower and monotelic paraclades (in the monotelic type). (Goebel
1931: 2 emend. Troll 1953: 40, 1964: 148).

terminal flower, a flower formed by the apex of and terminating the principal
axis or racemose branches.

thyrse, adj. thyrsic, an indefinite complex inflorescence with cymose branches
on a dominating main axis.

thyrsoid, similar to a thyrse, but with a terminal flower, see section 2.1.4
(Čelakovský 1892: 5, 1893: 46 and Briggs and Johnson 1979: 248).

triad, partial inflorescence with a terminal and two lateral flowers (Troll 1957:
237).

truncate synflorescence, Ger. "Rumpfsynfloreszenz", a synflorescence which
has reduced its terminal flower or its main florescence and thus consists of the
enrichment zone only, see section 2.2.10 (Troll 1960: 116; 1964: 157).

truncation[*], see section 2.2.10 (Troll 1964: 157).

zone of enrichment, see enrichment zone.

zone of inhibition, see inhibition zone.

zone of innovation, see innovation zone.

1   Morphology of flowers

1.1   Definition and morphological derivation of the flower, phylogenetic aspects

Investigations of the comparative morphology of the flower and its organs as well as the comparative morphology of higher plants, especially in their fundamental beginnings, reach back to the end of the 18th and the beginning of the 19th century. J. W. von Goethe's "Versuch die Metamorphose der Pflanzen zu erklären" (Attempt to explain the metamorphosis of plants, 1790) is the starting point. "The hidden relationship of the different external parts of plants, such as the leaves, the calyx, the corolla and the filaments, which develop one after the other and as if out of each other", of course in Goethe's own words "long since known by scientists", was here set out at length, and convincingly, for the first time. The insight that the flowering parts of a plant might be understood as "modified leaves" had already been expressed by C. F. Wolff in 1768.

The recognition that the important floral organs – the sepals and petals, the stamens and the carpels – are leaf organs, is equivalent to the observation that the flower bears characters of the shoot. This was illustrated by Goethe by, among other things, the "example of a rose grown through", a malformation in which the shoot returns to vegetative growth and the formation of pinnate leaves after developing floral organs, sepals and petals "of which some bear some vestige of the anthers" (Fig. 1).

When we speak of the flower as a shoot, we must of course bear in mind that:
1. the floral organs, similarly to a vegetative leaf rosette, are densely crowded upon one another, and that
2. the growing point of the flower, unlike that of a leaf rosette as in an *Echeveria* (Fig. 2 I, II), terminates its growth by the development of the innermost, i.e. the uppermost, floral organs. Moreover we already know that
3. the leaf organs of the flower have in some cases undergone a far-reaching change of form. Hence fundamentally different formations of successive floral organs are to be distinguished. In a perfect hermaphrodite flower (Fig. 2 III) these are:

1. The organs of the **perianth**, often distinguished as **calyx** and **corolla**,
2. The **stamens**, known collectively as the **androecium**; each stamen is com-

posed of a shorter or a longer, usually filiform portion, the **filament**, and at the end a sessile **anther**, which in the typical case exhibits four **pollen sacs** with **male spores** (IV).

3. The **carpels**, known collectively as the **gynoecium**, which surround the **ovules** (V).

The sequence in which the individual organ formations develop always remains the same. However, we already know from unisexual or from green, inconspicuous flowers, that these different formations of floral organs are not encountered in a similar form in all flowers, but that some or almost all can be rudimentary or completely lacking. A universally valid definition of the flower must therefore read: **a section of a shoot, or a branch resembling a short shoot, which bears leaf organs which serve for sexual reproduction and which are transformed accordingly**.

The stamens and the carpels may be regarded as the most important organs of the flower. We designate these also as microsporophylls and macro- (mega-) sporophylls, and hence equate them with the sporophylls of the ferns and their allies (Pteridophyta) and the gymnosperms (Gymnospermae). We are justified in this ever since Wilhelm Hofmeister (1851) established the fundamental correspondance between the life cycles of the pteridophytes, the gymnosperms and

Fig. 1 The rose "grown through", after Goethe, complete phyllody of the five sepals (from Troll).

the flowering plants, i.e. the angiosperms, and proved that the sporophyte of the ferns and allied plants (Pteridophyta) corresponds to the vegetative phase of the seed plants (Spermatophyta: Gymnospermae and Angiospermae).

In many of the ferns in the strict sense, the Filicatae, the leaves which serve for assimilation (trophophylls) are indeed scarcely distinguishable from the

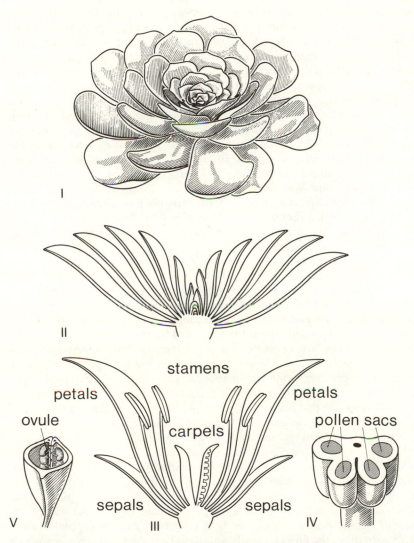

Fig. 2 The structure of an angiosperm flower. I, II vegetative leaf rosette in general view (I) and in axial cross-section (II). III schematic axial cross-section through a flower, IV schematic horizontal cross-section of an opened anther of a stamen, V schematic horizontal cross-section of an open carpel of *Colutea arborescens* (I, II, V from Troll).

sporophylls which also bear the sporangia. Still less in these plants is the development of the sporophyll confined to a definite and specially demarcated portion of a shoot. Where a clear difference between the sterile and fertile leaves is encountered, as for example in the Hard Fern (*Blechnum spicant*) or in the Ostrich Fern (*Matteuccia struthiopteris*), the vegetative growing point often produces in each year a number of sterile and fertile "fronds".

There are also quite a number of species of Clubmoss, as in the members of the genus *Huperzia* such as the Fir Clubmoss (*H. selago*) where in each year the flowering shoot forms a number of needle-shaped trophophylls and sporophylls. The latter are only distinguished from the trophophylls by the fact that they bear a large kidney-shaped sporangium on the upper side at the base. Whereas in this case there is no demarcation between the sporophyll and the vegetative region, and the vegetative growth of the shoot tip continues after the development of the sporophylls, the shoots of other Clubmoss species terminate their growth with a cone-shaped sporophyll-bearing structure or strobilus which is clearly differentiated from the vegetative region. An example of this is the Interrupted Clubmoss (*Lycopodium annotinum*). In the Stag's-horn Clubmoss (*Lycopodium clavatum*, Fig. 3 III, IV) these strobili are even elevated above the main system of creeping shoots on an upright, elongate and sparsely leafy "podium". If we make use of our previously formulated definition of the flower, then we must describe these strobili as "primitive flowers". In the species of Horsetail (*Equisetum*, Fig. 3 V), the remaining modern representatives of the Horsetail order (Equisetatae), these primitive flowers stand out even more clearly from the form of the stem leaves by the completely contrasting peltate shape of the sporophylls.

In the above examples of fern-like plants, which were mentioned for the sake of comparison, we were dealing entirely with homosporous plants. The structure of the angiosperm flower can be better understood if it is compared with the "primitive flowers" of *Selaginella* (Fig. 3 I, II), which is also classified with the Clubmosses (Lycopodiatae). This pteridophyte is heterosporous, i.e. the spores are heteroecious and are readily separated by their sizes into micro- and macro-spores. They are developed in special micro- and macro-sporangia which in *Selaginella* are placed singly on the base of the sporophylls, which we can therefore name micro- and macro-sporophylls respectively. Hence the sporophyll-bearing structures of *Selaginella* frequently appear to be like "hermaphrodite flowers", in that many microsporophylls follow on top of a number of basal macrosporophylls. In drooping "flowers" this spatial relationship can also be inverted. These "flowers" may also be set apart from the vegetative part of the plant by a more or less conspicuous "podium".

In view of the advanced level of development which *Selaginella* has reached among the pteridophytes with its "hermaphrodite flowers", and the fact that fossil representatives of this group from the Carboniferous era even reached the stage of seed production, it is quite surprising that, throughout the recent groups of gymnosperms at least, we find only unisexual sporophyll-bearing structures. In these, the flowers usually appear as cone-like structures, for which a female cone of the Cycad genus *Encephalarthos* (Cycadatae) and a male flower of the Yew (*Taxus baccata*) are convenient examples. However, in these seed-bearing plants we still fail to see any obvious perianth, which is so conspicuous and characteristic a feature of the flowers of the angiosperms. However, the male flowers in their juvenile stage are occasionally seen to be protected by scale-like "leaves" which precede the sporophylls, as occur similarly in the Horsetails and the Clubmosses. The male flowers of the Scots Pine *(Pinus sylvestris*, Fig. 3 VIII) and of the Yew (*Taxus baccata*, Fig. 3 IX) are good examples. In the Gnetales the flowers, which are obviously highly regressive, remain persistently surrounded by such bracts.

The protective function which is exercised by these leaf organs which precede the sporophylls in the above cases must also have been the first function of the perianth in the angiosperms. This must also be seen, undoubtedly, as a form of protection against flower-seeking animals. The development of brightly coloured display equipment and with this the implementation of the function of attraction has obviously been carried out as an adaptation to flower-seeking and -pollinating insects, and since the lower Cretaceous era and above all, in the Tertiary, a constant interplay with the evolution of the insect orders has led to the explosive development of an astonishing variety of forms in the flowering plants. The exercise by the perianth of not only a function of protection but also one of display determines the general appearance of the flower in the typical case to such an extent that in spite of the fact that an inconspicuous perianth or even (obviously always secondarily) flowers without a perianth tend to occur frequently, the perianth has to be regarded as the essential component of the angiosperm flower.

This perianth appears in different forms, namely as

a) a simple perianth (**perigon**), which consists of more or less similar segments (**tepals**), which can be large and conspicuously coloured as in the tulip (**corolloid** or **petaloid**), or small, inconspicuous and greenish or brownish coloured as in rushes (**prophylloid**);

b) double (compound, **heterochlamydeous**) perianth (Fig. 2 III) which shows a differentiation into a **calyx** which serves more of a protective function, and a **corolla**, whose lobes are more adapted to the attraction of pollinators by virtue of their size and bright colouring. Hence one speaks of sepals and petals, respectively.

It should be mentioned here that not only the members of the same set of floral organs can be joined together laterally but also that the members of different sets of floral organs may be united. In this way the sepals may be combined into a calyx which is campanulate or has some other shape, or the petals united into a sympetalous corolla, and the stamens may be joined to the corolla to a greater or lesser degree. Such organs are often united from their moment of origin onwards, i.e. their fusion is **congenital**. They may also grow together later on, when the fusion is said to be **postgenital**. The way in which such organs fuse often gives rise to important criteria for the recognition of systematic relationships, as does the fusion of the petals in the "sympetalous" group of families, or the fusion of the carpels into a homogeneous ovary, which is of particular importance.

From all that has been said so far, it follows that we regard the hermaphrodite flower as a *uniaxial* sporophyll-bearing structure with micro- and macrosporophylls. Such a flower is called a **euanthium** and, correspondingly, this

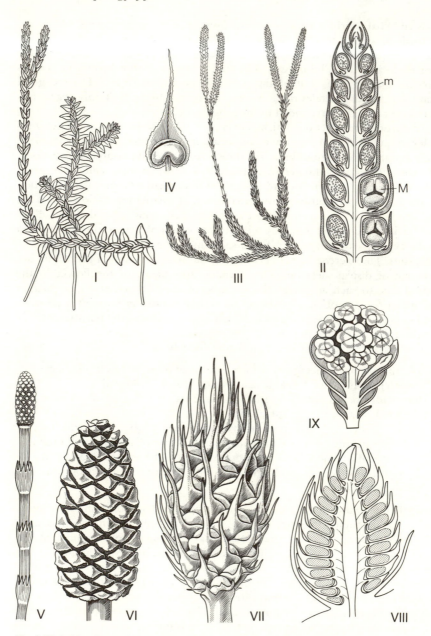

Fig. 3 "Primitive flowers". I, II *Selaginella helvetica*, I branching shoot with two young strobili, II strobilus in axial cross-section, M macro-, m microsporangia. III, IV *Lycopodium clavatum*, III branched shoot with strobili, IV sporophyll, upper side. V *Equisetum arvense*, fertile shoot with ripe cone. VI *Encephalarthos altensteinii*, male flower. VII *Macrozamia* sp., female flower. VIII *Pinus*, male flower in axial cross-section. IX *Taxus baccata*, male flower; the portion of the shoot which precedes the sporophyll in axial cross-section. After Luerssen in Rabenhorst (I), Troll (II, V, VI, VIII), Schenck (III, IV), Schuster (VII) and Richard (IX), partially adapted.

approach to the subject is called **euanthium theory** (compare Fig. 4 I-I' with II-II').

There is another point of view which is opposed to the above, which maintains that the angiosperm flower may have originated in a complex system of axes, similar to an "inflorescence" with numerous male and female flowers without perianths. This would only be consistent with terminal micro- and macrosporangia. The angiosperm flower would accordingly be understood as a **pseudanthium**.

This way of looking at the problem is termed **pseudanthium theory**. This theory goes back to Wettstein (1901/1935), and has undergone all kinds of modifications in the meantime. The best known of these derived theories is the hypothesis put forward by Lam (1948 et seq.), according to which **phyllosporous** and **stachysporous** flowers must be distinguished. For some cases it is assumed that the sporangia were positioned on the margins of the sporophylls, i.e. they originated from a leaf, then we have the expression **phyllosporous**. In other plants the sporangia are said not to have arisen from a leaf organ, but directly from the shoot axis. In fact the term **"stachysporous"** means that the sporangia are arranged in spikes on the axis. This is argued particularly for those cases in which the connection between the sporangia and the leaf structures which bear them is no longer immediately recognisable. This is quite commonly the case with ovules, and is the case in what is called central or basal placentation, which we will later recognise to be highly derived cases of the position of the ovules in the carpels. Some of these attempts at explanation, such as the **"gonophyll theory"** of Melville (1960, 1962, 1963) strike one as rather complicated. A gonophyll is a leaf organ with a fertile axial body, the "sporangiophore", as an epiphyllous outgrowth. There is also the **"anthocorm theory"** of Neumayer (1924) and Meeuse (1972, 1975a,b). These efforts represent speculative constructions which try to derive the structure and flower properties of recent angiosperms from fossil finds which are of rather dubious interpretation, e.g. of relatives of the seed ferns (also see further H. H. Thomas, 1931).

In addition there have been frequent attempts to identify those structures and organs which might be thought equivalent with the help of similarities in the branching of the system of vascular bundles (as by Henslow, as early as 1883, and more recently mainly by Eames, 1931, 1951). This may have been because of the suggestive influence of the impressive comparative studies of the skele-

Fig. 4 Diagrams of the derivation of an angiosperm flower (II, II'), by the euanthium theory (I-II) and by the pseudanthium theory (I'-II'). Pollen sacs dotted, ovules dark (I, II referring to Arber and Parkin, I', II' to R. von Wettstein's derivation of *Ephedra*; from Strasburger).

tons of vertebrate animals and their meaning for the elucidation of form and ancestral relationships (e.g. compare Eames 1929: "The skeleton of plants is in some respects – among these, conservatism in change – like the skeleton of animals." p. 423). Hence, without justification, a conservative role for the vascular bundle systems is assumed and the completely different function and mode of development of the vascular strands is left out of consideration. Agnes Arber (1933, p. 234) remarks on this already "that the phylogenetic speculations which have been based on the alleged 'conservatism' of the vascular strands, can no longer be accepted, and that we must cease to treat morphological interpretation as a puzzle-game elaborated by Nature, towards the solution of which she has thoughtfully provided neat little anatomical clues". Also Schmid (1972) warns against an uncritical application of the results of analysing vascular strands: "The only ultimate test is correlation with other lines of evidence" (compare also the discussion between Rohweder 1967, 1972 and van Heel, 1969). In particular the leaflike nature of individual floral organs is repeatedly challenged. In this way, for example, Payer (1857), Thompson (1929) and Hagerup (1936, 1939) explained the carpels or the placentae which they enclose (in the Personatae) as axial structures. ("The placentae are direct continuations of the floral axis", Hagerup, 1939, p. 36.) Gregoire (1931, 1935, 1938) thought that the homology of the flower with a vegetative shoot had been disproved on the basis of histological investigations. It is argued that the apical meristem of the flower (as also that of the inflorescence) shows no differentiation into tunica and corpus and by this character differs from the apical meristem of a vegetative stem (on this point see also Plantefol, 1948 and Buvat, 1952).

Gregoire therefore maintains that the floral organs may be "sui generis", and are definitely not to be made homologous with other organs of the plant. The alterations which are shown in the structure of the apical meristem during the transition to the reproductive region do, however, fit without further argument into our modern view of the zonation of the shoot tip (cf. Rohweder, 1963).

1.2   Topological and symmetry properties of the flower

The number and arrangement of the floral organs, as well as their form, determine to a great extent the overall appearance of the flower. Partly by their constancy, and partly by the appearance of a more or less continuous sequence of variation they provide important criteria for systematics.

In just the same way as the leaves in the vegetative region, the floral organs may be arranged either in **helices** (**acyclic**) or in **whorls** (**cyclic**), and likewise within each of the component structures. Especially, if the organs (still) occur on a more or less elongated axis and the number of organs in the individual structure is (still) not fixed and there may even (still) be intermediates appearing between the structures, then the helical arrangement is seen as primitive, which is entirely consistent with the view of the derivation of the flower and its interpretation as a "metamorphic shoot". In fact these kinds of arrangement are most frequently encountered in members of the Magnoliales and the Dilleniales, which also show primitive traits in other characters. On the other hand we must not overlook the fact that, in other members of these groups which are likewise regarded as being relatively primitive, the occurrence of whorls of two or three is the rule.

It is not uncommon to find a mixture of cyclic and acyclic arrangements occurring in one or other of the structures within the flower – such flowers are said to be **hemicyclic**.

1.2.1   Floral diagrams

The best way to present and compare the number and the topological properties of the floral organs is to sketch an outline of the flower, a **floral diagram**. In these the individual elements of a flower are projected onto a plane in their relative positions by means of schematic cross-sectional drawings. A flower-sketch of this sort may be obtained by preparing a cross-section through a young flower bud, as was done for the tulip in Fig. 5 I. A flower may also be prepared so as to register all the organs with their exact lateral and radial spacing. When dealing with whorls the leaves, each according to its proper position, may be inserted into a sequence of concentric circles, each of the circles symbolising the appropriate node. When dealing with a helical arrangement of floral organs, each symbol is included on a spiral which corresponds to the genetic spiral. Circles or spiral lines are of course usually left out of the final printed diagram, as in the examples of acyclic flowers of *Magnolia stellata* and *Calycanthus* reproduced in Fig. 14 II, IV.

The cross-section of a tulip flower bud which is reproduced in Fig. 5 I shows two successive whorls with three perigon (perianth) segments each, within which stand six stamens belonging to two whorls of three each. The three carpels of the gynoecium are united in their basal parts which enclose the ovules, thus forming an entire ovary with three locules.

A diagram such as this which reproduces unaltered the topological relationships of the organs in the flower is called an **empirical diagram**. We must distinguish the empirical diagram from the **theoretical diagram**. The most convenient way of explaining this difference and the meaning of a theoretical diagram is to comment on the example of the *Iridaceae* which is reproduced in Fig. 5 II. Furthermore, this diagram is also valid for the crocus, which closely resembles the tulip in the external form of the flower. The diagram, derived empirically in a similar way, also shows a perianth with two whorls of three. After this, however, there come only three stamens, and not six as in the tulip. The fertile portions of the three carpels have also grown together into a single ovary with three locules.

For flowers with a whorled arrangement of organs there are two rules:

1. The **alternation rule** states that the organs of two successive whorls **alternate** with one another in such a way that they "fill up the gaps", so to speak.
2. The rule of **equidistance** states that the angle of divergence between the organs of one and the same whorl is always the same; the organs therefore appear at an equal distance from each other.

Exceptions to these rules will always need to be specially explained. In the floral diagram of the Iridaceae the locules of the ovary and likewise the three carpels with the three stamens should alternate with the preceding whorl. Since this is not the case it may be supposed that an inner ring of stamens is secondarily lost. This assumption is confirmed by the results of other comparative studies. In rare and exceptional cases individual members of the missing ring of stamens do appear. Accordingly we have to indicate the exceptional stamens in

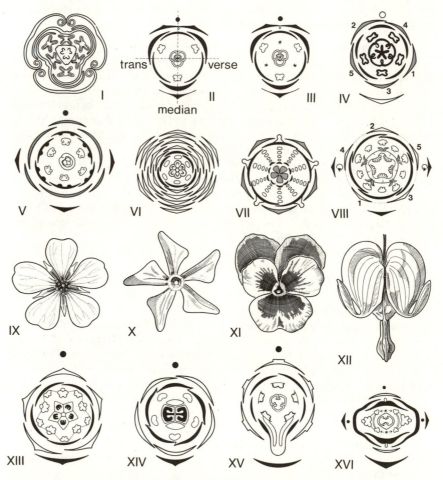

Fig. 5 Floral diagrams. I *Tulipa gesneriana*, cross-section through a flower bud, schematic diagram, empirical. II, III *Iris*, empirical (II) and theoretical diagram, in diagram II the median and transverse planes of the flower are given; IV *Aldrovanda vesiculosa*, the arabic numbers indicate the aestivation and the "genetic spiral" of the calyx; V *Berberis vulgaris*, VI *Sychnosepalum paraense*, VII *Aquilegia vulgaris*; VIII *Carica papaya*, IX and XIII *Geranium* (*pratense*), flower seen from above (IX) and floral diagram (XIII); X and XIV *Vinca minor* similarly, XI and XV *Viola tricolor* similarly; XII and XVI *Dicentra spectabilis*, flower in horizontal view (XII) and in a diagram (XVI). After Troll (I, IX–XII) and Eichler (the remainder), partially adapted.

the Iridaceae diagram with corresponding symbols, as has happened in Fig. 5 III, in order to make clear the structural relationships between the tulip or the Liliaceae and the Iridaceae. A diagram such as this which does not immediately reflect the actual facts, but which includes an **interpretation**, is called a **theoretical diagram**. The lateral fusion of floral organs and other characters may also be illustrated in a floral diagram (Fig. 5 VIII, XIV, XVI).

In the majority of flowers all the floral organs are arranged in whorls. However, the calyx very often forms an exception. In particular this applies to the 5-**merous** calyx (i.e. with 5 sepals), which very often exhibits a 2/5-spiral (Fig. 5 IV) not only in the position of the sepals but also at times in their shape. This arrangement is termed **quincuncial**, or else we speak of the **quincunx** of the calyx.

In cyclic flowers (and with regard to the calyx, in hemicyclic also) we very often find a structure in which there is one circle each of sepals and petals, two whorls of stamens and a whorl of carpels (Fig. 5 VIII). Flowers which are built up of five whorls like this are called **pentacyclic**. **Tetracyclic** flowers with only one circle of stamens are also common (Fig. 5 IV, XIV). One also speaks of a **diplostemonous** flower with two whorls of stamens, and of a **haplostemonous** flower with one whorl. The number of whorls of one kind of organ can also be very much higher. So for example the calyx of the Barberry (*Berberis vulgaris*, Fig. 5 V) consists of two whorls of three, and of three whorls in the closely related *Mahonia*. In the genus *Epimedium* which also belongs to the Berberidaceae there are five whorls of two, one after the other; in *Nandina domestica* (also Berberidaceae) there are seven 3-merous whorls and in *Sychnosepalum paraense* (Menispermaceae, Fig. 5 VI) nine 5-merous whorls! There is a similar situation with the whorls of stamens in the *Lauraceae* where one often finds four 3-merous whorls of fertile stamens with an additional whorl of rudimentary sterile ones, or in Columbine (*Aquilegia vulgaris*) with 8 or 10 fertile whorls plus two sterile whorls of five stamens each (Fig. 5 VII). Examples with several successive whorls of carpels are the Pomegranate (*Punica granatum*, Fig. 19 V) and the Navel Orange (*Citrus sinensis* var.).

The rule about the alternation of successive whorls is often broken in the region of the androecium in the sense that the stamens in the outermost ring are placed in front of the petals. For more details about this phenomenon of **obdiplostemony** see section 1.5.11 and Fig. 5 XIII.

The number of members in the successive whorls of each kind of floral organ may be the same, as for example in the Columbine (Fig. 5 VII), but is not necessarily so. The gynoecium in particular often comprises very many fewer members than the other sets of floral organs, as for example in the Berberidaceae (Fig. 5 V) where we regularly find only one carpel. Flowers with **isomerous** whorls – i.e. bearing the same number of parts in each whorl – are said to be **isocyclic** (**eucyclic**), and those with unequal numbers in each (**heteromerous**) whorl are said to be **heterocyclic**. One also speaks of **polymerous** whorls (or of polymery) when the number of members increases, and of **oligomerous** or even of **monomerous** whorls when the number of parts decreases or is reduced to one, respectively.

In a sequence of heteromerous whorls it is often necessary to recognise whether it is a case of typical heteromery, or whether this has come about secondarily through the loss or gain of individual organs. Obviously in this and in other cases the study of the development of the flower can give important information.

In the case of such "secondarily" derived heteromery the rule holds good that the floral diagram shows the normal alternation of the whorls (although of course displacements between the individual members may appear). For "typical" heteromery the following rule applies, according to Eichler (1875, p. 12). In general the whorls are so placed that the floral symmetry is disturbed as little as possible, by arranging an alternation of neighbouring whorls in as nearly as possible a uniform manner. The new lay-out, it may be said, should arise with the least possible disturbance of neighbouring structures, whether this disturbance is caused by increasing the size of some organs, or by a competitive inhibition. Thus the new organs originate "where there is the most room" (also known as Hofmeister's rule).

As it is often a matter of presenting diagrams of flowers which occur laterally on a main axis, this is taken care of by introducing the **principal axis** and the **leaf** which subtends the flower into the diagram, and the convention is that this leaf is introduced under the flower and the main axis above it.

The following representation is used in order to give the orientation of a floral diagram. The plane which passes through the principal axis and the subtending leaf and which divides the diagram into two halves which are mirror images of each other is called the **median plane** (central plane), or the **median** for short. The plane which stands at right angles to this and which divides the flower into its upper and lower halves is called the **transverse plane** (**transverse** for short). When the organs occur in the region of the median plane they are termed **medial**, and **transversal** when they are in the transverse plane. **Above** or **behind** means facing the main axis, and **below** or **in front of** means facing the leaf.

In the presentation of a floral diagram consideration must also be given to the **prophylls**, in so far as these have developed or at least appear in related plants, so that they then have to be brought into the theoretical diagram. It must be taken for granted that the sequence of the leaves on lateral shoots very often begins with one or two leaves which are distinguishable in their shape and arrangement from the other leaf organs, i.e. the prophylls. Inasmuch as we are dealing with two prophylls, as is the case among the majority of dicotyledonous plants, these are usually found in a transverse position, that is to say lateral to the median plane which is defined both by the symmetry plane of the subtending leaf, and also by the longitudinal axes of the main and lateral shoots. They may both be inserted at the same level to form a whorl of two, or there may also be a more or less definite internode developed between them. In so far as this may be ascertainable, the oldest prophyll will be termed the **α-prophyll**, and the youngest the **β-prophyll**. The successive internodes from the bottom upwards are called the **hypopodium, mesopodium** and **epipodium** (Fig. 112 II). In the monocotyledons on the other hand the prophylls usually occur singly and are often inserted not in the transverse plane but on that side of the lateral flowering shoot which faces the main axis. In these so-called "**addorsed prophylls**" one is quite often dealing with the product of fusion between two leaf organs.

1.2.2 Aestivation in the presence or absence of prophylls

There are also certain rules for the positioning of the floral organs in relation to the subtending leaf and the principal axis which have undoubted descriptive and systematic value. In certain cases the topological properties may provide information about prophylls which were originally present but later aborted Just a few examples are presented here in order to explain the basic principle; for further particular instances the remarks made by Eichler (1875, p. 25 et seq.) may be consulted. A distinction is to be made between:

1. Position of the floral organs along with one or two prophylls;
    prophyllate aestivation, and
2. Position of the floral organs without prophylls; eprophyllate aestivation.

1. **Prophyllate aestivation**. With a *single* prophyll and a helically arranged 3-lobed calyx (or perigon) the first lobe always lies approximately opposite the prophyll.

With an addorsed prophyll and a 3-merous flower the first segment of the flower is placed medially in front (Fig. 6 I, II e.g. Iridaceae). With the prophyll placed laterally the whole cycle is displaced through a corresponding angle (Fig. 6 III, IV); in the case taken in diagram IV of a prophyll which diverges from the subtending leaf by 90° (whether primarily or secondarily) one of the segments enters the median, so that either the second leaf comes into position in front of the principal axis (*Lilium*) or the third segment lies above the subtending leaf: *Hemerocallis, Scilla* (Fig. 6 V). In the aestivation of a genuine whorl with a single prophyll, for which only cases with an addorsed prophyll are possible, one segment is always found in the median position either at the back or at the front.

Let us consider a calyx in a 2/5 helical arrangement (dicotyledons only) with a single lateral prophyll adjacent to it, where this prophyll is genuinely unique and not just the result of the suppression of a second prophyll. There are then two possibilities. Firstly, that the prophyll is placed at a position at 2/5ths of the way round from the first (outermost) calyx segment, whereby in a manner of speaking the genetic spiral runs around "behind" between the flower and the main axis, which is termed **opistodromous** ("running round the back", Fig. 6 VII). Secondly, the first calyx segment stands diametrically opposite the prophyll, whereby the genetic spiral passes round the "front", which is termed **emprostodromous** ("running round the front", Fig. 6 VI).

For the case of prophyllate aestivation with *two* prophylls we will only consider the attachment of a calyx with the 2/5 arrangement; here as a rule one of the calyx segments is placed in the median and the β-prophyll is placed obliquely with respect to the first calyx segment. In this case too, the genetic spiral may run diagonally round the back of the flower from the β-prophyll, giving opistodromous flowers (Fig. 6 VIII; Lobeliaceae). This is distinguished from the emprostodromous spiral where the first calyx segment is directed obliquely towards the front and the second at the back in the median (Fig. 6 IX). The emprostodromous flower is the usual case among dicotyledonous plants with 5-merous flowers. An arrangement comparable to that in diagram VIII is found in many leguminous plants (X). Here the first calyx segment does not lie diagonally across from the β- prophyll, but in the median at the front (emprostodromous).

2. **Eprophyllate aestivation.** When the prophylls are missing the rule which typically applies is that the first two floral segments are placed wherever possible as if they were prophylls, whereupon the arrangement of the remaining segments follows suit. If one or both of the prophylls is absent merely because they have been suppressed, then the attachment of the flower is usually found to be unchanged by their absence, so that we have a situation which is equivalent to that with the typical number of prophylls. Hence a calyx which is arranged with two segments in the median usually indicates that the prophylls which were originally present have regressed.

1.2.3   Vernation and aestivation

Already during the discussion of the attachment of the outer floral organs to the prophylls it became apparent that the genetic succession of the floral organs is often recognisably similar to that of the leaf buds, in that the margins of the older leaves overlap those of the younger. This **aestivation (prefloration)** thus allows inferences to be made about the development sequence of the floral organs, particularly with the 2/5-phyllotaxy of the successive segments of the quincuncial calyx, although this is by no means always the case (on this point see, among others: Reinsch 1927 and Schoute 1935). Above all the **aestivation of the petals** in its diverse forms is often characteristic of large groups of related taxa. The first character to observe is whether the margins of the organs:

1	2	3
do not reach each other	only touch	or pass over each other

in which case one distinguishes the following kinds of aestivation:

| open (apert) | valvate | imbricate. |

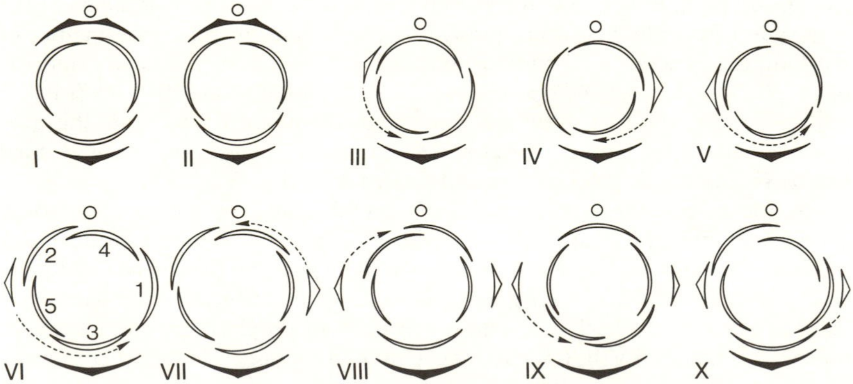

Fig. 6 Prophyllate aestivation of the floral organs, I, II 3-merous flower with addorsed prophyll, III, IV 3-merous perianth and lateral insertion of the prophyll, in V displacement of the 3rd perianth segment to the median in front, emprostodromous attachment (arrow). VI-X Attachment of a 5-merous perianth to a lateral prophyll (VI emprostodromous, VII opistodromous) and to two lateral prophylls (VIII-X), in many leguminous plants the 1st calyx segment drops to the median in front (X), further explanation in text. After Eichler, somewhat adapted.

For example, the aestivation in the Rhamnaceae is open, and valvate in the Cornaceae and in Lilac (*Syringa*).

The **imbricate** condition may be further classified in different types:

1. quincuncial (Fig. 7 I; common in calyces; *Saxifraga granulata*, corolla)
2. contort (twisted)
   left-handed (Fig. 7 IV; *Vinca, Oxalis acetosella* etc.) right-handed (Fig. 7 V; Convolvulaceae, Gentianaceae). The direction of rotation may change within one inflorescence (Malvaceae).
3. cochleate (irregularly helical), in this type of aestivation, which occurs commonly in 5-merous corollas, there is always one segment which lies entirely outside, one which lies completely inside, and the other three with one margin in and the other out. There are then two types to be distinguished; a) the innermost segment can be immediately adjacent to the outermost, which is called **proximal-cochleate** or **paratact** (Fig. 7 VI; *Drosera, Jasminum, Pittosporum*), or b) the innermost and outermost can be separated by other segments: **distal-cochleate** or **apotact** (*Primula*). Two important forms with the innermost segment in the median position are:
   ascending cochleate (Fig. 7 II, VIII; Caesalpiniaceae),
   descending cochleate (Fig. 7 III, VII; Papilionaceae).

Let us consider once more the cross-section through the bud of a tulip which is reproduced in Fig. 5 I. From this we can establish the fact that, apart from the valvate outer and the imbricate inner perianth whorls, each of the individual floral organs exhibits its own special bearing in the flower bud, just as do the

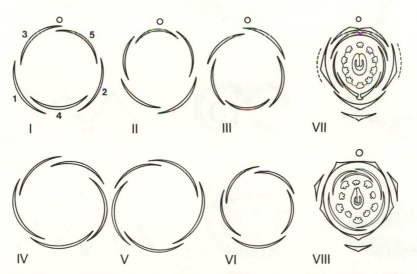

Fig. 7 Types of imbricate aestivation. I quincuncial, II ascending cochleate, III descending cochleate, IV, V contort, left-handed (IV), right-handed (V), VI proximal-cochleate (paratact). VII flower diagram of *Vicia*, Papilionaceae (descending cochleate), VIII flower diagram of *Cercis*, Caesalpiniaceae (ascending cochleate). After Eichler.

leaf organs of a vegetative bud. Here in the tulip the margins of the outer tepals are rolled inwards (involute), and in other cases they are rolled outwards (revolute), or once (or more) folded, and so on. These phenomena are summed up under the name of **vernation**. Some of the forms of vernation which appear in flowers and leaf buds have been brought together in Fig. 8. It is not rare to find that the way that the floral organs open is more or less strongly correlated with their vernation. When this type of inhibition of development occurs, although often only partial, and if this is linked with the preferred lateral and longitudinal modes of growth of the other parts of the flower, then very complex flower forms may develop as a result, as for example in the "window flowers" of *Ceropegia* species (see section 1.4.3.7).

1.2.4   Symmetry properties

The flower diagrams often offer a regular representation of the equivalent or similar structural elements, in a definite order (sometimes more clearly than does the flower itself), so that **symmetry** is recognisable.

In order to express these "ordered representations of equivalent or similar

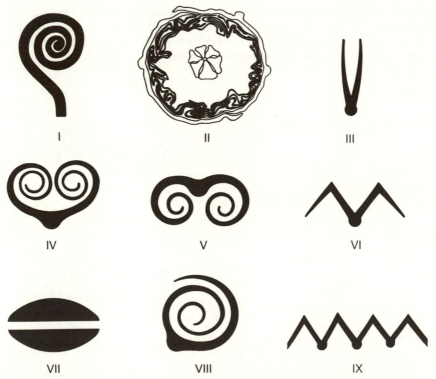

Fig. 8 Types of vernation. I circinate (as in many stamens), II crumpled (corrugate) (*Papaver*, also *Cistus* etc.), III conduplicate, IV involute (*Syringa*), V revolute, VI reduplicate, VII planate (flat, Mesembryanthemaceae), VIII convolute, IX plicate.

structural elements" we may use the so-called **symmetry operations**, namely **translation**, **rotation** and **reflection**. Correspondingly one distinguishes between translational, rotational and reflection symmetry. **Translational symmetry** is produced when similar elements are repeated along a straight line, which is called the axis of translation. The corresponding symmetry operation here consists in shifting the equivalent elements of the pattern along the axis of translation until they come into coincidence again. In the flower translational symmetry is shown in the repetition of individual floral organs along the floral axis. However, this longitudinal symmetry in the flower is of less interest than lateral symmetry in the form of rotation and reflection. **Rotational symmetry** is produced when equivalent elements can be brought into coincidence by turning them about an axis of rotation of the configuration in question. In floral diagrams, which are of course two-dimensional, this axis appears as a point, the "centre of symmetry". Hence for example in the flower of the Lesser Periwinkle (*Vinca minor*) the floral diagram repeats itself in a rotation of 72° (Fig. 5 X, XIV). In **reflection symmetry** the two building elements behave like a picture and its mirror image. They can therefore be brought back into coincidence by being "turned over" about a "reflection axis", or by reflection at a "**mirror-**" or "**symmetry-plane**" which includes the reflection axis and stands vertically at right angles to the configuration in question. In the flower diagram of the tulip (Fig. 5 I) there are three such symmetry planes, and in that of the columbine (Fig. 5 VII) there are five, so that these flowers are described as three- or five-fold symmetric, respectively. In these two examples rotation by 120° or 72° respectively could be used as well as reflection; not so in the flower of *Vinca minor*, whose petals likewise repeat regularly at an angle of 72° but which are formed asymmetrically, so that rotation is the only possible symmetry operation. In **acyclic** flowers the operations of simple rotation (or reflection) cannot be applied and so the only way to bring the elements into coincidence is to combine rotation with translation along the floral axis.

By the symmetry properties we distinguish:

1. **Radially symmetric, regular** or **actinomorphic** flowers; they usually exhibit more than two planes of symmetry (tulip, columbine etc.; see Fig. 5 I, IV, VI to VIII, IX and XIII);
2. **Bisymmetric (bilateral)** flowers; they possess only two planes of symmetry, which divide the flowers in two different ways, such as the flowers of *Dicentra spectabilis* (Fig. 5 XII, XVI);
3. A **zygomorphic (dorsiventral**, monosymmetrical) flower structure is by far the most characteristic for plagiotropic lateral flowers; in these there is only one plane of symmetry, because the upper and lower side of the flower are differently constructed, such as in the violet (Fig. 5 XI, XV) or in bilabiate flowers (Fig.42 I). Quite commonly the zygomorphy consists only of a deviation from the radial arrangement in the orientation of the perianth segments, or in curvature in the stamens and the style, which is usually due to the influence of gravity. A well-known example of this is in the flower of the Rosebay Willowherb (*Epilobium angustifolium*, Fig. 9 I). Occasionally this is combined with minor differences in the way that the perianth segments are laid out, as in *Gladiolus byzantinus* (Fig. 9 III), or with differences

in the size of the petals, as in the Burning Bush, *Dictamnus albus* (Rutaceae, Fig. 9 VII), and in *Antholyza aethiopica* (Iridaceae, Fig. 9 V), in which at times the medial perianth segment – the lower in *Dictamnus*, the upper in *Antholyza* – is greatly enlarged. All these features commonly appear in flowers which are held horizontally and they tend to disappear by reversion to radial symmetry when the influence of gravity is removed (II, IV, VI) – perhaps the ancestral habit of the flower may have been erect. In the Polemoniaceae the genus *Loeselia* differs from the related genera (with the exception of *Bonplandia*) by zygomorphy in the flower and the stamens. If one positions the inflorescence of *L. glandulosa* so that the young flowers come into an upright habit, then these remain completely radially symmetric in the orientation of the corolla lobes and the stamens.

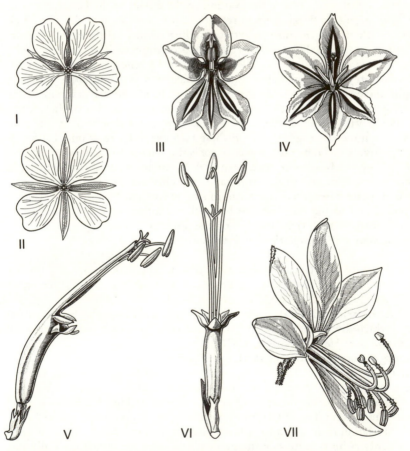

Fig. 9 Geotropically induced floral symmetry. I, II *Epilobium angustifolium*, II with the elimination of the influence of gravity with the help of a clinostat, in I and II, stamens and style cut away. III, IV *Gladiolus byzantinus*, IV flower which is approaching radial symmetry through an erect habit. V, VI *Antholyza aethiopica*, VI flower developed after the removal of gravity. VII *Dictamnus albus*. After Vöchting (I, II), Troll (III-V, VII) and Haeckel (VI).

4. **Asymmetrical** flowers, which do not demonstrate symmetry under any of the standard operations; from amongst these one must distinguish between primarily asymmetric flowers and those whose asymmetry came about from reduction, multiplication or transformation of cyclically arranged organs. Completely asymmetric flowers are relatively scarce; they are found for example in the Valerian (*Centranthus*, Fig. 43 IV) and in Indian Shot (*Canna indica*, Fig. 74 I, II).

In some cases it may not be possible to prepare a flower diagram. In such cases the numerical characters, the manner of arrangement, and if need be the fusion of a number of the floral organs as well as the symmetry properties can be summarised in a **flower formula**. In order to represent the symmetry properties one makes use of special symbols (℮ spiral, ⋆ radial, ·|· bisymmetric, ↓ zygomorphic). Then the number of organs in each of the formations and rows are given (P = perigon, K = calyx, C = corolla, A = androecium, G = gynoecium) and the number of fused organs is indicated in brackets ( ). The formula for the Evening Primrose illustrated in the diagram in Fig. 19 IV then runs: K4 C4 A4 + 4 G ($\overline{4}$). In this the expression A4 + 4 means that the stamens occur in two whorls of four. The ovary in this case is inferior, which is shown by a line over the number of carpels. Had it been superior, this would have been shown by a line underneath the number of carpels (_), and if it were semi-inferior the line would have been omitted, and only the number given. The sign ∞ means that the organ appeared with a large and indefinite number of parts. It must be admitted, nevertheless, that some kinds of information which are presented in a flower diagram, such as that for the Papilionaceae (Fig. 7 VII) cannot be expressed completely in a flower formula.

The flower diagram is not a sufficient means of giving a complete picture of the structure of a flower, and the form and arrangement of its organs, especially those of the ovary. It needs to be supplemented by outline diagrams and this means that we next have to turn our attention in particular to the form of the floral axis, or receptacle.

1.3 The receptacle

1.3.1 Forms of the receptacle

The form of the flower is highly dependent on the structure of the receptacle, even though this may not always be obvious externally. The receptacle is also called the **floral axis**, or it is sometimes called the **torus**, which may be translated as "swelling". These expressions *per se* imply that, although in the majority of cases the receptacle is greatly reduced, it is frequently thickened in a capitate form or broadened into a definitely disc-like shape. If there is a moderate lengthening of the axis, as is frequently encountered in the region of stamens and free carpels, if these are developed in large number, the receptacle takes on a spherical or cylindrical shape. This causes the conical shape of some flowers as referred to above for which many examples can be found in the Magnoliales, and in the Mousetail, *Myosurus minimus* (Fig. 11 VI), and in general this is judged to be a primitive character. By contrast we have to regard the flattened or discoid receptacle as a derived character. This is even more strikingly the case if the receptacle grows up and around the carpels which remain in position at the tip of the axis, raising the other floral organs in a ring in the process, and surrounding the gynoecium (perhaps also the androecium) in a cup-shaped structure, or even fusing with the gynoecium to make the ovary "inferior". Consequently, as is well known, three types are distinguished according to the position of the gynoecium:

1. Flowers of the common type, in which the perianth segments and the androecium are inserted below the gynoecium, are called **hypogynous**. Their ovary is superior (Fig. 10 I);
2. Flowers in which the receptacle is cup-shaped and surrounds the gynoecium are called **perigynous**. Here the androecium will, like the perianth, be raised in most cases, so that all these organs appear inserted on the edge of the **hypanthium**, or so-called receptacle (Fig. 10 II); here the ovary is surrounded by the cup-shaped axis although not fused with it, and remains superior;
3. The gynoecium is said to be inferior if the floral axis has extended above the region of the insertion of the carpels to such a degree that the ovary is finally enclosed in and united with a cup-shaped axis, leaving only the styles or the tips of the carpels free and bearing the calyx, corolla and stamens on its upper margin (Fig. 10 III). The flower is then **epigynous**.

There are of course other possible interpretations of the very diverse structural properties of the gynoecium, of which a rather sketchy account has just been given, and we shall discuss some of these further. In what follows, the possible forms of the receptacle, with different structural properties of the gynoecium, are by no means exhausted.

It is not uncommon to find very different types of construction of the receptacle in the region of different formations of organs. Thus, for example, in the Wood Avens (*Geum urbanum*, Fig. 11 I) the receptacle is club-shaped in the region of the gynoecium, which consists of numerous free carpels, but below this is expanded like a disc into the form of a shallow funnel, on whose margin the sepals, petals and stamens are inserted.

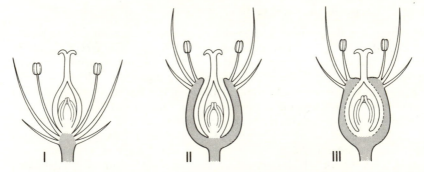

Fig. 10 Hypogynous (I), perigynous (II) and epigynous (III) flowers in axial cross-section. After Troll.

It should be mentioned that Hillmann (1910) thought that the expanded discoid receptacle in *Geum* and the corresponding dish- or cup-shaped structures of other members of the Rosaceae, in contrast to the hypanthium in *Rosa*, had to be interpreted as "a fusion product of leaf structures" on the basis of the vascular bundle system. See the previous comments on the value of vascular anatomy, section 1.1.

In some very peculiar flowers the floral axis is elongated, but only in certain internodes, or even between organs of particular formations, while all the other sections of the axis remain of normal size. So for example in *Lychnis flos-jovis*

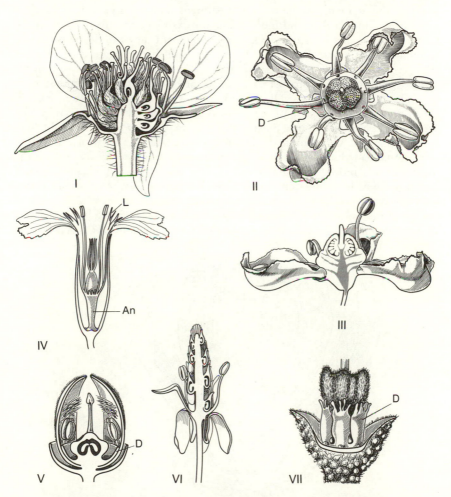

Fig. 11 Structure of the receptacle. I *Geum urbanum*, axial cross-section through a flower; II, III *Ruta graveolens*, flower from above (II) and in longitudinal section, D disc; IV *Lychnis flos-jovis*, An anthophore, L coronal scales (ligulae); V *Cathedra rubricaulis* (Olacaceae), flower in axial section, with dish-shaped disc; VI *Myosurus minimus*, flower in axial section; VII *Spiranthera odoratissima* (Rutaceae). IV after Sachs, V–VII after Engler.

there is an extension of the axis between the calyx and the corolla, the **antho-phore** (Fig. 11 IV, An). In a similar way elongated internodes can develop between the corolla and the androecium (termed the **androgynophore**) or between the androecium and the gynoecium (**gynophore**). An abundance of different forms is found in the Capparidaceae (Fig. 13 I to VI), where they can reach a length of 30 cm (*Cleome siculifera*). However, one also finds them in many members of the related Resedaceae or in the Sterculiaceae (Fig. 13 VII to IX).

It often happens that by outgrowth certain thickenings of the floral axis become more or less prominent, either between the insertions of the petals and those of the stamens, or between the stamens and the gynoecium. Such a growth is termed a **disc** and mostly functions as a nectary (for details on the anatomy of the nectary see E. Frei, 1955). The form they can develop may be cushion-like (pulvinate, Fig. 11 II, III), discoid, dish-shaped (crateriform, Fig. 11 V), urceolate or even as an elongated tube on one side of the floral axis, as in the genus *Cadaba* of the Capparidaceae, where these structures are mounted on a stalk-like androphore (Fig. 13 II, III). It is not rare for these axial growths to fill up all the space at the base of the stamens, and then a disc with a crenate or grooved appearance results. Finally the disc may grow out into palmately divided (Fig. 11 VII), filiform, scale-like or capitate glands, which are often very similar to reduced stamens, all the more so since these may also assume the functions of glands. A decision about the morphological significance of such structures, which can be quite important in order to answer questions in systematics, is not possible without further comparative research.

One of the finest examples of a complex receptacle is offered by the Passion Flower (*Passiflora*). An axial cross-section through such a flower is reproduced

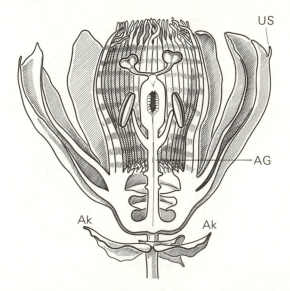

Fig. 12 *Passiflora alata*, flower in axial section, Ak epicalyx, AG androgynophore, US unifacial processes at the tips of the sepals. After Engler.

in Fig. 12. A three-leaved epicalyx (Ak, in Fig. 12) follows above the two prophylls, then five sepals with hooded tips which are provided with small unifacial excurrent points (US) and in this case a corolla with many petals. The calyx and corolla rest on a cup-shaped receptacular structure. The floral axis

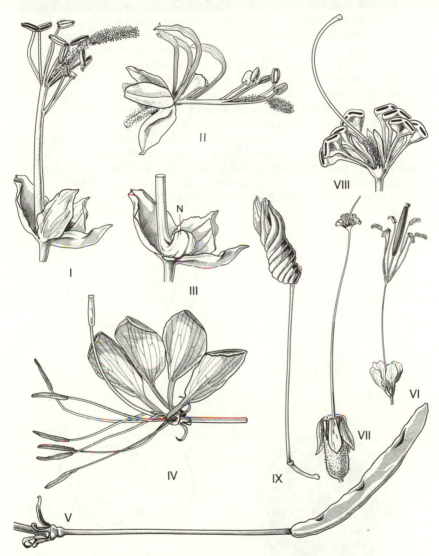

Fig. 13 Gynophores and androgynophores. I, III *Cadaba aphylla*, I flower after anthesis (petals fallen off). III Base of the flower more highly magnified, with a sepal removed to make the nectary (N) visible. II *Cadaba kirkii*, flower with tubular nectary, N. IV, V *Cleome spinosa*, flower (IV) and young fruit on a long gynophore (V). VI *Gynandropsis gynandra*, flower with androgynophore. VII-IX *Helicteres ovata*, VII flower with androgynophore, VIII the same with apical portion magnified, IX fruiting carpels (spirally coiled and free!) on a long (andro-)gynophore. VI after Warming, VII-IX after Wettstein.

continues from out of the base of this cup in the form of a stalk, on which are borne the five stamens, and then supports the three-locular ovary which is separated from the stamens by a short internode. We are once more dealing with an androgyno-phore. Next to the petals on the edge of the receptacular cup we observe a fringe of varying width which is composed of long filiform structures, which are often splendidly coloured like a cockade. These filiform structures represent nothing more than outgrowths of the floral axis formed from the upper edge of the receptacular cup. On the one hand this follows from ontogenetical and compara-tive studies (Puri 1948), and on the other hand it is also self-evident from the fact that on the inner side of the receptacular cup, next to the long filiform fringe come even more shorter structures, still somewhat fringe-like, which finish up in a solid ring projecting inwards and forming a kind of collar about the inner wall of the receptacular cup. Up to thirty such rings may be formed.

Passion flowers were introduced into Europe from South America about 350 years ago and have been cultivated here ever since. Their strange flowers have given rise to all kinds of interpretations. The name "Passion flower" originates from the belief that "all the attributes of the Passion of Christ" could be recognised in the parts of the flower (see Ferrari, "De florum culturae", 1633).

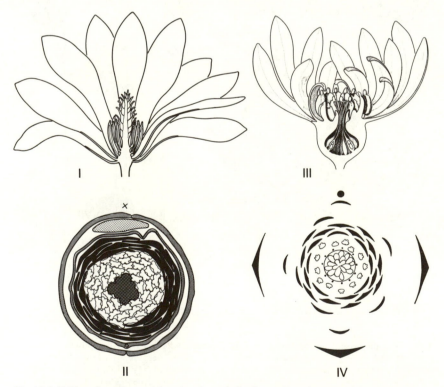

Fig. 14 I, II *Magnolia stellata*, axial section through the flower (I) and empirical flower diagram (II), × position of the main axis. III, IV *Calycanthus floridus*, axial section of flower (III) and flower diagram (IV). I after Troll, III after Baillon, IV after Eichler.

We find a very distinctive form of the receptacle in the families of the Magnoliales and Ranunculales, which are already characterised by an array of primitive features. The flower of *Magnolia stellata* (Fig. 14 I, II) will serve as a typical example, in which the organs of the individual formations are very

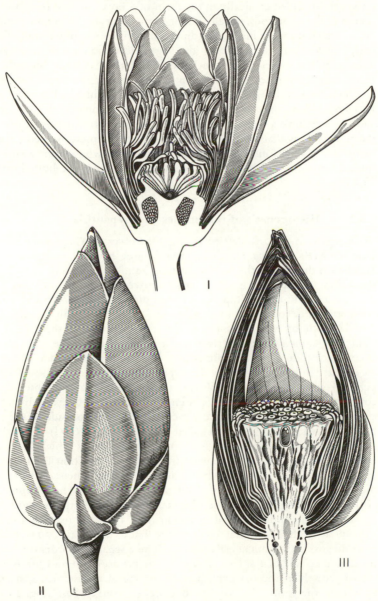

Fig. 15 I *Nymphaea alba*, axial cross-section, after Troll. II, III *Nelumbo nucifera*, II flower bud, shown in cross-section in III.

numerous and inserted on a long conical receptacle in a totally acyclic arrangement, in contrast to the similarly acyclic flower of *Calycanthus floridus* (Fig. 14 III, IV) with its hollow urceolate receptacle – an example of the heterobathmic combination of primitive and derived characters. In *Nymphaea alba* (Fig. 15 I) the receptacle has expanded around the numerous free carpels, which are now, however, arranged cyclically in the form of a bowl, arising from the insertion region of the numerous, spirally arranged members of the perianth and androecium, which now are inserted at different levels on the outer wall of an apparently entire ovary so that only the stigmatic region is left free. (As is well known, in the ripe fruit of the Yellow Waterlily, *Nuphar lutea*, the layer of tissue which simulates a coenocarpium falls away again from the carpels, see section 1.6.13 and Fig. 93 V.) The receptacle is quite different in the Lotus Flowers (*Nelumbo nucifera*, *N.lutea*) which also belong to the Nymphaeaceae and likewise have a completely acyclic arrangement of the floral organs. Here the receptacle widens above the insertion of the densely packed perianth segments and stamens into an obconical pedestal, in whose gently curved upper surface the numerous one-seeded nutlets develop from the buried carpels (Fig. 15 III).

1.3.2   Histogenesis of receptacles and hypanthia

With regard to the urceolate receptacle the flowers of *Calycanthus* resemble the Dog Rose (*Rosa canina*, Fig. 16 I). In both cases there are numerous free carpels attached at the base of the receptacular cup, and only the styles project through the orifice at the top. In *Rosa canina*, however, we do not find a perigon constructed of numerous tepals inserted at different levels on the receptacle, but the perianth is composed of a 5-merous, quincuncial calyx and a 5-merous corolla, both of which are placed on the upper lip of the cup along with numerous stamens. The difference in the arrangement of the perianth segments becomes clearest when the fruit is ripe. Whereas the outer side of the *Calycanthus* fruit exhibits the now widely spread-out scars of the spirally arranged perianth segments (Fig. 16 IV, V), the outer side of the rose hip remains completely smooth (Fig. 16 II). The enclosure of the gynoecium by the receptacle is particularly evident in many cacti, in which the outside of the receptacle bears reduced axillary buds, so-called "areoles" (Fig. 16 VI). These may even show subtending leaves in the form of scale-like bracts which form a gradual transition to the prophylloid tepals of a perigon (see section 1.4.4).

The histogenetic processes which lead to the development of the receptacle in the rose were investigated by Rauh and Reznik (1951), see also O. Zeller (1975). According to this the development of the floral organs begins as soon as the still undifferentiated floral meristem (Fig. 17 I) has reached a certain size. The initials of the individual sepals are the first to take shape, one after the other according to their quincuncial sequence. These sepal primordia run downwards on the outer side of the base of the flower primordium and lift as a result of active growth of the marginal part of the apical meristem, so that this now appears in the form of a bowl with a slightly convex base (Fig. 17 II). What happens next is that first the petals emerge simultaneously from the meristematic "rim of the dish" and then stamens begin to appear successively, progressing

towards the flower centre; and finally the primordia of the carpels arise succes-
sively from the convex basal part (Fig. 17 III). At the same time as the floral
organs develop, active anticlinal cell divisions take place in the receptacle at the
boundary between the primary cortex and pith in the region of the pith, which
lead to a more marked expansion of the receptacle (Fig. 18 VII, VIII). The rows
of cells ("periclinal rows") which result from this active cell division advance
along a curved path into the rim of the dish. Finally, by the activity of a residue
of meristematic cells remaining in the rim of the bowl, the receptacle itself
increases in height and with it the calyx, corolla and stamens are raised, as
indicated in Fig. 17 III, IV with horizontal lines. From what we know of the
development of the receptacle it is probably true to say that the outer surface of
the cup-shaped primordium and later of the receptacle is already more or less
covered with the downward-directed bases of sepals.* From consideration of
the form of the receptacle and fruit of *Calycanthus* species (Fig. 16 III to V) it is

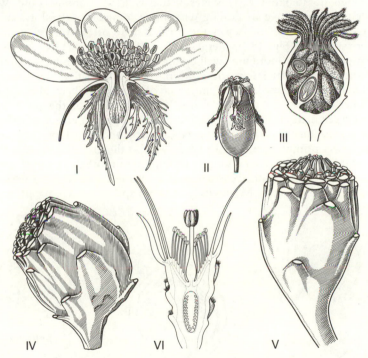

Fig. 16 I, II *Rosa canina*, I Flower in axial section, II Fruit. III *Calycanthus occidentalis*, axial
section of fruit, IV, V *Calycanthus floridus*, fruits. VI *Opuntia humifusa* (*O. rafinesqui*) axial section
of fruit. II after Warming (modified), III after Engler, VI after Buxbaum.

* This possibility was actually excluded by Leinfellner (1954) on account of the detailed agreement
(in *Rosa*, *Cydonia*, *Manettia* and *Philadelphus*) between the forms of the (free) bracts and the sepals
which are inserted in the receptacular cup. Although for *Rosa* this comparison does not appear
convincing, nevertheless in *Begonia* (*B. heracleifolia*) the complete agreement between the perianth
segments of the male flowers and those of the female flowers with their inferior ovary is very much
clearer.

not surprising that its development proceeds similarly to that of *Rosa* (Rauh and Reznik 1951, Dengler 1972).

The floral axis may also be elongated in a tube- or cup-shaped form above and outside the ovary in epigynous flowers. By this means, that kind of floral section which is called a "hypanthium" is built, which is for example more or less conspicuous in the flowers of the Onagraceae (Fig. 19 III, VI), for which it is characteristic. This tubular structure may reach a length of 10 to 14 cm in some species of Evening Primrose (e.g. *Oenothera missouriensis*), and often appears in a colour similar to that of the corolla, as in many species of *Oenothera* and *Fuchsia*, so that it is often mistaken at first glance for part of the corolla. However, we find the sepals and the stamens as well as the petals inserted at the upper margin of this tube. Naturally this tube has been explained as a fusion product of the calyx, petals and stamens and called the "calyx-tube". Also in this case the study of the floral development in *Epilobium*, *Oenothera* and *Fuchsia*, which pass through homologous early stages (Pankow 1966 Bunniger and Weberling 1968), shows that we are dealing with a development of the floral axis. Before the development of the sepals the earliest stage of the meristem, a so-called meristem plug (Fig. 18 I), merges into a "circular wall" stage, in which the margins of the primordium swell up through cell divisions in deep-lying layers of the apical meristem (marked × in Fig. 18 I) so that the structure soon becomes bowl-shaped. After initiation of the sepals here also appear rows of meristematic cells running from the central part of the receptacle up to the curved margins (Fig. 18 II, III, IX). These rows of meristematic cells mainly provide for the formation of the petals, stamens and carpels on the inner wall of the flower primordium (Fig. 18 IV to VI). As the growth proceeds the tips of the carpels which are emerging from the margins of the bowl-shaped flower primordium towards the centre close in over the centre to form what is now a clearly inferior ovary. Meanwhile, in those species which are provided with

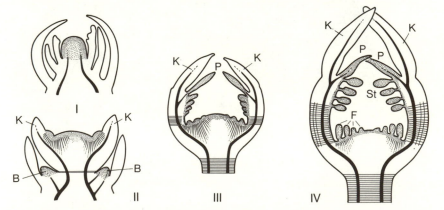

Fig. 17 Rosa rugosa, flower development. Axial sections through floral primordia in different stages of development, schematic. Apical meristem and petal primordia (P), stamens (St) and carpels (F) dotted. The "periclinal rows" which result from active division of young cells are shown with diagonal lines. Intercalary meristems which form the receptacle and the pedicel shown by shading of horizontal lines, K calyx, B primordia of lateral flowers. After Rauh and Reznik, somewhat modified.

one, the growth of a more or less extended hypanthium starts again from residual meristem laid down in the receptacular cup below the insertion of the petals and stamens (Fig. 19 I, II). The relation of the floral axis to the receptacular cup can also be proved by histogenetic research in the Myrtaceae and Thymelaeaceae (Bunniger 1972; see also Leins 1965, Mayr 1969) as well as the

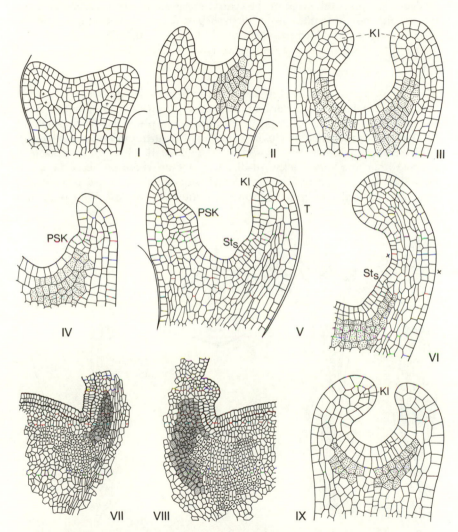

Fig. 18 I–VII Flower development in Onagraceae. I–VI, IX Axial sections through flower primordia (taken in the median of the bracts) in different successive stages of development for *Epilobium hirsutum* (I–VI) and *Oenothera fruticosa* (IX), KI calyx initials, PSK petal-stamen complex, Sts primordia of the episepalous stamens. VII, VIII *Rosa rugosa*, axial sections through two flower primordia of different ages (before stage III in Fig. 17), sections from the marginal zone, active cell divisions, which lead to the expansion of the receptacle and the growth of the receptacular cup. I–VI, IX after Bunniger and Weberling, VII, VIII after Rauh and Reznik.

Punicaceae (Mayr 1969) and in the corresponding structures of many Saxifragaceae (*Tellima*, Klopfer 1968b, 1969a; *Astilbe*, Eckert 1966). In *Astilbe* "the
growth of the receptacle clearly begins in the rib meristem" (Eckert).

On the other hand the typically tubular section between the perianth and the
whorls of stamens in the flowers of the Lythraceae ("perianth tube") can be
understood as a congenital fusion of the calyx and corolla (Mayr 1969). The
evidence for this is "the lack of the central meristem in the histogenesis of the
tube, which resembles the raising of the sepals from the dermatogen and the
subdermatogen and which may be termed "leaf-like" (p. 252).

1.3.3   Axial spurs

Many spur-like appendages of flowers can also be considered as structures of the
floral axis; certainly we must distinguish carefully between axial spurs and spur-
like extensions of the sepals or petals. A superb example of the development of
an axial spur is presented by the flower of the Nasturtium (*Tropaeolum majus*),
which is reproduced in medial section in Fig. 20 II. Here the receptacle is
hollowed out in a long, backward-directed tube in whose tip nectar accumu-
lates. This tube originates on the (adaxial) side which is turned towards the
principal axis and is originally formed by intensive intercalary growth in a

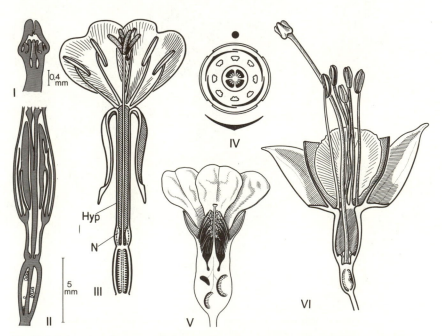

Fig. 19 I, II *Oenothera fruticosa*, medial section through a younger and an older flower bud,
III *Oenothera biennis*, longitudinal section of flower, IV flower diagram of *Oenothera*, V *Punica
granatum*, axial section through the flower, VI *Fuchsia magellanica*, longitudinal section of flower;
Hyp hypanthium, N nectary. I, II after Bunniger and Weberling, III after Firbas, IV after
Eichler, V after Warming, modified.

highly localised zone below the adaxial stamens (Mair 1977). The formation of this spur takes place at a very late stage of the floral development. The flowers of *Pelargonium peltatum* similarly possess an axial spur. Hardly anything of this is to be seen externally however (Fig. 20 IV, V), because it is not formed as an independent structure, but is combined with the flower stalk over its entire length, a property which can be seen by comparing median sections II and III with the schematic IIa and IIIa in Fig. 19. If a cross-section of the pedicel is taken (Fig. 20 IV), the hollow tube of the spur which is cut through at the same time is immediately recognised, and a nectary is found at the base. What are generally taken to be sepal or petal spurs in the Vochysiaceae (see Scholz in Engler 1964, Warburg 1913/22 Vol. 2), with its prominently and obliquely zygomorphic flowers, are another case of axial spurs. In the comparable and obliquely zygomorphic flowers of the Chrysobalanaceae (Fig. 20 VI) on the

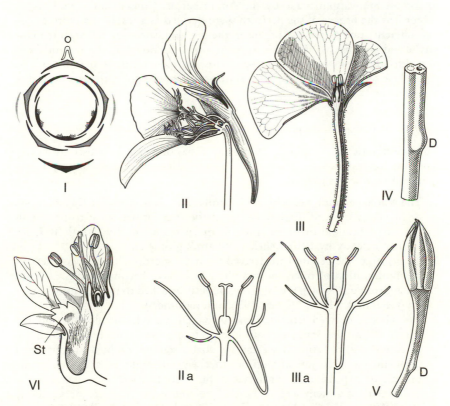

Fig. 20 I, II *Tropaeolum majus*. I Flower diagram (androecium and gynoecium omitted); II flower, in medial cross-section; II-V *Pelargonium peltatum*, III flower, medial cross-section, showing the hollow of the spur; IV base of a pedicel, with the cut spur tube showing above, and at D a swelling at the lower end of the spur tube, similarly in the bud which is about to open in V; IIa, IIIa diagrammatic explanation of the spur structure in *Tropaeolum* and *Pelargonium*. VI *Hirtella butayei* (Chrysobalanaceae), longitudinal flower section, St staminodes. I, II, IIa, IIIa, IV, V after Troll, VI after Engler.

other hand what is found is just a very one-sided and urceolate or tubular hollowing-out of the receptacle, where the gynoecium with its single carpel appears in a more or less elevated position on the margin of the cup.

1.4   The perianth

1.4.1   General

We have already stated at the beginning (section 1.1) that the stamens and the carpels, i.e. the sporophylls, are the essential components of the flower, and that the "primitive" flowers which occur in the pteridophytes and gymnosperms are composed of these elements alone. The creation of the perianth, together with the enclosure of the ovules in a secure protective cover – the single carpel, or an ovary formed by the fusion of several carpels – represent the two most important innovations made by the Angiosperms. Consequently we have to accept that the organs of the perianth were derived as a result of a corresponding differentiation of the leaf organs in the intermediate stage between trophophylls and sporophylls. We already mentioned (section 1.1) the fact that this perianth appears in two contrasting forms; as a **perigon** with more or less equivalent segments, or as a "**differentiated**" **perianth** with separate calyx and corolla.

1.4.1.1   *"Origin" of the perianth*

The question of the "origin of the perianth" has occupied botanists since early days, and has been answered very differently by different authors. The opposing points of view are as follows:

1. The whole perianth, i.e. calyx and corolla, is derived from the foliage leaves or bracts by upward displacement and the transformation of trophophylls towards a petaloid or prophylloid form (as essentially by Prantl 1887; see further Velenovský 1910, Glück 1919, among others);
2. the organs of the perianth are derived from the sporophyll-bearing structures by modification of microsporophylls which have become sterile, i.e from the androecium, which is in any case directly attached to the perianth further up (Čelakovský 1900, Němejc 1956, among others);
3. the calyx is derived from the bracts which precede the flower, and the corolla from the stamens; this interpretation was already proposed by Goethe (1790, p. 120): "If we can say that a stamen might be a contracted petal, then we can just as well say that a petal might be a stamen in the process of expansion; a sepal might be a contraction or a stage in the refinement of a stem leaf..." Further references for this are de Candolle (1813, 1817), Nägeli (1884), Drude (1887) and above all Troll (1928a);
4. the calyx is derived from the bracts, and the corolla from a different origin; here for example the points of view represented by Mattfeld (1838) for the Caryophyllaceae and other groups, and by Roth (1959, 1962a) for *Primula* and *Armeria* must be mentioned, in which the petals are considered as stipules or "dorsal leaf outgrowths" but "in no way as independent leaf organs"

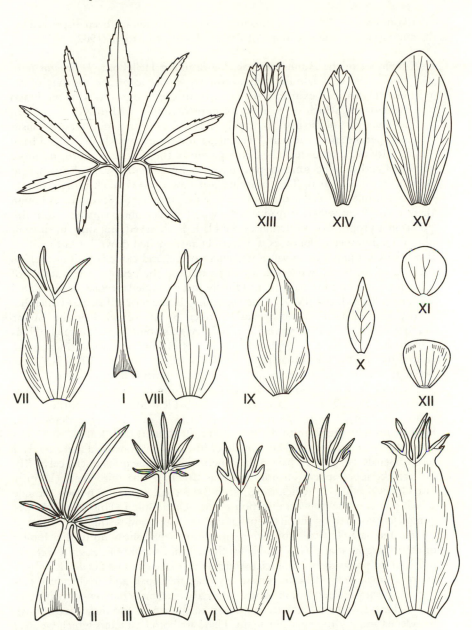

Fig. 21 I–XII *Helleborus foetidus*, leaf series. I foliage leaf, II–X bracts, XI, XII perianth segments (sepals). XIII–XV *Helleborus niger*, bracts (XIII, XIV) and corolloid sepal (after Troll).

(Roth 1962a p. 229). In the meantime this explanation has been disproved by, among others, the researches of Eckert (1966) and Sattler (1962).

A member of the Ranunculaceae, the Stinking Hellebore (*Helleborus foetidus*), will serve as a familiar and instructive example of the morphological relationships that can exist in a sequence from the foliage leaves to the bracts and the perianth, which we want to examine here (Fig. 21 I to XII).

The lower part of the plant is densely leafy. The lower stem leaves (I) consist of a long petiole and a pedately lobed blade and have a weakly developed basal sheath. Moving up the stem the development of the petioles is more and more restricted and they become shorter, until finally they disappear altogether, as shown by the leaf in II. The sheathing leaf base is now much more strongly developed, but the blade still shows the same number of divisions, although these are now less widely spread and the whole is very much smaller. From this point on a progressive reduction of the blade is observed, not only in size but also in the number of lobes, as if it were being "melted down". On the other hand the sheathing base expands more and more, and extends to join itself to the base of the blade, while still leaving a trace of the petiole. The number of lobes is next reduced to 3 or 2, until finally just a simple lanceolate and pointed leaf organ is produced (IV–X). The continuity in the series of shapes which has been preserved until now is abruptly broken by the rounded outline of the nearly cuneate perianth segments (XI, XII). At this point one occasionally finds intermediates between the highest of the bracts and the perianth segments; in a related species, the Christmas Rose (*Helleborus niger*, Fig. 21 XIII to XV), the bracts are commonly connected to the perianth segments by transitional forms.

The way in which some of the sepals from the flowers of the Dog Rose (*Rosacanina*, Fig. 22) are divided is still somewhat leaf-like. In a sense they show a close relationship with the pinnate upper leaves, but this segmentation has more to do with the form of the margins of the 'Unterblatt' (leaf-basis), as was observed by Leinfellner (1954a). On closer inspection it is realised that in only 2 of the 5 sepals, i.e. the first pair in the genetic spiral of the quincuncial calyx (II), do leaf-like lobes appear on both margins (III, IV). The third sepal (V) only bears lobes on the side which is closest to the second sepal, but not on the side which points in the upward direction of the genetic spiral. This latter side was completely covered by the first sepal in the bud stage, and is still partly covered by it at the base. The two following, innermost sepals show no further lobing of their margins at all (VI, VII). Hence the leaf-like nature of the sepals recedes more and more as the quincuncial sequence is followed. The petals of the rose are of a much more delicate consistency than the sepals and are brightly coloured, i.e. they are petaloid in nature. They are extremely narrow at their point of attachment to the receptacle, in contrast to the sepals, and they have a much broadened and slightly emarginate tip. These are both characters which we shall come across very frequently in the petals of the dicotyledons.

Sometimes a whole series of leaf-like organs may be found between the androecium and the petals of roses, intermediate in form between stamens and petals (Fig. 22 VIII to XI). This occurs occasionally in the Dog Rose and frequently in cultivated, double varieties. The stamens (XI) are, as usual, divided into a filiform stalk, the **filament** and a thickened terminal portion, the

anther. The latter comprises two halves called the **thecae** which are mirror images of each other and enclose the pollen sacs, i.e. two microsporangia. In the above-mentioned transitional forms between the stamens and the petals, there is often only one theca which is fully developed and fertile, while the other has become sterile and broadened out and runs down the filament in a wing-like shape (X). This wing-like structure has something of the colouring and consistency of a petal, and can also include both thecae (Fig. 22 IX). In such

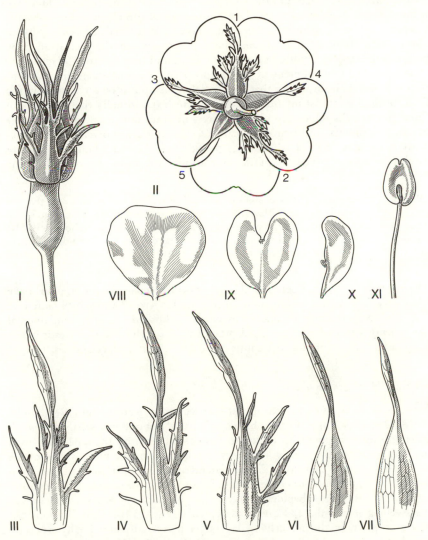

Fig. 22 Rosa canina. I Flower bud, II flower seen from below, III-VII sepals in their natural sequence (II 1–5); VIII petal, IX, X petaloid intermediate forms of the stamens; XI stamen (after Troll).

organs the out-growth of these wings is restricted by the influence of the anthers, whose vestigial callous-like remains are often still visible at the base of a more or less distinct sinus in the upper margin. This sinus is seen to disappear gradually as the petaloid organs further outward are examined.

What is the significance of the above observations, along with numerous others which have not been explicitly presented here? Are they evidence that e.g. the corolla originated phylogenetically by a petaloid transformation of the stamens, or did the calyx or the perigon, or parts of the same, originate by transformation of bracts and their shifting towards the sporophylls? Here we must refer back to the assertion made at the head of this chapter, that we have to accept that the organs of the perianth have been formed secondarily by an appropriate differentiation among the foliar organs in the transition zone between the trophophylls and the sporophylls, and that this took place at the time when the development of the Angiosperms began. Given this assumption it is not at all surprising that a close relationship can be recognised in the form of the innermost members of the perianth (i.e. usually the petals) with those sporophylls which are placed nearest i.e. the stamens, whilst the outer perianth segments, and hence the calyx too, generally show closer affinities with the bracts. Also Troll (1928a, p. 198) has warned against premature phylogenetic interpretation of the transitional forms in the region of the flower:

"When we are speaking of the 'origin' of the petals, that does not mean to say that the petals were 'originally' stamens. Nevertheless, they are *morphologically* identical to them. The question of the origin of the petals only refers to the purely morphological relationship between the petals and the neighbouring floral organs, hence the problem: are the petals nearer to the stamens or to the bracts in the way that these appear in the perianth as the calyx?"

The relationships of the petals (and partly also the tepals) to the stamens have been very thoroughly investigated during the last two decades by Leinfellner (1954a et seq.). In some instances this work established far more fundamental agreements than had been expected previously. In such circumstances it is also possible to venture phylogenetic interpretations in certain special cases.

1.4.1.2   *Histogenesis of the perianth*

A thorough analysis of the histogenetic processes by Kaussmann (1941), Boke (1948), Tepfer (1953) and others has revealed that not only the sepals, the petals and the perigon segments, but also the stamens, all resemble the foliage leaves in the way in which they grow. This is true not only for the initialisation of the organs but also for their longitudinal, lateral and superficial growth. Just as in foliage leaves, the initiation of sepals and petals is first visible in median sections, where the cells of the subepidermal layer elongate themselves in the direction of the swelling primordia and divide periclinally (Fig. 23 Ib). Cell divisions can also take place in the underlying layers. A periclinally-anticlinally dividing initial cell in the subepidermal layer (Fig. 23 IIb to IIIb, IV and section 1.5.2) is very soon observed, from which cell multiplication and longitudinal growth first proceeds. A short while later its activity is continued by a bilaterally dividing subepidermal apical cell, which produces new cells alternately on either side, which then themselves divide further (V); towards the end of the subapical growth this initial cell often produces new cells on one side only (V to VIb). In this way the mesophyll is built up, while the epidermis keeps pace with it by anticlinal divisions. Later on the dividing activity passes over from a subepidermal initial cell to an epidermal one, which likewise begins to divide bilaterally (VIb), until finally the further expansion of the leaf primordium continues by intercalary growth at its base or through cell elongation. The lateral development of the sepals and petals begins as usual from subepidermal marginal cells, and can be associated with a more or less generalised active division of the marginal epidermal cells (VII to IX). Initially a periclinal-anticlinal mode of division of the marginal subepidermal cells is observed here too (cf. VII, left-hand side).

A consequence of the scheme of marginal growth outlined above is **open venation**, which is believed to be a general characteristic of perianth segments. The procambial tissue can only be initiated in those parts of the perianth primordia, which have a mesophyll of at least two cells in thickness. However, the bilateral division of the marginal subepidermal initials often gives way at an early stage to unilateral division and finally even is continued to a more or less high degree by marginal epidermal cells. Hence the individual veins terminate with free ends at some distance from the margin (Fig. 24), without – as is usual in foliage leaves – joining up together again in a sort of network (Fig. 24 II). The open venation of the perianth segments is accordingly a sign of restricted mesophyll development, just as it is in the bracts and cataphylls (see Müller-Hoefs 1944). Since the mesophyll of herbaceous calyces is that much more strongly developed, the open nervature is less often apparent there. On the other hand, more or less conspicuous hyaline margins in both calyx lobes and bracts, resulting from more or less extensive epidermal growth of the margin, are quite commonly observed.

We have already pointed out that in many instances the petals have a very narrow point of attachment, and so differ from the sepals which are inserted on the floral axis with a wider base. The nervature of the sepals and petals is consistent with this: the sepals are usually trilacunar with three or more leaf trace bundles, and the petals unilacunar and provided with one bundle (Glück 1919, von Gumppenberg 1924, Kaussmann 1941 & 1963, Puri 1947 & 1951, Hiepko 1965, etc.). The broad-based perigon segments of monocotyledons are similarly often provided with three or more leaf trace bundles (Fig. 24 I, III). At the base of those perianth segments which are provided with one (or three) leaf trace bundles one often observes a more or less pronounced sympodial branching of the incoming vascular bundles (Fig. 24 VI, VII). In other respects the layout of the veins often very clearly reflects the properties of the longitudinal and lateral growth of the perianth segments. This is shown in the example investigated by von Gumppenberg (1924) of the large "standard" petal (see section 1.4.3.4) of the flower of the Bush Vetch, *Vicia sepium* (Fig. 24 VIII to X). In younger stages of development (VIII, IX) the way in which the development of the veins is directed towards the tip emphasises the vigorous growth there. Similarly the early onset of lateral growth (IX) is later continued in the distal region, while the proximal part elongates into a claw (X).

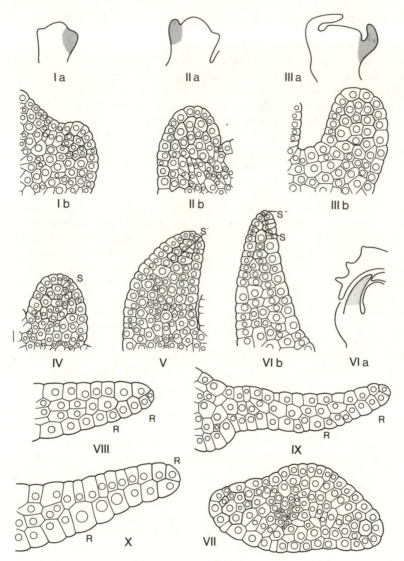

Fig. 23 Histogenesis of the sepals and petals. I-VI, VIII, X *Cleome gigantea*, VII, IX *Passiflora violacea*. I-III general views of axial sections through floral primordia in different stages of sepal development (Ia-IIIa) and detailed drawings (Ib-IIIb) of the shaded areas in Ia-IIIa. IV-VI apical growth of the petals, IV periclinal-anticlinal activity of the division of the initial cell S, V of a bilaterally dividing apical cell S′, VI unilateral division of the subepidermal (subapical) cell S and bilateral division of the epidermal (apical) cell S″ in an older petal primordium (shaded section in VIa). VII, VIII, X marginal growth in petals of various ages, R subepidermal, R′ epidermal marginal cells; IX cross-section through a sepal primordium. After Kaussmann, partly adapted.

1.4.2   Calyx and epicalyx

1.4.2.1   Forms of the epicalyx

We have already mentioned several times that the sepals often still exhibit a
more or less leaf-like texture, which is quite consistent with their protective

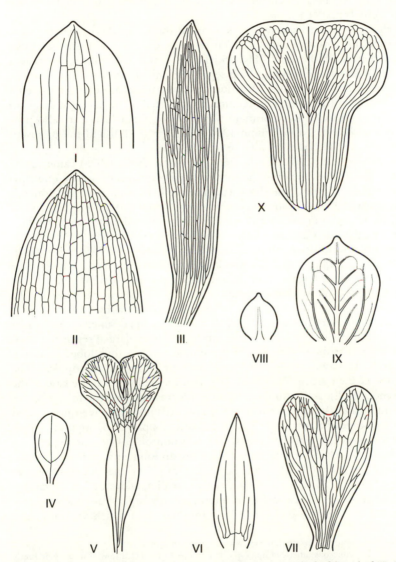

Fig. 24 Venation of petals. I, II *Clivia nobilis*, apex of a tepal (I) and of a foliage leaf (II); III
Canna indica, tepal; IV-VII petals of *Saxifraga hederacea* (IV), *Silene longiflora* (V), *Sedum acre* (VI)
and *Malva sylvestris* (VII); VIII-X standard of *Vicia sepium* in various stages of development
(after Kaussmann I-IV, VI, VII, Glück V and von Gumppenberg VIII-X).

function. They probably have a certain value for assimilation, which is to the advantage of the flower or the developing fruit. After all, the simple fact that they have a leafy texture is usually considered to be sufficient evidence of their foliar nature. In addition to this sepals are occasionally found which by their shape make it possible to establish more thoroughly the homology of certain parts with those of foliar leaves. This is especially true when the sepals are provided with stipules, which usually make their appearance collectively as a **stipular epicalyx**.

This character is particularly well-known in the Potentilleae, to which the strawberries and cinquefoils belong. These generally have five smaller lobes which alternate with the sepals (Fig. 25 III, IV). It is usually taken for granted that both the stipules of neighbouring sepals are united into a single "epicalyx leaf", as are interfoliar stipules. Occasionally two stipules are found between a pair of sepals (★ in III). This interpretation is supported by the fact that this kind of "stipular epicalyx" only appears in plants whose foliar leaves are also provided with stipules (cf. also Domin 1914). Bugnon (1929) held the opposing point of view that the epicalyx lobes might be bracts.[*]

While investigating deformed flowers[†] of *Geum*, Bolle (1935) came to the conclusion that in this case only the three outermost members of the quincuncial calyx possess stipules. The third sepal only has them on that side which is directed towards both the older sepals, similarly to the way in which the sepal margins are arranged in the Dog Rose. Hence although the five stipules which form an epicalyx alternate regularly with the five sepals, they are not to be understood as interfoliar stipules. The form of the calyx of *Rhodotypos scandens* (= *R. kerrioides*, Fig. 26 I to V) is especially informative, as already researched by Domin (1914) and again more recently by Schaeppi (1953b). This shrub, belonging to the Rosoideae-Kerrieae, occupies a special position in the Rosaceae on account of the decussate arrangement of its leaves. Similarly, in its four-lobed calyx two lower, outer sepals and two upper, inner sepals can be recognised. Usually only the two outer and somewhat enlarged sepals bear two subulate stipules at their base, so that the whole forms a four-lobed epicalyx. More rarely the two inner sepals are also provided with stipules. The comparison of these sepals (Fig. 26 III, IV) with the bud scales at the base of the seasonal shoot (I, II), and with the transitional forms between the foliage leaves and the bud scales (V) which occasionally appear, shows that we must indeed consider the somewhat narrower basal lobes of the sepals to be stipules. This is equivalent to saying, moreover, that the sepal is homologous to the hyperphyll ('Oberblatt'), which of course here shows no division into stalk and blade. Particularly interesting in this context is the calyx of the yellow-flowered *Fragaria indica* (Fig. 25 II). In this species five smaller, more or less inwardly-directed, broad-based, lanceolate lobes alternate with larger, patent and widely spreading, very narrowly-based, regularly toothed lobes which are not unlike

---

[*] For *Rhodotypos* Bugnon agrees nevertheless with Domin's interpretation.
[†] If, as above, the numerous other examples of "deformities" or "aberrations" can provide useful indications for the morphological interpretation of organs, then it is also generally true that in the evaluation of developmental aberrations one must proceed extremely critically. A summary of both the older and the more recent literature on the subject is found in Meyer (1966) as well as Dupuy (1963).

the bracts in shape. Domin (1914) took these larger lobes, probably on account of their prominence, as an epicalyx formed of stipules, although comparison with the epicalyx of a frondescent flower of *Potentilla aurea* which he also illustrates (Fig. 25 I), in wich the sepals are developed "in the form of green leaflets", suggests that this is another, extreme, case of a reverted sepal leaf-blade.*

At this point, exceptionally, we can go further than the general statement that the foliar organs of the flower and the foliage leaves are homologous with each other, and also homologise individual parts of these organs, a statement for which otherwise no adequate basis of comparison is given.

Among the "stipular" epicalyces we must distinguish between those which are composed of prophylls and those which are made up of a number of bracts. In the lateral flowers of the Hedge Bindweed (*Calystegia sepium*) there is an epicalyx formed from two inflated prophylls. Because of the shortness of the epipodium these are inserted directly beneath the flower and completely enclose the calyx proper, whereas in the Field Bindweed (*Convolvulus arvensis*) the prophylls are only weakly developed and separated from the flower by a long epipodium. *Thunbergia alata* (Acanthaceae), for example, behaves similarly to the Hedge Bindweed. A "genuine" epicalyx composed of several bracts is characteristic of, amongst others, the Passifloraceae (Fig. 12) and the Malvaceae (Fig. 25 V), where it consists of three to many, inconspicuous to strongly developed bracts. In *Hibiscus costatus* these bracts are peltate (Baum 1952b).

In the Dipsacaceae, and similarly in the closely related Morinaceae, a constant family character is provided by the epicalyx, which is closely associated with the flower and constructed from a whorl of prophylls with a further whorl of bract-like leaves. This can play an important role in the wind-dispersal of the mature fruit. The similarly related genus *Triplostegia* (Triplostegiaceae) even possesses a double epicalyx.

The so-called "involucre" is comparable to an epicalyx, as in *Hepatica* where it is positioned just below the flower. This is a case of a false whorl of three bracts (Fig. 27 IV), which is comparable to the three pinnate stem leaves, which in the Wood Anemone (*Anemone nemorosa*) are arranged in a false whorl in the middle of the peduncle, and which also appear in a similar form in the Yellow Anemone (*A. ranunculoides*), in the Pasqueflower (*Pulsatilla*) and in the Winter Aconite (*Eranthis*). In *Anemone hortensis* (V), these can often be placed far below the flower in forms intermediate between the pinnate stem leaves and the lanceolate bracts (Fig. 27 VII to XI), but can also be brought up close beneath the flower by the shortening of the following internode (VI). In many text-books an arrangement based on the above examples in a morphological sequence serves as a model for the origin of the calyx from the bracts. Hiepko (1965) was able to show, however, that the perianth of the genera *Anemone*, *Hepatica* and *Pulsatilla* is really a perigon (see further Hiepko 1975 and the literature cited there).

---

* Kania (1973) regards the epicalyx of the Rosaceae as a mere excrescence, since it "first appears after the sepal primordia become visible" (which however agrees with the usual properties of the stipules), and because it shows no prolepsis. The prolepsis of the stipules is not however an abso-lutely reliable criterion, especially not in the floral region where the growth relationships are often highly modified.

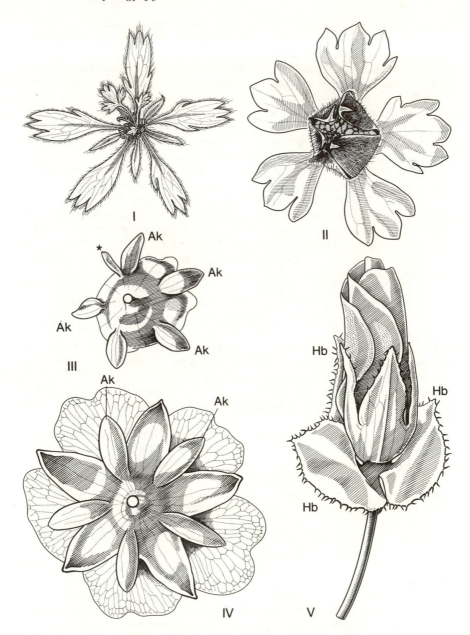

Fig. 25 Epicalyces. I *Potentilla aurea*, frondescent flower with a prominently leafy calyx (after Domin); II *Fragaria indica*, flower after anthesis; III, IV *Potentilla tabernaemontani*, flower without (III) and with (IV) petals seen from below, showing the calyx and epicalyx lobes (Ak), the latter "doubled" at ★. V *Malope trifida*, flower bud with calyx and "genuine" epicalyx of 3 leaves (Hb).

1.4.2.2   Form and function of the sepals

In as far as the sepals perform a protective function, this applies principally before the flowers open, but they can also continue to serve in this way during anthesis. The sepals are then often not fully reflexed during the flowering period, but only open partially. In many Cruciferae they even remain fully erect. Often the sepals drop off as soon as the flowers open (caducous), or they are not shed until after flowering (deciduous). By comparison with the foliage

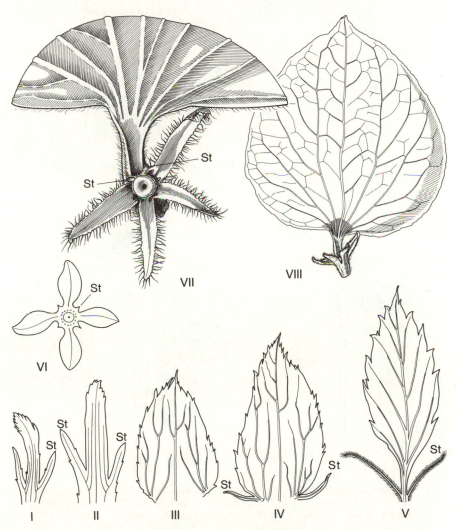

Fig. 26 I–V *Rhodotypos scandens (R. kerrioides)*, bud scales (I, II), inner (III) and outer (IV) sepals, transitional leaf between scale leaf and normal leaf (V), after Schaeppi. VI *Manettia inflata*, calyx of a flower after anthesis (after Leinfellner). VII, VIII *Mussaenda* cf. *erythrophylla*, calyx with a lobe formed like a leaf and coloured like a corolla, enlarged in VII. St Stipules.

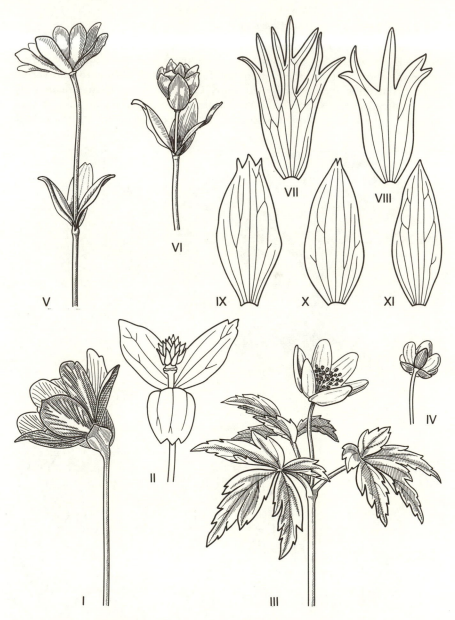

Fig. 27 Involucre. I–VI Flowers or flowering shoots of *Anemone transylvanica* (I), *A. angulosa* (II, after flowering), *A. nemorosa* (III), *A. hepatica* (IV), *A. hortensis* (V, VI). VII–XI different forms of the involucral leaves of *A. hortensis*, IX–XI belonging to the same involucre (all after Troll).

leaves the anatomical structure of the sepals shows a reduction of the mesophyll, which becomes that much more noticeable as the sepals become less leaf-like in form.

In the passion flower (Fig. 12) we have already seen an example of a hood-shaped sepal which bears what, by its outward appearance, seems to be a small process or "corniculation" on the outer side. This was interpreted by Troll (1932b, p. 171) and Baum (1951a) as a unifacial rudiment of the primordium of the 'Oberblatt' of the leaf (hyperphyll). Correspondingly, the flat section of the sepal is interpreted as the 'Unterblatt' of the leaf (leaf basis), whose upper margin, which runs transversely across the leaf, has an outgrowth as in a median stipule, so that the rudimentary 'Oberblatt' (hyperphyll) appears to be borne on its dorsal side.

This interpretation can of course only be confirmed if correspondingly constructed more or less foliaceous stem leaves or bracts show a transition to sepals of this kind, and definite homology of the individual parts can be established. Sepal processes of the kind described here "can of course by their structure and their position within the sepal correspond just as well to the completely reduced and undifferentiated Oberblatt as to the excurrent tip of this Oberblatt" (Baum-Leinfellner 1953, p. 593). The excurrent tips known from foliage leaves often exhibit unifacial, point-like terminal sections, whose growth is hastened on before the development of the corresponding lamina. In this case it is the cross zone at the boundary between the bifacial blade and the unifacial excurrent tip, whose growth can bring about the hooded shape. This character was established by Leinfellner (1952a) for the sepals of the Mesembryanthaceae. Here the sepals, as well as the foliage leaves, possess "a unifacial excurrent tip, which in the foliage leaves consists of an inconspicuous appendage to the bifacial Oberblatt, but which in the sepals produces the characteristic dorsal process of the hood" (p. 317).

Examples of the appearance of unifacial sepal processes and a more or less hood-shaped form of the sepals are known from a wide range of different families, as in the Onagraceae (*Oenothera, Godetia*), Geraniaceae (*Geranium, Erodium*), Caryophyllaceae, Vitaceae, Bignoniaceae (*Tecoma*), *Impatiens* species, *Ipomoea* species, and so on. Baum (1950, 1951a) refers to unifacial excurrent tips in all these cases. For the sepals of *Oenothera spectabilis, Impatiens balsamina* and other species in which several vascular bundles enter the tip, it can be shown by means of cross-sections that the arrangement of vascular bundles of the excurrent tips is completely in agreement with the existing view of any unifacial part of a leaf. This would of course also be true if the case of the unifacial point were seen not as an excurrent point, but as the equivalent of the whole Oberblatt. In the case of *Polygala myrtifolia* Leinfellner (1952b) was able to demonstrate the homology of the tip with the very inconspicuous excurrent tip of the foliage leaves.

Apart from the examples of unifacial excurrent tips Baum (1951a) also names many plants on whose sepals *bifacial* excurrent tips appear (*Cobaea scandens, Campanula pyramidalis, Trifolium repens, Alcea (Althaea) officinalis* etc.).

Further to this, Baum (1951a) and Leinfellner (1955a) even give examples of excurrent tips on the petals of polypetalous ('choripetalous' = with separate sepals) and sympetalous corollas, namely of unifacial ones (*Impatiens* species;

Cucurbitaceae: *Cucurbita, Cucumis, Lagenaria*; Convolvulaceae: *Quamoclit, Ipomoea*) and also bifacial ones (other Cucurbitaceae: *Bryonia, Echinocystis*; Solanaceae: *Datura*; Melastomataceae: *Bertolonia*; Plumbaginaceae: *Ceratostigma* etc.). Tepals may also possess unifacial excurrent points (Juncaceae; Liliaceae: *Lilium, Ornithogalum, Hemerocallis*; Amaryllidaceae: *Leucojum* etc.; Baum 1950).

1.4.2.3   Gamosepaly

In the context of the protective function of the calyx it must be realised that congenital fusion occurs much more often in the sepals than in the petals, although of course it may not always be very conspicuously expressed. This **gamosepaly**, which is often also observed in calyces with a quincuncial arrangement, is found to a greater or lesser degree in various families in which the petals are not united into a sympetalous corolla. This illustrates the fact that the systematic value of gamosepaly is not set too highly in comparison with that of sympetaly, although the fusion of the sepals is used as a character which clearly separates one of the subfamilies in the Caryophyllaceae from the rest.

Fundamentally gamosepaly can be compared with the gamophyllous development of the leaf bases in leaf whorls, which we meet with time and time again in the vegetative phase. However, this statement does not mean to say that the more or less well developed calyx tube which results from the congenital unification of the sepals corresponds to the parts of the 'Unterblatt' (leaf basis) of the foliage leaves, nor that the more or less well developed free lobes of the calyx margin (limb) correspond to its hyperphyll. We have already commented that often no adequate basis of comparison is given for recognising homologies in such detail.

The calyx tube can assume a great many different forms; for example it can be bowl-shaped (crateriform), bell-shaped (campanulate), tubular, inflated (ventricose), funnel-shaped (infundibuliform) or urn-shaped (urceolate). Moreover it is possible to distinguish between an actinomorphic and a zygmorphic calyx structure. In the latter case the calyx usually has a bilabiate form, in which the calyx lobes (usually in the ratio 2:3) forming the upper and lower lips often scarcely appear, or may be no longer recognisable at all.

1.4.2.4   Interlocking sepal margins

Apart from the gamosepaly which comes about through congenital unity, other, postgenital means have been observed by which both sepals and petals can become united. These consist of a more or less firm attachment between the margins of these organs during the bud stage.

In many flower buds the margins of the sepals and petals overlap, and this is sufficient in order to keep the young buds tightly closed. However, a quite special anatomical arrangement is often found which brings about a close and precise interlocking of the sepal or petal margins. This is what is called **interlocking of perianth segment margins by adhesion** ('Kontaktverschluß').

Sigmond (1930) distinguished two basic types of this through special attachment organs, namely 1. **dentonection** ('Verzahnung') and 2. **capillinection** ('Haarverschluß').

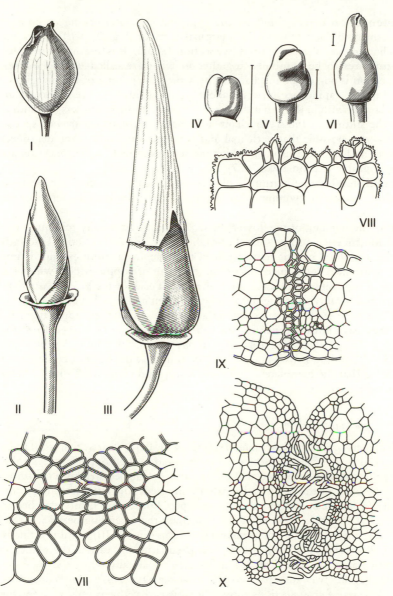

Fig. 28 I *Eomecon chionantha*, flower bud. II–VI *Eschscholzia californica*, II young bud after removal of the calyx, III flower before opening, where calyx has broken away at the base but still encloses petals, IV–VI stages in the flower development, IV after formation of the calyx, V, VI, intercalary growth demonstrated. VII–X adhesion-interlocking of petal margins (VII, VIII) or sepal margins (IX, X) VII *Galium odoratum*, VII *Vitis vinifera*; contact area of calyx; IX *Oenothera biennis*, X *Tilia platyphyllos*, valvate aestivation. VII–X after Sigmond.

Dentonection comes about by the engaging and interlocking of pointed papilliform epidermal cells on the marginal surfaces (Fig. 28 VII), or by the dovetailing of cuticular projections (see section 1.4.3.6). It is not uncommon for both structures to be combined together in what are called double sutures. Dentonection is one of the mechanisms by which the joining of the margins of the keel of the Papilionaceae comes about. Joins of this kind are by no means only to be met with in the perianth, but also in the joining of anthers (synandry) and in the sutures of the stylar channel (*Oenothera biennis, Campanula* species etc.). It is also possible that periclinal and anticlinal divisions of the dentonect cells occur and lead to a tissue in which the sutures caused by dentonection are no longer visible (Chrometzka 1967).

In **capillinection** the marginal adhesion of perianth segments comes about through the close intertwining of trichomes, as for example in the sepals of the lime tree (Fig. 28 X).

During the opening of the flower the sepals and petals usually relinquish this mutual attachment. In some special cases, however, the dentonection is so intimate that the floral organs remain joined together for a more extended time period, or are shed as one piece. Hence in the sepals of poppy species, which fall off early, it is possible to observe the epidermis of each of the broadly overlapping margins weakly adhering to the other. It often happens that at the time the flower opens the sepals are still coherent at their tips or by their upper margins, whilst they have already separated in their basal parts and been shed from the receptacle. The petals of *Vitis* are joined to one another by double dentonection (Fig. 28 VIII) and, as is well known, remain united at the tip, but come free at the base, so that the complete corolla is lifted up and thrown off like a cap by the expanding stamens.

1.4.2.5   Calyptrate calyces

The characters of *Eschscholzia californica* (Fig. 28 II, III) appear at first sight to be an advanced stage of the calyx characters observed in the Poppy species. In this instance the receptacle is broadened into a flat bowl-shaped structure on whose margin the bases of the petals and sepals are inserted. The latter also break free from the receptacle at the time the flowers open, but remain firmly attached to each other. The whole calyx is lifted upwards by the extending petals and rests on top of them for a while longer like a small cap (Fig. 28 III), until it is finally thrown off by the expanding petals. This peculiarity has earned the plant the vernacular name of "night cap". This character is usually explained by commenting that the "valvate sepals have fused at their margins" (e.g. Fedde 1936, p. 22). A series of cross-sections shows, however, that this is only true for the extreme tip of the calyx. It is at this point only that the sepals are separated from each other by closely interlocked epidermal tissues. The sepal tips are separated fully from each other a little further up. Over all its remaining surface the calyx has no trace of any joins or boundaries, but shows complete unbroken tissue. The study of the calyx development confirms this statement: at their formation the sepals appear clearly as two foliar organs, but a gamophyllous basal zone can already be recognised (Fig. 28 IV). It is this gamophyllous basal zone which subsequently grows out into a pear-shaped hood while the growth of both the calyx tips is greatly retarded (Fig. 28 V, VI). It thus displays no postgenital fusion, but genuine gamosepaly. An intensive intercalary growth of the congenitally unified basal zone of this calyx leads to the development of a so-called "calyptrate calyx". Sigmond (1930) had already noticed the "completely continuous nature" of the calyx of *Eschscholzia* and the lack of "dividing lines", whilst Baum (1951a) already remarks on the fact that the calyx of *Eschscholzia pulchella* is laid down as "a completely enclosed circular wall", and that the common basal zone of the sepals is greatly enlarged in the subsequent growth. The calyx of *Eomecon chionantha*, which also belongs to the *Papaveraceae*, behaves in a similar fashion (Fig. 28 I; Weberling, unpublished). According to Fedde (1936, p. 22), the sepals here "are fused into a helmet-shaped perianth, which splits open longitudinally". The fusion of both sepals into a calyptra has also been reported for the Himantandraceae. (For details on *Calycotome spinosa* see Goebel 1924, pp. 66/67.)

Calyptrate calyces similar to those of *Eschscholzia* occur in many members of the family Myrtaceae, especially the aptly-named genus *Eucalyptus* (see also Raciborski 1895). As our next example we shall take *Eucalyptus globulus* (Fig. 29 I, II). Here the 4-merous calyx is developed in the form of a thick, hard, lid, the so-called operculum, which comes completely away from the receptacle when the flower opens. Moreover, the four petals are also fused together. They stay behind as a short-lived inner cover which shrivels up during the expansion of the numerous, long, splendidly coloured filaments with which the stamens are provided. Hence we have here both a "calyptrate calyx" and a "calyptrate corolla" (Goebel 1924, p. 65 ff.) Apart from such bioperculate flowers, which according to Pryor and Knox (1971) are characteristic of the subgenus *Symphomyrtus*, there are also monoperculate flowers (subgenus *Monocalyptus*) in which the operculum is constructed from the calyx alone. Conversely, the so-called "inner operculum" (subgenus *Eudesmia*) can be constructed entirely from the whorl of petals while the calyx is greatly reduced, or there may be a less significant contribution from the sepals, of which then often only a pair is present, united with the inner operculum. However, the inception of individual calyx and corolla lobes at the beginning of the development of both the calyx and the corolla, which each arise from a common ring-shaped growing point, can be clearly seen in each case. The latter is greatly expanded during further development due to intercalary growth and finally comes to form by far the most significant part of the respective operculum. At the time the calyx opens only a narrow "opercular canal" remains, which can then become largely closed by further tissue growth. Whilst completely free petals and sepals can be found in the genus *Angophora* which is closely related to *Eucalyptus*, the sepals in other Myrtaceae (*Calyptranthes, Campomanesia, Psidium* species) are fused into a perianth which either tears in an irregular fashion when the flower opens, or splits off at the base; in *Syzygium aromaticum* (*Eugenia caryophyllata*), the clove tree, and *Eugenia* species it is the petals which are united into a cap and fall off together.

We can also observe the separation of the complete calyx in the form of an operculum in members of the Monimiaceae, as for example in the female flowers of *Mollinedia* species (Fig. 29 III, IV). On the other hand, in a species of the Bignoniaceae, *Enallagma cucurbitana* (Fig. 29 V, VI), "the calyx is fused into a perfectly spherical, smooth, fleshy structure, which shows no trace of any calyx teeth whatever" and which finally splits open irregularly when the flower opens. *Mitranthes langsdorffii*, a member of the Myrtaceae, behaves in a similar way (see Velenovský 1910, pp. 924/25).

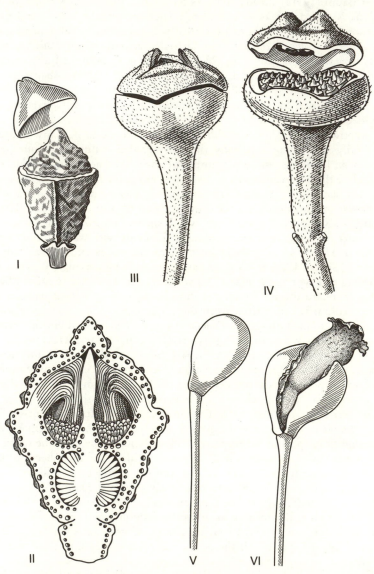

Fig. 29 I, II *Eucalyptus globulus*, general view of flower, with the outer operculum falling away (I) and in axial cross-section (II). III, IV *Mollinedia floribunda*, female flower with the operculum separating (III) and lifting off. V, VI *Enallagma cucurbitana*, closed (V) and open (VI) flower. I, II after Engler, III, IV original V, VI after Velenovský.

The peculiar phenomenon of a **water calyx** was first observed by von Treub (1889) in the buds of *Spathodea campanulata* (Bignoniaceae). In the closed state these are brimful of water liberated by hydathodes (water-secreting glandular trichomes) on the inner wall of the calyx. The corollas therefore develop, as it were, "in a bath of water". It is not clear, however, whether this has any biological significance. There do exist, moreover, entire inflorescences which develop in reservoirs of water, such as in the genera *Nidularium* and *Vriesea* of the Bromeliaceae, where they are often submerged in the water stored in a funnel of leaves. Further details of water calyces and similar phenomena are given by Koorders (1897) and Pascher (1960). These occur in very different families, such as the Bignoniaceae, Solanaceae, Scrophulariaceae, Verbenaceae etc., and also in *Aconitum variegatum* (Molisch 1920).

1.4.2.6 Change of function in sepals

In *addition to* its protective function, the calyx quite frequently serves other purposes as well. It is often seen that the calyx intensifies the display effect of the corolla by being brightly coloured. As an example of this we may quote the popular plant of flower borders, *Salvia splendens*, whose calyx, corolla and all the axes and bracts of the inflorescence are coloured bright red. In *Clerodendrum thomsoniae* (Verbenaceae), which is often cultivated in greenhouses or as a house plant, the greenish-white, inflated calyces contrast vividly with the dark red corolla. The spreading and deep red-coloured calyx of *Holmskioldia sanguinea* (Fig. 30 VII) undoubtedly serves for display; other well-known examples are *Molucella laevis* (Labiatae) and the genus *Petraea* of the Verbenaceae.

In not a few genera of at least five different tribes of the Rubiaceae individual sepals within the inflorescence are transformed into large white, or vivid yellow or red-coloured display organs, which considerably increase the conspicuousness of the inflorescence by being always placed around its periphery (see Weber 1955). These enlarged sepals are often differentiated into petiole-like and blade-like sections (Fig. 26 VIII), so that recognition of homology with the petiole and blade of the foliage leaves then seems to be justified, since rudimentary stipels are clearly visible at their base (Fig. 26 VII). Unequal growth of the sepals (heterosepaly) is also found in several other genera of the Rubiaceae (e.g. *Pentas*).

In the genus *Hydrangea* the display effect of the inflorescence is achieved by the petaloid development of the complete calyx, which normally only occurs in certain sterile flowers situated at the tips of higher-order branches. The corolla in these flowers, which are placed at the edge of the inflorescence, remains quite inconspicuous (Fig. 30 XI), but it also stays quite small in the fertile flowers, in which the calyx lobes are now only recognisable as tiny teeth (VIII to X). In the inflorescences of the "double" garden cultivars practically all the flowers are modified in the manner illustrated, with the result that the display effect of the inflorescence is emphasised even more, although the flowers are sterile.

In general therefore a transfer of the attraction function to the sepals is often observed when the corolla or some of its parts have either been much reduced or entirely suppressed (secondary apetaly). A step by step reduction of the petals and the adoption of their function by the sepals and even by the receptacle too is very well demonstrated by members of the Thymelaeaceae, of which three species of *Gnidia* may serve as an example (Fig. 30 I to III). The flowers of *Gnidia spicata* correspond basically to the flowers of the more familiar Spurge Laurels *(Daphne)*. In the Polygalaceae the flowers are reminiscent of the "butterfly flowers" of the Papilionaceae, but without the two "wing" petals of the papilionaceous flower. These have been "replaced" by the two corresponding transversally-placed sepals which have a petaloid form.

Occasionally the sepals also bear extranuptial nectaries or elaiophores, as often for example in the Malpighiaceae, where they can even be greatly transformed for this purpose (Fig. 30 IV to VI). In the Lime trees *(Tilia)* and in the Mallows *(Malva)* the nectaries occur on the upper side of the sepals. In species of *Impatiens* (Balsaminaceae, Fig. 31) they are borne in more or less deepened hollows in the sepals or even in a sepal spur, which is constructed from a median sepal and contributes to the zygomorphic form of such flowers to a considerable degree. These sepals occur in a great variety of forms and exhibit spurs of very different shapes and lengths (Fig. 31 V–VII).

The position which these spurred sepals take up in the unfolding flower comes about just before anthesis through a rotation of the pedicel by 180° (resupination, see section 1.4.3.4). This sepal is originally orientated towards the principal axis, as can be seen in the theoretical diagram in Fig. 31 III, which also indicates the position of both the anterior sepals for *Impatiens wallerana* (*I. sultani*), which are missing in many species (still present in *I. glandulifera*). The empirical diagram in IV shows the characters encountered in the open flower (I), whose calyx is reproduced separately in II.

Delphinium is different from *Impatiens* in that the production of nectar does not take place in a sepal spur. On the contrary, in this case two long-spurred nectar-leaves project into a long spur which is, once again, constructed from the median sepal, whereas in Monk's-hood *(Aconitum*, Fig. 47 II) two long-stalked nectaries are sheltered inside the single helmet-shaped sepal which corresponds to the spur of *Delphinium*.

The sepals of the Wild Pansy, *Viola tricolor* (Fig. 31 VIII), and other species of *Viola* bear at the base and at the back a short, broad and solid spur, which probably originates through the activity of a dorsal meristem (see Roth 1957a & b, also Goebel 1933, p. 1875; and further Troll 1949).

1.4.2.7   Persistent calyces

We have already drawn attention to the known fact that the sepals often fall off when flowering begins or after flowering. However, particularly in flowers with inferior or semi-inferior ovaries, and also in not a few flowers with superior ovaries, the calyx is preserved after flowering. Such persistent calyces often undergo even further development. This can consist mainly of a considerable enlargement (accrescent calyx) as in Primroses or in *Alectorolophus* and particularly so in the Solanaceae, in which it can reach an extreme degree, and in which "all possible modifications of calyx enlargement" are repeated, "which otherwise occur singly in individual families" (Pascher 1910a, p. 268).

In *Przewalskia tangutica* the calyx achieves a 14- to 17-fold increase in length and a 3000-fold increase in volume, according to Pascher. An enlarged fruiting

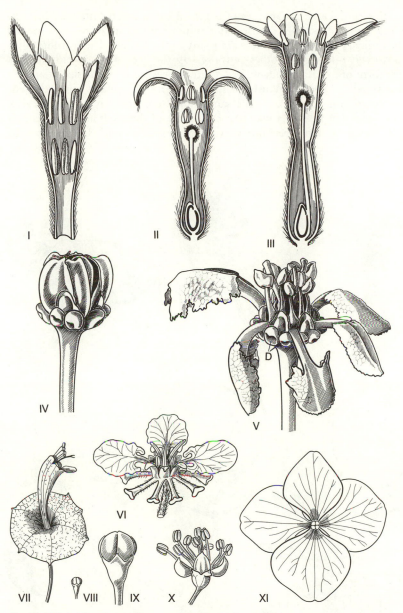

Fig. 30 I–III Axial sections of flowers of *Gnidia* species, showing a progressive reduction of the corolla, I *G. stuhlmannii*, II *G. spicata*, III *G. carinata*. IV, V *Malpighia punicifolia*, bud (IV) and open flower (V) with paired glands (D) on the underside of the sepals. VI *Dinemandra glauca*, flowers with stalked nectary glands on the sepals. VII *Holmskioldia sanguinea*, zygomorphic flower with spreading, salverform calyx. VIII–XI *Hydrangea hortensis*, VIII–X fertile flowers in bud (VIII, IX) and open (X), the flower in VIII is on the same scale as that in XI. I–III after Gilg, VI after Engler, VII after Velenovský, VIII–X after Troll.

calyx of this kind with a surface area of between 200 and 300 cm² if it is leafy in nature may at times acquire even greater importance as an assimilation organ than the ordinary leaves, which here are many times smaller. The fruiting calyx here, however, fulfils yet another purpose. By the end of its development it takes the form of a "rather thin-walled, ellipsoid vesicle with the size of a hen's or goose's egg". "The previously delicate network of veins is strengthened, becomes tough and strong". At the same time "the similarly enlarged calyx teeth bend together in such a way that the escape of the seeds after this frontal closure of the calyx orifice becomes very much more difficult". The majority of

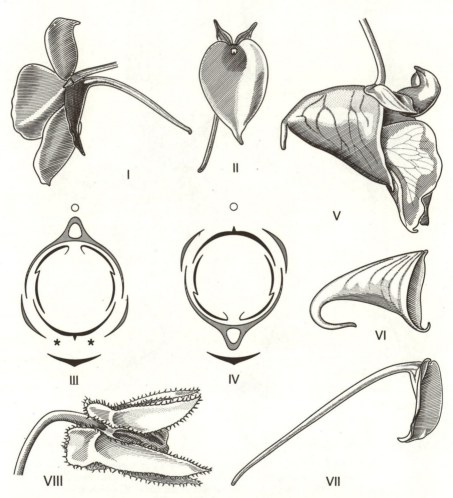

Fig. 31 Sepal spurs in *Impatiens* species (I-VII) and *Viola* (VIII). I-IV *Impatiens wallerana*, flower in side view, II calyx viewed from above, III, IV floral diagram before (III) and after (IV) resupination. V *I. glandulifera* (*I. roylei*) side view of flower; VI, VII spurred sepals of *I. nolitangere* and *I. wallerana*; VIII *Viola tricolor* ssp. *arvensis*, calyx at the fruiting stage of the flower. I-VII after Troll.

the seeds, which fall out of the circumscissile capsule into the empty space of the calyx, are finally released in quite a different fashion. The sections of tissue between the rope-like, thickened veins of the fruiting calyx dry out more and more, become ever thinner and more papery and at last break away completely from the reticulate calyx veins, "so that all that remains is the lattice of the vein network". Since the fruiting pedicel withers, the fruit breaks gently away, and is rolled and whirled across wide stretches of the steppe (of North China) by the wind, as light as a balloon, so that its seeds can be scattered far and wide (Pascher 1910b). Similar "lattice calyces" are also developed by several other steppe-dwelling species of the Solanaceae.

The bladder-like fruiting calyces of the Chinese Lantern (*Physalis*, Fig. 32 IV, V) contribute to the dispersal of the fruit in another way. This time they are conspicuously coloured and can therefore serve to attract animals, which distribute the seed by consuming the fruit. Bladder-like, inflated calyces which permit the fruit to float and to be transported by flowing water are found in *Hernandia ovigera* (Hernandiaceae) according to Ridley (1930, cited by Stopp 1952); these are termed floating fruits.

The postfloral enlargement of the calyx of the Solanaceae contrasts with another function in other genera, as in *Anisodus* and *Atropanthe*, where the calyx is already greatly enlarged before anthesis, and "serves to protect the small, undeveloped flower buds" (Pascher 1910a).

The development of fleshiness in the calyx while the fruit is ripening can also contribute to distribution of fruits by animals (endozoochory), as Stopp (1952) has described for the Labiate *Hoslundia decumbens* (Fig. 33 XI). Persistent calyces can assist with distribution by animals by developing hooks, thorns or glandular hairs, which readily fasten the fruit to the coats of passing animals (burrs or sticky fruits). Examples of this are *Phryma leptostachya* (Phrymaceae), Madwort (*Asperugo procumbens*, Fig. 33 VI) and the Leadwort, *Plumbago capensis* (= *P. auriculata*, Fig. 32 I, II). See also the commentary by Stopp (1952, 1958).

Graphic examples of the development of a persistent calyx for distribution by wind (anemochory) are the winged calyces of the Dipterocarpaceae (notice the name!), in which some or all of the sepals elongate to form wings as the fruit ripens (Fig. 32 VI), or the shuttlecock-like calyx of many members of the Plumbaginaceae (*Armeria*, Thrift; *Limonium*, Fig. 32 III), and the genus *Triplaris* of the Polygonaceae. Also many genera of the Rubiaceae (*Otiophora*, *Otomeria*, *Cruckshanksia*) have enlarged, persistent calyces to aid wind dispersal, and in *Jacksia*, *Alberta*, *Nematostylis* and *Gaillonia* these calyces first begin to enlarge as the fruit ripens.

The variety of different forms of pappus developed in the Compositae is also very familiar (see especially Lund 1874). It can be developed as a crest of simple or pinnate bristles or a "parachute on a stalk" (Fig. 33 I). In the genus *Centranthus* (Valerianaceae) and in many *Valeriana* species this pappus is recognisable during anthesis as an inrolled ring-like swelling which first expands as the fruit ripens (Fig. 33 IV, V). In *Valerianella* species it may occur only as a ring, or as various kinds of calyx adapted to wind dispersal, such as in the form of a crown (*V. coronata*, Fig. 33 II), or as a spreading star (*V. discoidea*), or as an inflated calyx (*V. vesicaria*, Fig. 33 III). In other cases the calyx develops hooked bristles for dispersion by animals (epizoochory, see section 3.4.2). In

V. uncinata (Fig. 33 VII), this is connected with an unequal development in different parts of the infructescence; especially noteworthy is that in *Fedia*, where it leads to conspicuous heterocarpy (Fig. 33 VIII to X; see Weberling 1970).

There are further cases of persistent calyces in the Boraginaceae, Labiatae and the Verbenaceae, which "exhibit dehiscence phenomena at the time the seed ripens, which appear to be analogous to what occurs in the pericarps of capsules" (Stopp 1958a). This is particularly true for those calyces, e.g. in *Aeolanthus*, which can be compared with a circumscissile capsule (pyxidium).

1.4.3   The corolla

The petals are distinguished from the sepals in general by their more delicate texture and typically also by the development of a considerably greater surface area. This is equally true whether we are dealing with a polypetalous (choripetalous) corolla or a sympetalous one. Moreover they are as a rule white (because of numerous intercellular spaces) or variously coloured, and constitute the most conspicuous part of the flower, evidently representing what might be called "display equipment". We have of course already established that this function

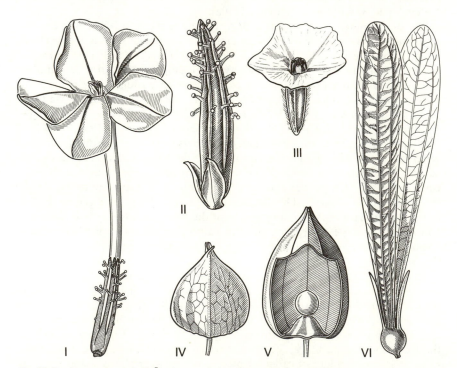

Fig. 32 Persistent calyces. I, II *Plumbago auriculata*, flower (I), fruit (II). III *Limonium mucronatum*, shuttlecock-like fruit with enlarged calyx. IV & V *Physalis alkekengi*, fruit with inflated, reddish-orange calyx, in V inside of calyx seen by cutting out a window. VI *Anisoptera curtisii* (Dipterocarpaceae), fruit. IV, V after Troll, VI after Gilg.

Fig. 33 Persistent calyces. I *Tragopogon orientalis*, long-beaked achene, calyx in the form of a pappus with pinnately-branched rays; II-V achenes of *Valerianella coronata* (II) and *V. vesicaria* (III) with crown-shaped and inflated calyx, IV, V *Valeriana tuberosa* with pappus still folded up (IV) and undeveloped (V). VI *Asperugo procumbens*, calyx developed as a sticky burr; VII *Valerianella uncinata*, achene with calyx bearing hooks, VIII-IX *Fedia cornucopiae*, VIII Longitudinal section through the hypopodium of a dichasially-branched partial inflorescence with the fruiting primary flower (P), which is fused with the hypopodium but still bears calyx teeth, IX achene from the monopodially-branched section, X Achene from *Fedia caput-bovis* with a 3-hooked calyx. XI *Hoslundia decumbens*, fruit with fleshy sepals. VI & XI after Stopp, VII after K. Müller.

can occasionally be taken over by the sepals, which are then modified accordingly. However, in such cases it is still true to say that the sepals are attached to the floral axis by a broad base, whereas the petals, in polypetalous flowers at least, are often characterised by being very narrow at the point of attachment. Nevertheless, there are many exceptions to this, especially among monocotyledonous plants.

1.4.3.1   *Anatomical structure and colouring of the petals, flower scent*

In keeping with their more delicate texture, the perianth segments often have an anatomical structure which differs rather more from that of the foliage leaves than that of the sepals. The division of the mesophyll into palisade and spongy parenchyma is usually lacking (Fig. 34 I, II). The cells of the mesophyll are usually rounded in cross-section, or, if the intercellular spaces are very extensive, they can be irregularly branched with several arm-like outgrowths (Fig. 34 II). The epidermal cells on the upper side are often elongated to form papillae, which accounts for the velvety upper surface of many petals. The anticlinal cell walls are often highly undulate, and can produce ridge-shaped processes as a result of intensified and localised centripetal growth. These can either be solid (Fig. 34 III) or appear as more or less clearly defined folds (Fig. 34 IV). According to Kaussmann (1963, p. 485), this folding of the membrane may be regarded as a mechanical strengthening of the cell walls, which would not otherwise be so rigid. As an extension of this phenomenon intercellular spaces even appear in the upper epidermis of the petals of certain plants (Hiller 1884), in contrast to the unbroken pattern of interlocked cells in the structure of a normal and characteristic epidermis. They are usually covered with a cuticle on the outer side of the epidermis, but correspond on the inward side with the intercellular spaces of the mesophyll (Fig. 34 V). In special cases, such as in the pitfall flowers of *Ceropegia* species (Vogel 1961), and also in the spathe which encloses the inflorescence of Lords and Ladies (*Arum* species, Knoll 1923, and see section 3.3.4.2c) genuine perforations appear in the epidermis on the inner surface of the organ.

By way of contrast to the velvety upper surface of many petals we have the phenomenon of **glossiness**; this comes about when the outer surface of the epidermis is not only entirely smooth, but covered with a layer of fatty oils (e.g. petals of *Ranunculus* species, orchid flowers, *Dendrobium*).

Stomata are generally absent from the epidermis of petals, which once again demonstrates the non-leaflike character of these foliar organs. This is not to say, however, that petals may not be provided with chlorophyll. In fact, this is quite often the case, especially in the early stages of the development of the corolla. In such a condition petals or tepals possess a pale green coloration, such as in the Hyacinth, whose flowers later fade and turn white. This is because the chlorophyll content is not increased as the organs enlarge and unfold, and may perhaps even be destroyed, so that the light is then reflected from the numerous intercellular spaces which the tissues contain. Leucoplasts are often found in petals instead of chloroplasts, and these often originate secondarily from chloroplasts; the Hyacinth is also an example of this. In red- or blue-flowered Hyacinths the colour is due to anthocyanins dissolved in the cell sap, which give their colour

to the flower after the disappearance of the chlorophyll. The same is true for the
Tulip, in which the colour of the young flower buds is similarly brought about
by chlorophyll. In this case, however, there is plastid pigment (a carotenoid)
dissolved in the cell sap besides the anthocyanin, which also contributes to the
colour of the flower.

The coloration of flowers can thus be brought about in different ways, of
which three can be distinguished: 1. the so-called chymochromes, i.e. colours
dissolved in the cell sap, namely the flavone- or flavonol-glycosides, the anthocyanins (including their colour-giving complexes of aluminium and iron) and
the nitrogen-containing betalains, which are characteristic of the Centrospermae (with the exception of the *Caryophyllaceae*); 2. the plasmochromes, i.e.
yellow and red carotenoids which exist in the plastids, especially the chromo-
plasts; these are responsible for various yellows, orange-yellows or orange-reds
which also produce various mixed colours in conjunction with chymochromes
such as the blue anthocyanins, e.g. the brown colour in Wild Pansy; 3. by
pigment deposited in the cell walls. The intensity of the colours can be increased

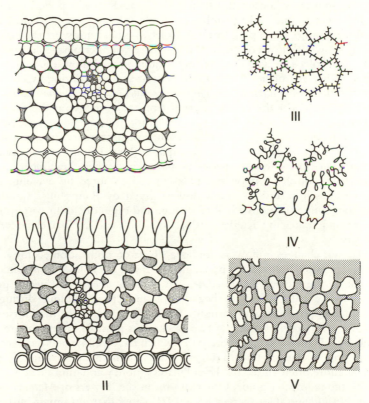

Fig. 34 Anatomy of petals and tepals. I, II Cross-sections through petals of *Tulipa kaufmanniana*
(tepal) and *Viola tricolor* f. *maxima*. III–V View of the epidermis of petals of *Pelargonium* sp. (III),
Anchusa officinalis (IV) and *Linum usitatissimum* (V). After Kaussmann (I, II) and Hiller (III–V
adapted, after Kaussmann).

(as in water-colours) if the light incident on the petal is reflected by air-bearing mesophyll after passing through the coloured layers, and hence passes back through these layers again. This "cat's-eye" effect can be further increased by reflection of the light from cells which are rich in starch granules.

Plasmochrome colours also make their first appearance after the disappearance of the chlorophyll, as can be clearly observed in the flowers of *Forsythia* in the spring. Indeed, the fatty acids which are liberated by the decomposition of the chloroplasts often combine with the xanthophylls which are also present in the chloroplasts to produce the so-called wax colours, which occur widely as pigments, as for example in *Physalis* and *Helenium*, among others. There is a correspondence between processes such as this which depend on decomposition for the production of pigments with the processes which take place in the leaves when they turn yellow in autumn. This makes it clear that in general the blossoming of the flower, perceived as the climax of the development of the plant, is closely linked with processes of decomposition (autolysis). For evidence of this the comprehensive survey of Matiles (1977) on the investigation of processes of synthesis and decomposition during and after anthesis in an *Ipomaea* flower should be consulted. Often a loss of acids over a period of time in the cells of petals coloured by anthocyanin, i.e. a change of pH-value, leads to a **colour change** from pink through violet to blue. Lungwort (*Pulmonaria officinalis*) is a well-known example of this, where younger pink-coloured and older blue- to violet-coloured flowers can be found in the inflorescence all at the same time; other members of the Boraginaceae are also distinguished by similar colour changes (*Lithospermum purpureo-caeruleum, Symphytum, Echium*), and in *Myosotis versicolor* the colour changes from pale yellow at the time the buds open to dark blue in old flowers. In *Lathyrus vernus* the flowers are first red and then turn bluish-green. The colour change of flowers of *Hibiscus mutabilis* was described at an early date by Rumphius (Herbarium amboinense, Utrecht 1743). The flowers of this species are pure white when they open in the morning, pale reddish by noon and pink- or red-coloured by evening; hence the name "flos horarius" (= hour flower) given by Rumphius. Significant and thorough investigations of colour change and types of colour pattern in flowers were undertaken by Kugler (1936, 1952 etc.) and by Vogel (1950), among others.

Colour changes in flowers can have effects on pollination ecology, as shown in the example investigated by Kugler (1936) of the colour change in the nectar guide of the Horse Chestnut (*Aesculus hippocastanum*). The nectar guides (see below) are pale yellow to begin with and change colour during anthesis through yellowish-red to carmine. Since the "candles" of the Horse Chestnut continue flowering for one or two weeks, it is possible to find different flowers side by side on the same inflorescence which have yellow and red nectar guides. The pollen- and nectar-providing flowers with the yellow nectar guide are much more often visited by the pollinators (honey and humble bees) than those with the red nectar guide. Observations on the flowers of *Malvaviscus arboreus* var. *penduliflorus* (Gottsberger 1967, 1971) show that humming-birds also react to colour changes in flowers.

It is not in every case that colour changes are caused by alteration of the pH-value; it may also happen that the synthesis of different pigments proceeds

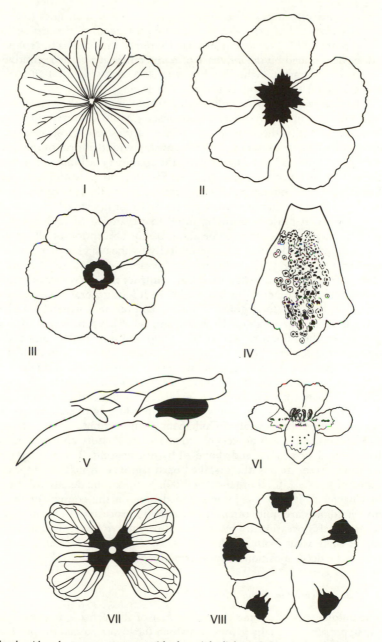

Fig. 35 Floral guide colour patterns as seen with ultra-violet light, in I *Geranium pratense* (linear patterns), II *Cistus villosus* (basal blotches), III *Androsace chamaejasme* (basal ring), IV *Digitalis purpurea* (speckled on the upper side of the lower lip), V *Linaria vulgaris* (dorsiventral blotch, on the palate), VI *Tozzia alpina* (spotted), VII *Raphanus raphanistrum* (basal blotch and linear patterns combined), VIII *Nemophila maculata* (marginal markings). After H. Kugler.

during particular stages of anthesis to give different intensities (see Harborne 1976). Flower colours can also be altered by temperature changes, as for example in Stork's-bill (*Erodium*) or in Lilac (*Syringa*); in Petunias the colour pattern can be determined by the influence of temperature at a certain sensitive stage of flower development (Schröder 1934; see also: Suessenguth 1936, 1938, Lüke 1943, Seybold 1954, Kisser & Hauer 1955, Reznik 1956).

The total range of flower colours runs from red to the ultra-violet which, as is well-known, can be perceived by many insects. Multi-coloured corollas are quite frequent, in which the arrangement of the various colours, shades and patterns can usually be seen to have a close relationship with the form of the flower. The different colour shades and particularly the coloured markings and patterns point the way to the sources of nectar (see Fig. 35), albeit in quite different ways. These **floral guides**, or **nectar guides** in nectar-secreting flowers, or pollen guides in flowers without nectar, provide optical signals for insects which visit and pollinate flowers. It must be borne in mind that many of these floral patterns are invisible to man, since they come about by the absorption of ultra-violet; they can, however, be detected photographically by the use of certain filters (Fig. 35, see further Kugler 1963, 1970 and the literature cited there). In this case also changes in the characteristic pattern can influence the behaviour of insect visitors (see for example Jones & Buchmann 1974). It is of general biological interest to point out that the floral guides are often laid out in a generalised plan which extends over some or all of the petals, and what happens is that the individual patterns on these floral organs are combined into an overall design as the flower opens. In the flowers of Nasturtium (*Tropaeolum majus*) the pattern of lines which point inward towards the nectar-bearing spur even overlaps from the petals onto those surfaces of the sepals which are visible between the petals. This is reminiscent of the patterns sometimes seen on butterfly wings, which develop separately on each of the four wings, and which only produce an overall pattern when the wings open out (the so-called Oudemans phenomenon).

The methods which flowers use to attract pollinators optically are enhanced by chemical attractions, and above all by flower scent. This is often discharged by certain scent glands, the so-called **osmophores**, which can appear on very different kinds of floral organs (Fig. 186). Some of the details will be discussed in Chapter 3. It will just be mentioned here, in the context of the fact that flower colours are quite often determined by processes of decomposition, that scent can also occasionally be attributed to autolytic processes, as for example the hay-like scent from the flowers of Sweet Clover (*Melilotus*) is due to the enzymatic decomposition of coumarin glycosides.

1.4.3.2   Development and life span

The unfolding of the petals or tepals is first of all a process of growth, in which the upper surface usually grows more than the lower, so that the petals are more or less curved outwards. In the flowers of the Tulip or of the Crocus it is well known that temperature-dependent growth movements produce repeated opening and closing. Moreover, turgor movements can also play a part in this. The petals of many species of *Silene* roll up inwards during the day, which is made possible by a loss of turgor on the upper surface of the petals and a corresponding support by the vascular bundles which are shifted to the lower side. Goebel (1924, p. 75) already pointed out that the inrolling of the petals does not take place if there is an adequate supply of water via atmospheric humidity. It is important that such movements must be related to the general symmetry of the flower, or in the case of the heads of the Compositae, with the symmetry of the inflorescence. This is particularly true for the many zygomorphic flowers, in which active or passive articulation zones occur, such as in the labellum of the Orchidaceae (see Goebel 1924, p. 79 ff.).

The life span of petals or tepals is variable. They often drop off after a shorter or a longer time without even withering (*Rosa, Geranium, Linum, Aquilegia, Canna*), or they fall in a withered or dried state, in which case they drop off at the point of insertion (*Lilium, Tulipa*, Cruciferae), or somewhat further up, leaving a basal part behind (*Alcea, Datura, Nicotiana*). The separation may be effected by a special abscission zone, in which the abscission layers are usually only slightly differentiated and where no cell division activity can be recognised. The walls of these cells can become impregnated with fatty substances to form an abscission scar. Similar processes can also be observed in the shedding of the stamens or the style. (In plants with unisexual flowers, complete male flowers are often shed after anthesis, as are the catkins in catkin-bearing plants; the same often happens to the unfertilised flowers of an inflorescence). It is not unusual for the petals to remain in a dried state until the fruit ripens, so that in the process they surround the fruit to a greater or lesser degree, as for example in species of Clover. The tepals of Langwort (*Veratrum*) or of Herb Paris (*Paris*) become greener after anthesis and finally form part of the fruit.

We have already discussed the fact that in some cases the sepals may become fleshy, especially in the context of the ripening of the fruit, and can play a part in its dispersal by animals. A change of function of this kind can also occasionally be found in petals, as for example in the genus *Coriaria* (Coriariaceae, Fig. 37 VII to X). The five small fleshy petals of the greenish flowers are keeled on the inner surface and after flowering they grow larger and force their way in between the free carpels so that a berry-like compound fruit is formed.* A further example is provided by the inconspicuous perigon of the Mulberry (*Morus*, Moraceae), which consists of two whorls of two segments. These surround the gynoecium which encloses just a single seed and likewise become fleshy and juicy as the fruit ripens; as is well-known the complete inflorescence gives rise to an infructescence which is known as the "mulberry", which superficially resembles a blackberry.

1.4.3.3   Types of petals

The outline shapes of petals are so varied that a thorough account of them would be just a useless enumeration. However, it is worth mentioning that in flowers with free petals the narrowed base of the petals where they are attached to the floral axis can be elongated into an attenuated, proximal section which is termed the **claw**. Frequently, as for example in many Pinks (Caryophyllaceae-Silenoideae, Fig. 36 I to III, VII to IX, XII), the narrowed section, which is

---

* Distribution by birds (blackbirds) has been reported from Italy.

usually enclosed by the calyx, is sharply differentiated from the freely spreading terminal section, so that one can distinguish between the **claw** (unguis) and the **blade** (lamina). During the ontogeny of petals which are constructed in this way, the blade usually develops far faster than the claw, which then elongates more vigorously as the flower opens (Fig. 36 VIII, IX). The fact that there also exist petals with a broad basal portion (cap-shaped in the example) and a relatively narrow distal portion is shown by *Guazuma ulmifolia* (Sterculiaceae, Fig. 36 V). Here the basal portion develops faster than the distal (Leinfellner 1960b). In the Silenoideae and a few other taxa, a small, scale-like and often bifid appendage or **coronal scale** often develops at the boundary between the claw and the blade, arising from the upper side of the petal (Fig. 36 II, III). This appendage is sometimes interpreted as a stipular structure (Glück 1919, and others). However, when dealing with so strongly modified an organ as the petal there is in this instance as little sufficient basis available for the recognition of such a precise homology as there is in the example given by Glück (1919), where the teeth at the base of the standard of *Scorpiurus subvillosus* are interpreted as "stipular denticulation" (Fig. 36 XIII).* The coronal scales of a flower taken as a whole can form a so-called **corona** (Fig. 11 IV).

We have already mentioned (section 1.4.1.1) the phenomenon of emarginate or bifid petals, which is of frequent occurrence (Fig. 22 II). This can be observed not only in the case of free petals but also at the tips of the lobes of sympetalous corollas, and can doubtless be regarded as evidence of the close structural relationship which exists between the petals and the stamens, which is to be discussed more fully later on. Troll (1928a, p. 184) drew attention to the fact that in radially symmetric flowers, if the petals are emarginate, they are all equally so. However, in bisymmetric flowers it is mostly only the larger and more developed petals which are emarginate, as for example in the standard of the Papilionaceae (see next section) or in the largest abaxial petal in *Viola* (Fig. 5 XI).

Occasionally, as in the Pink family (Caryophyllaceae), or as in the Common Whitlowgrass (*Erophila verna*), the petals are not just emarginate but are more deeply cleft, or in other species are even split practically all the way to the base (Fig. 36 IV), so that at a casual glance, the flower appears to have twice the true number of petals. As is shown by the petals of Ragged Robin (*Lychnis floscuculi*, Fig. 36 II), it is possible for both lobes which have originated by a central division to be further subdivided. In *Dianthus superbus* (Fig. 36 VII) and the species of the genus *Schizopetalon* (Cruciferae, Fig. 36 X, XI) we find examples of pinnately divided petals; also a more or less regular denticulation of the upper margin (Fig. 36 IX), or of the complete margin (Fig. 30 V), or an irregularly fimbriate margin, are not uncommon.

1.4.3.4   Zygomorphic flowers with free petals

As already mentioned, a monosymmetric (zygomorphic) flower structure can come about as a result of unequal development of the petals of a radial or bilateral flower. This is very clearly shown in the flower of the Candytuft

---

* Weber (1980) has attempted to find precise homologies for the individual tepal lobes in the perigon of the Zingiberaceae.

(*Iberis*, Fig. 108). The racemose inflorescence of this member of the Cruciferae appears to be umbellate at its apex, because the internodes of the inflorescence axis do not begin to elongate before the flowers in that part have finished their anthesis. In addition the pedicels lengthen so much before and during anthesis that the flowers are arranged in a slightly domed "umbel". The resemblance of

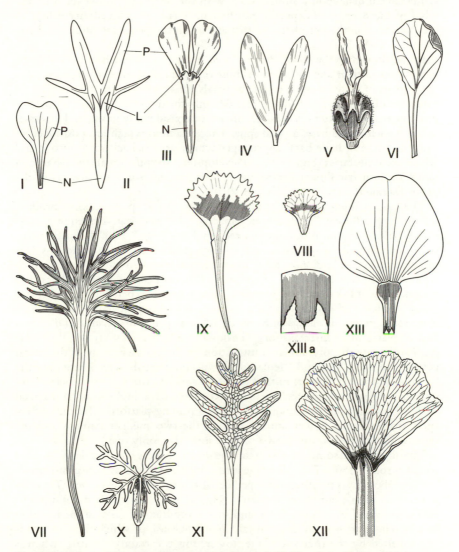

Fig. 36 Petal shapes. I *Tunica saxifraga*, II *Lychnis flos-cuculi*, III *Silene dioica*, N claw, P blade, L coronal scale, IV *Stellaria media*, V *Guazuma ulmifolia*, VI *Erucastrum gallicum*, VII *Dianthus superbus* subsp. *speciosus*, VIII, IX *Dianthus barbatus*, young (VIII) and adult petal (IX), X, XI *Schizopetalon walkeri*, flower (X) and petal (XI), XII *Lychnis coronaria*, XIII *Scorpiurus subvillosus*, XIIIa base of petal, enlarged. I-IV, VIII, IX after Troll, X after O. E. Schulz, XIII, XIIIa after Glück.

the flowering part of the inflorescence to an umbel is further emphasised by the fact that, although the four petals are equivalent in other respects, the two lower ones which are directed outwards towards the periphery of the inflorescence become greatly enlarged during anthesis compared with the two inwardly-directed upper ones. The peripheral flowers in the double umbels of many Umbelliferae behave in a similar way, with the difference, however, that the lobes of the deeply bifid petals are often unequally developed, according to their position on the periphery, so that as a result the petals are often asymmetric (Fig. 37 IV, V).

In the flower of the Wild Pansy (*Viola tricolor* ssp. *arvensis*, Fig. 5 XI) the petals differ in their alignment from those of a radially symmetric flower. Both of the pairs of upward-directed lateral petals are asymmetrically formed by an extension of the downward-directed side; only the downward pointing petal is bisymmetric. The latter exceeds the other four considerably in size and possesses a deep pouch at the base, a petal spur. Those stamens which are positioned in front of this, and have basal appendages which project backwards into it and which act as nectaries (Fig. 38 VII). Development of spurs on all the petals of a radially symmetric flower is rare, but is found in the 5-merous flowers of the genus *Halenia* (Gentianaceae, Fig. 38 VIII).

The most striking example of a highly zygomorphic, polypetalous corolla is that of the pea flower (*Papilionaceae*, Fabaceae). The cochleate-descending aestivation has already been discussed in section 1.2.3 (Fig. 7 VII). The rear petal, which overlaps all the others, is both the largest and the most conspicuous. It is known as the "standard" (*vexillum*); this is because it is often the only petal which is directed upwards and spreads out flat as the flower opens, whilst the other petals remain more or less folded up (Fig. 37 II). The lower two petals are united into a "keel" (*carina*), which encloses the stamens and the gynoecium, and which commonly functions along with the two lateral petals or "wings" (*alae*) as a landing platform for insect visitors. The flowers of Milkwort (*Polygala*, Polygalaceae) are of superficially similar construction. In these, however, both the lateral petals have been lost, and "replaced" by petaloid sepals which resemble the "wings" of the pea flower; instead of the "standard" there are two petals, and instead of the two petals which form the "keel" there is a single boat-shaped petal with a fimbriate appendage which serves as a landing platform. (Further differences from the pea flower are the lack of the two median stamens and the gynoecium which consists of two carpels, instead of only one.)

The flowers of some members of the Papilionaceae are regularly twisted round through 180°. This phenomenon, which is known as **resupination**, is brought about by a twisting of the pedicel or of the base of the flower. *Trifolium resupinatum* is an example of the latter (Fig. 37 III). The standard is turned to face downwards by this operation and so as a result could take on the function of a landing platform. In the tropical *Clitoria ternatea*, in which the standard is also inverted by the rotation of the flower, this is certainly the case, whereas according to Goebel (1924, p. 257) *Trifolium resupinatum* is autogamous. Resupination of the flower by 180° is also known in many species of the Acanthaceae, Lobeliaceae, Melianthaceae, for many of the Balsaminaceae (Fig. 31) and above all in the Orchidaceae. It is well known in the Orchidaceae that this comes about by a twisting of the inferior gynoecium (Fig. 74 VII, IX).

The plane of symmetry of zygomorphic flowers is not always placed vertically from the outset. In Fumitory (*Fumaria*) and in *Corydalis*, (Fig. 38 IV)
which like the Bleeding Heart (*Dicentra spectabilis*, Fig. 38 I, III; Fig. 5 XII, XIV)
belong to the Papaveraceae-Fumarioideae, only one of the transversely placed

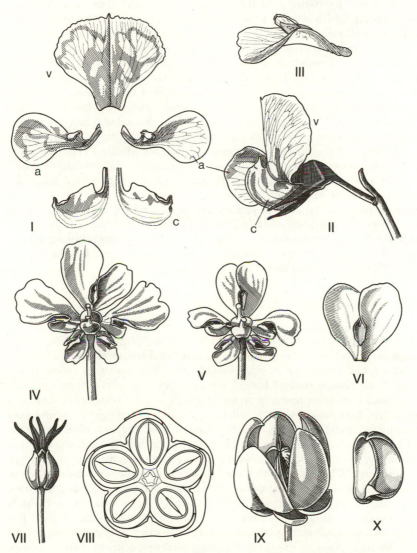

Fig. 37 I, II *Pisum sativum*, I Members of the corolla: v standard (vexillum), a wing (ala), c keel
(carina), II complete flower in side view with one wing removed. III *Trifolium resupinatum*,
lateral view of flower. IV-VI *Daucus carota*, flower with partially 'favoured' petals, seen from
above (IV, V), and a single completely 'favoured' petal (VI). VII-X *Coriaria japonica*, VII side
view of flower, VIII, IX young fruit in cross-section (VIII) and in side view (IX), X a single
petal dissected from the young fruit. I-II and VII-X after Troll, III after Goebel, modified.

petals is provided with a spur (Fig. 38 IV), whereas otherwise the basic plan of the flower is completely equivalent to that of *Dicentra* (Fig. 5 XVI). Further, it is only on this side that a nectar gland is provided at the base of the stamen which is placed in front of the spur (Fig. 38 IV). This kind of floral symmetry is called **transverse zygomorphy**. Shortly before anthesis this flower undergoes a rotation of 90° by twisting of the pedicel, so that the spur as well as the opposing petal, which is formed for use as a landing platform, take up a suitable position for pollination (Fig. 38 V, VI). The two members of the inner whorl of petals in *Fumaria, Corydalis* and *Dicentra* differ from the two outer petals in other respects besides the way in which the spur is constructed. This is particularly true of *Dicentra* (Fig. 38 II), where the petal is clearly differentiated longitudinally into a proximal and a distal section, divided by a constriction which acts as a joint. The distal parts of these petals, which hang well down out of the flower and are weakly united at their tips, enclose the stamens and style like a hood. The articulated zone makes it possible for the hood to be pushed to one side by visitors to the flower (for technical details see Müller 1954). The same kind of division can also be seen in the outer petals, where the terminal section is narrowed and reflexed during anthesis, and is clearly demarcated from the saccate and inflated basal part.

Besides transverse zygomorphy, diagonal zygomorphy is also known, and is characteristic of many members of the Sapindaceae and also of the Horse Chestnut (*Aesculus hippocastanum*). In these the original symmetry plane departs by only a small angle from the vertical, and here also it is turned to the vertical position shortly before anthesis.

1.4.3.5   Sympetaly

The form of the corolla is markedly affected when the petals are united and the corolla is sympetalous. As was discussed in section 1.4.2.5, this fusion of the petals is in general not a consequence of a postgenital recombination of initially separate organs, but rather they are united from their initiation onwards, i.e. they are fused **congenitally**.* Indeed, for Troll (1937, p. 29) the formation of a sympetalous corolla represents a typical example of the (congenital) fusion of organs: "As their development is followed, it can be observed that at the very beginning free petal primordia appear, whose number corresponds to the number of lobes in the mature flower. However, they do not continue to grow individually, so that when they have reached a certain size, they begin to merge at the margins. What we see is that when the primordia begin to enlarge out of the meristematic tissue of the growing point, they widen laterally at their bases, so that the initially independent primordia, *whose free sections moreover remain independent*, are from now on positioned on the edge of a ring-shaped zone. It is principally this which now grows on to form the mature, sympetalous corolla tube. The coalescence of the petals therefore already occurs at their initiation, *simultaneously with their creation*, and it is this that is termed '*congenital*' fusion." Accordingly what occurs is not only a fusion of the marginal meristems (Hagemann 1970) but also growth by means of an intercalary, basal meristem. The

---

* In particular see Čelakovský 1884a.

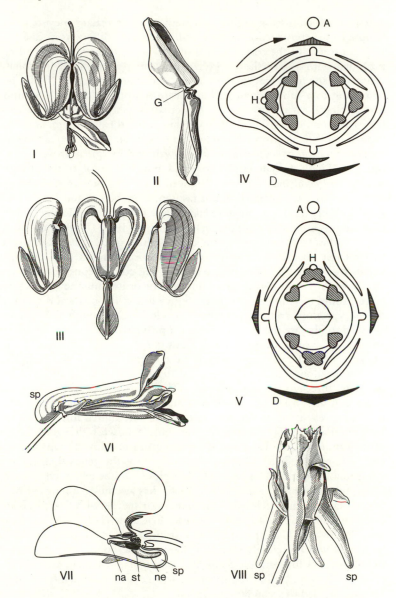

Fig. 38 I–III *Dicentra spectabilis*, I flower in side view, with the hood-like tip of the inner petals pulled aside so that the stamens and style emerge, II inner petal with joint G, III flower with the outermost petals removed, IV–VI *Corydalis lutea*, IV, V diagram of the flower before (IV) and after (V) resupination, H stamen with nectary, A principal axis, D subtending leaf, VI lateral view of flower, sp spur. VII *Viola odorata*, flower in median cross-section. na stigma, st stamen, ne nectary. VIII *Halenia umbellata*, side view of flower. I–VI after Troll, VII after Firbas.

extent to which fused marginal meristems and intercalary meristems each contribute to the construction of sympetalous corollas differs from case to case.

The growth of the rotate corolla of *Solanum dulcamara*, which admittedly is provided with a very short corolla tube, can be accounted for solely by the activity of the fused marginal meristem, according to Sattler (1977). However, according to a more recent communication from Sattler (5th Symposium of Morphology, Anatomy and Plant Systematics, Ghent 1979), the earlier findings have to be revised. Nishino (1976) writing about the development of the corolla tube of *Pharbitis nil* (Convolvulaceae) remarks that: "The corolla tube is initiated by co-operation of interprimordial growth and marginal growth of petal primordia." Singh and Jain (1979) observed marginal fusion of the calyx and petal bases in *Adenocalymna alliaceum* (Bignoniaceae) and preferential growth of the common basal zone created in this fashion. Gerstberger and Leins (1978) made similar observations on *Physalis philadelphica* (Solanaceae). In this context attention should also be drawn to the earlier studies of Boke (1948) and Cusick (1966).

In contrast to gamosepaly in the calyx, which often only provides a constant and reliable character for certain genera or subfamilies (e.g. the Silenoideae versus the Alsinoideae), the sympetaly of the corolla constitutes an important character for family classification on account of its constancy within even very large groups of families. This finds expression in the classical division between the polypetalous and sympetalous dicotyledons, although this should not be allowed to obscure the fact that amongst polypetalous families (or orders) there are various exceptions with sympetalous corollas, and that not a few members of sympetalous families have polypetalous corollas. Hence, for example, among the polypetalous Violales the petals of the Achariaceae are united into a campanulate, three- to five-lobed corolla. The genus *Sympetaleia* (Loasaceae), as its name indicates, has petals which are united throughout most of their length. In the Caricaceae the petals of the male flowers are united into a campanulate or long-tubular corolla, whereas in the female flowers they are free or quite shortly united at the base, and the hermaphrodite flowers which occasionally occur are intermediate. Within the Saxifragaceae the genus *Roussea* differs in its sympetalous corolla. This series of examples could be extended further.

On the other hand, Suessenguth (1938), has assembled a long list of "sympetalous" families in which individual genera or species with free petals are found, including the Diapensiaceae, Pyrolaceae, Ericaceae (*Ledum*), Styracaceae, Oleaceae, Rubiaceae, Campanulaceae and Stylidiaceae. Also members of the genus *Synthyris*, which is closely related to *Veronica* (Scrophulariaceae), may be mentioned here, as for example *Synthyris ranunculina*, in which, according to Hartl (1965), the considerable reduction of the corolla tube which is observed in *Veronica* has led to secondary polypetaly.

1.4.3.6   False sympetaly

Sometimes sympetaly appears to occur when the petals become united after their separate initiation (postgenitally), and this is known as **false sympetaly**. An example of this is *Correa speciosa* (Rutaceae), which was regarded by Wettstein (1935, p. 824) as a possible connecting link between the Choripetalae and the Sympetalae on account of its apparently sympetalous corolla. As was

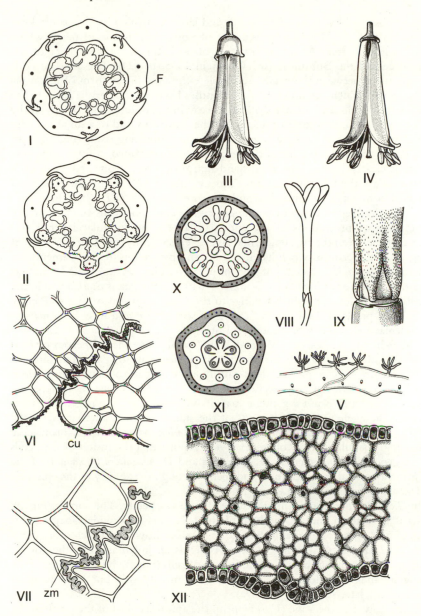

Fig. 39 False sympetaly. I, II *Lonchostoma pentandrum*, cross-sections through the flower in the region of the fusion between petals and filaments (I) and above (II), F filaments of the outer stamens. III–VII *Correa speciosa*, III, IV flowers with (III) and without (IV) calyx. V–VII Cross-sections through the connecting tissue between two petals, V general view, VI part of V (enlarged), VII part of VI (enlarged again), cu cuticle, zm toothed cuticular layer. VIII–XII *Oxalis tubiflora;* VIII flower shortly before the complete opening of the petals, IX base of corolla after the calyx has been removed; X, XI Cross-sections through a young flower in the zone of complete fusion of the petals (XI) and in the transitional zone further up (X), XII Cross-section in the completely fused zone, enlarged, inner side of tube facing downwards. After Leinfellner (I, II) and Hartl.

already observed by Troll (1937, p. 29) and later confirmed by the painstaking researches of Hartl (1957a), this is a case of postgenital fusion of petals which were originally free. A similar situation also obtains in other genera of the Diosmeae and the Boronieae (see Wettstein 1935, p. 824). The areas of connecting tissue between the petals are visible externally in the form of inconspicuous ridges, which run down from the sinuses between the corolla lobes to the base of the corolla and show a shallow furrow in the middle (Fig. 39 III). If the calyx is removed, it becomes evident that these furrows lead to a slit at the base of the corolla (IV). A cross-section through the connecting tissue (Fig. 39 V to VII) shows that the integrity of the petals is also maintained in this region: the suture is clearly visible. The right-hand margin (as seen from above) of each petal overlaps the left-hand margin of its neighbour (V), and so the epidermal layers of both petals are enmeshed and cemented together (VI, VII). The dentonection is doubled; it depends on the projecting epidermal cells and also on the dovetailing of projections of the cuticular layer. The cuticular layer is still detectable between the two epidermal layers. This is really just a further stage of progress in the process which has already been observed in the phenomenon of what is known as interlocking of sepal or petal margins (section 1.4.2.4). Similar phenomena were confirmed by Winkler (1940) for the genus *Stackhousia* in the family Stackhousiaceae which belongs to the Celastrales. According to Reiche (1891), the petals in *Tupa salicifolia* (*Lobelia* sect. *Tupa*) and in *Selliera radicans* (Goodeniaceae) also exhibit postgenital fusion.

The exhaustive researches of Leinfellner (1964b) into the properties of *Lonchostoma* and some related genera of the Bruniaceae are of particular interest. These members of a polypetalous family also appear to have sympetalous corolla tubes. Series of cross-sections through the flowers (Fig. 39 I, II) show, however, that this is not a genuine case of sympetaly, but a postgenital fusion through the agency of the filaments of the stamens which alternate with the petals. According to Leinfellner the genus *Pachyphytum* in the Crassulaceae behaves in a similar way. Rohweder (1969) reports that the fusion of the petals in the genera *Zebrina*, *Coleotrype*, *Setcreasea* and *Dichorisandra* (*hexandra*) of the Commelinaceae also comes about indirectly through the agency of the episepalous stamens (but directly in *Cyanotis*).

In *Oxalis tubiflora* there is another case of false sympetaly. The corolla (Fig. 39 VIII) is salverform, and at the base of the tube there are widely-gaping slits (Fig. 39 IX). The postgenital fusion of the petals is, however, much more complete here than in *Correa* and is only visible as an indentation in the corolla surface and as an inconspicuous double row of cells (Fig. 39 XII). A study of the ontogeny makes it clear that the petals are initially free but become fused together at a later stage (Hartl 1957a).

In the flower of the Vine (*Vitis vinifera*) the apical sections of the petals are so firmly fastened together in their valvate type of aestivation by dentonection of the epidermal cells and the cuticle (Raciborski 1895, Sigmond 1930; see Fig. 28 VIII) that the petals cannot open. In fact, quite the reverse, since they become detached at the base instead, so that the entire corolla is lifted up and thrown off by the expansion of the filaments of the stamens. According to Goebel (1924, p. 65), the rejection of the corolla in *Vitis vulpina* is assisted by an increased tension in the tissue on its inner surface, which would lead to the opening of the

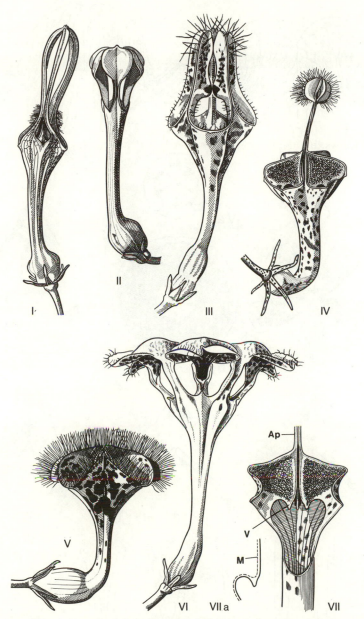

Fig. 40 Flowers of *Ceropegia* species. I *C. barkleyi*, II *C. medoënsis*, III *C. nilotica* × *sandersonii*, IV *C. distincta* var. *haygarthii*, V *C. elegans*, VI *C. sandersonii*, VII *C. distincta*, var. *haygarthii* funnel region of the flower in bud, cut open, VIIa outline of the ventral wing, with the path of the central vein M drawn in, Ap stalk of the appendage, V ventral process. I, II, IV and V after Troll, the rest after Vogel.

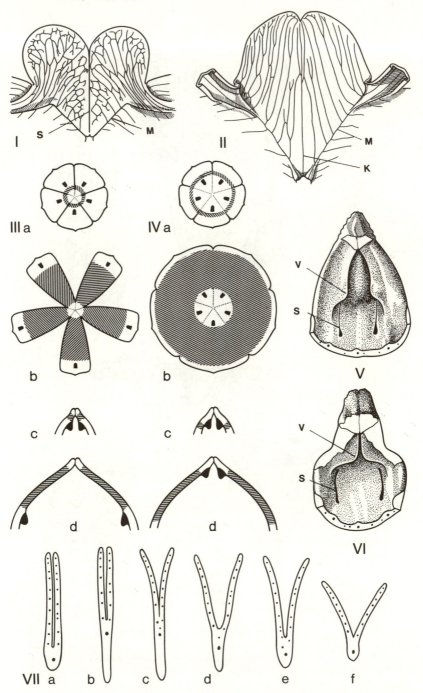

Fig. 41 I, II Construction and venation of the central dome in the flower of *Ceropegia rendalii* (I) and *C. sandersonii* (II), showing a sector of the dome with two claws. III, IV schematic

petals, if they were not held together at their tips, and so this just causes a pronounced outward curvature of the bases of the corolla.

1.4.3.7 Window flowers

In the Rampions (*Phyteuma*, Campanulaceae), at the beginning of anthesis the tips of the five corolla segments remain attached to one another, forming a narrow beak which encloses the stamens. This is then later pierced by the expanding style. First the corolla is opened just below the beak as the petal margins split apart to form five "windows". Only after this do the tips of the petals separate from each other (see Goebel 1924, p. 72).

It is also usually the case in the sympetalous flowers in species of *Ceropegia* that the corolla lobes are so firmly attached by their tips as a result of postgenital fusion, that these do not separate during anthesis. This **inhibited opening of the flower** is one of the reasons for the occurrence of these specialised flower types, which function biologically as myiophilous (fly-pollinated) pitfall-trap flowers (Vogel 1961).

In this case too the tips of the corolla lobes only partially separate along their lateral margins in such a way that five openings or windows are created. These are therefore termed "window flowers" (Goebel 1924, p. 69).* The margins of the lobes may curve outwards to a greater or lesser degree (Fig. 40 I), and by differentiation into a broadened terminal section and a proximal claw may even form an umbrella-like dome which is spread out over the mouth of the corolla tube (Fig. 40 II, VI). In many species (*C. dichotoma*) the margins of the corolla lobes are rolled outwards, so that the windows are opened wide; the widened and trumpet-like limb of the corolla of *C. elegans* shows the same feature (Fig. 40 V; see further Troll 1950a, 1951). In many cases, however, as for example in *C. nilotica* × *sandersonii* (Fig. 40 III) the lobes are so far "intended for window construction right from the start" since the main part of their surface area is already orientated radially as a result of markedly unifacial development. The form of the horizontal cross-sections (Fig. 41 VII) resembles cross-sections through the basal part of an Iris leaf, with the difference that in *Ceropegia* it is an instance of a reduction of the morphologically lower side of a leaf, and not of the upper surface.

In *Ceropegia distincta* var. *haygarthii* these sections are folded inwards almost at right angles and form five septa which divide the funnel of the corolla from

Caption for Fig. 41 (*cont.*). comparison of the development of distal corolla sections of *Ceropegia sandersonii* and the related species without a dome (III), zone of the greatest surface growth shaded. Ventral process (uvula) black; a, view of the primordial apical zone, the petal sections which later separate are delimited by solid lines, definitively fused sections by dotted lines; b, view of the expanded corollas; c, d, longitudinal sections through the apical region of the (c) young and (d) expanded corollas. V, VI Apical sections of the young flower of *Ceropegia nilotica* × *sandersonii* (V) and *C. sandersonii* (VI) opened by removing two segments, V ventral processes or primordia of the uvula, s slits (later windows), surfaces of intersection shown in white. VII a–f *Ceropegia nilotica* × *sandersonii*, series of cross-sections through a petal tip from (a) the apical to (f) the basal region, with the segment margins in the upward direction and the ventral process downwards. After Vogel.

* The expression "window flower" has been taken up from Goebel (cf. 1924, p. 69 et seq.) and goes back to W. Hooker (Botanical Magazine 1845, Pl.4161), who used it to describe the properties of the flower of *Masdevallia fenestrata* (Orchidaceae).

above and meet in the centre of the flower. Each of these septa bears a central downward-pointing conical process (Fig. 40 VII, VIIa). Vogel (1960) found that these ventral processes in *Ceropegia sandersonii* (Fig. 41 VI) were fused into a uniform conical body, the so-called "uvula". The situation is like that in *Oxalis tubiflora*, as previously illustrated, where there is no recognisable sign of suture left by the postgenital fusion of the petals. Indeed, according to Troll (1928, p. 370), this is true for many species in which there is postgenital fusion of the tips of corolla lobes. As in the examples of interlocking of sepal and petal margins (section 1.4.2.4), it is true here also that the contact comes about "in the first place through the touching and enmeshing of the adjacent but still embryonic epidermal layers. The thickening of the cell walls does not take place, however, and the cuticle is never established. The fusion (symphysis) is therefore so thorough, and the original boundaries are so completely effaced, that in the end the suture is no longer recognisable." (Vogel 1960, p. 403).

In most species this complete fusion of tissues is confined to the furthermost tips of the lobes, but in *C. sandersonii* it reaches below the ventral processes, which are already developed at an early stage. In the former cases, further growth proceeds in a zone above the ventral processes, which later separate from each other, so that these processes are finally situated at some distance from the apex (Fig. 41 IIIa to d). In *C. sandersonii*, however, the zone of greatest expansion lies below the ventral processes, which is to say in the region of symphysis, while the ventral processes remain fixed in their embryonic relative position and fuse firmly together into the uvula. The central dome of the flower of *C. sandersonii* is therefore produced by an *"enormous enlargement of the genuinely united apical region"*, which *"is able to expand itself as a single unit by vigorous cell division"*. The tissue which originates in this way "is completely 'secondarily and congenitally' united as a result of intercalary growth, thus a suture between the organs can not be assumed, even theoretically" (Vogel 1960 p. 407). This is reflected in the venation of the lobes of the dome, which must be attributed to two corolla lobes: where one would expect to find a suture there in fact appear commisural nerves leading from the centre of the dome (Fig. 41 IIk), running along what would be accepted in theory as the dividing lines between two neighbouring petals, as is often the case in sympetalous corollas. The lobes of the dome of *C. sandersonii* are therefore demonstrably unified structures (Fig. 41 II), in complete contrast to those of *C. rendallii* (Fig. 41 I).In the latter a suture (S) separates the lobes of the dome, i.e. the nerves of the two halves clearly originate from the median veins (M) of the neighbouring corolla lobes.

The above differences are already recognisable in the flower buds of *C. nilotica* × *sandersonii* (Fig. 41 V) and *C. sandersonii* when they are only 2 mm or 1.2 mm long, respectively. In the latter the close proximity of the ventral ribs (V) which later grow out into conical processes results in their subsequent total fusion and at the same time creates the compact tip which later becomes the "umbilicus" which is situated on the apex of the dome.

1.4.3.8   Labiate and personate flowers, coronal scales

In many sympetalous families the corollas have a more or less clearly two-lipped form. In this case the generally 5-merous limb of the corolla is mostly arranged in such a way that two corolla lobes form the upper lip, and three the lower. This is true in the familiar case of the Labiatae, in which the several stamens and the style (temporarily at least) are placed under the protective roof of the upper lip, whilst the lower lip, which is usually provided with a nectar guide, serves as a landing platform. The pollen is therefore rubbed off onto the backs of insects which visit the flower and is correspondingly carried across to the styles of other flowers. This kind of pollination is said to be **nototribal**. The allocation of two corolla lobes to the construction of the upper lip and three to the lower is frequent, but is by no means always obvious externally. The lower lip is often strongly divided, as for example into two smaller lateral lobes and a larger, conspicuously divided and lobed central portion; it can even appear in a five-lobed form as in Germander (*Teucrium*), since the two lobes of the upper lip are united with the lower lip. The flowers of the Bugle (*Ajuga*, Fig. 42 I), however, represent an example in which the upper lip is greatly reduced. In Basil (*Ocimum basilicum*) the opposite is seen: the lateral lobes of the lower lip are bent upwards towards the upper lip, so that the lower lip appears to consist of one lobe only (the calyx on the other hand exhibits a four-lobed lower lip whereas the upper lip consists of a single lobe only!). *Plectranthus glaucocalyx* behaves in a similar way. In both flowers the style and stamens are not placed below the upper lip, but are adjacent to the lower and are curved upwards slightly; the pollination must consequently be **sternotribal**, and take place via the underside of the insects visiting the flower. In the bird-pollinated genus *Columnea* of the Gesneriaceae (Fig. 43 VII) the two lateral lobes which originally belonged to the lower lip are likewise transferred to the upper in the open flower. Generally, the 2:3 ratio of lobes which is so common among the Labiatae is by no means the only one. In the labiate flowers which are found in the genus *Lonicera* of the Caprifoliaceae (Fig. 42 VIII) the upper lip is formed from four lobes while the lower lip has only one. In many members of the Lobeliaceae, as for example *Lobelia* and *Centropogon*, the lobes occur in the ratio 3:2, however, the original upper lip later appears to be the lower, and vice versa, as a result of resupination (*Lobelia*) or of the flower being bent over (*Centropogon*). The genus *Strobilanthes* (Acanthaceae) behaves rather differently. The corolla of *S. isophyllus*, which initially has an ascending position, bends over backwards through an angle of more than 90° as the flower opens, so that the lower lip points upwards and the upper lip downwards (Fig. 42 IX).

The corollas of certain members of the Acanthaceae in the tribe of the Justicieae and the sub-tribe of the Odontoneminae are distinguished by another special structure, i.e. the style is borne in a central fold of the upper lip, the so-called rugula (Fig. 42 II to IV). As shown in Fig. 42 V to VII in *Beloperone*, this structure takes the form of two ridges or folds, one running on either side of the median, whereas in *Jacobinia* it evidently consists of no more than a furrow running the length of the underside of the upper lip (see Troll 1951a).

Most of the personate flowers (Scrophulariaceae) are arranged in the ratio 2:3, for which the Snapdragon and the Toadflax are good examples (Fig. 43 II, V). An exception is the genus *Nemesia*, in which the upper lip has four lobes and the lower lip only one. In the Snapdragon (*Antirrhinum*), the Toadflax (*Linaria*), and in a series of other members of the Scrophulariaceae, the entrance to the corolla tube is closed by a "mask" which is the reason why the family (or order) name *Personatae* was given. It is a matter of a curvature of the lower lip in the region of the common basal parts of the corolla lobes. As is shown by median cross-sections through the flowers of *Linaria* (Fig. 43 VI) and above all of *Antirrhinum* (Fig. 43 III), in both cases the lower lip is so to speak "tucked in" from below, which has the effect that it touches the upper lip and the entrance to the tube is closed. The curvature of the lower lip corresponds to the formation of a "palate", caused by the fact that the slight downward curve of the upper lip leads to a sudden upward and backward curvature of the upper lip apex. In the "ladies' slipper" flowers of the genus *Calceolaria* (see section 3.3.4.2) on the other hand, the lower lip is curved from within and greatly inflated. The opening of a personate flower is generally made possible by a (springy) articulated zone at the base of the lower lip (Fig. 43 I G). Not only in *Antirrhinum* but also in *Linaria* a nectary is found on the ventral side of the corolla tube at the base of the gynoecium, the secretion from which collects in the saccate base of the tube in Snapdragon, and in the tip of a long spur which extends from the base of the corolla in Toadflax. The style and the stamens, whose anthers open

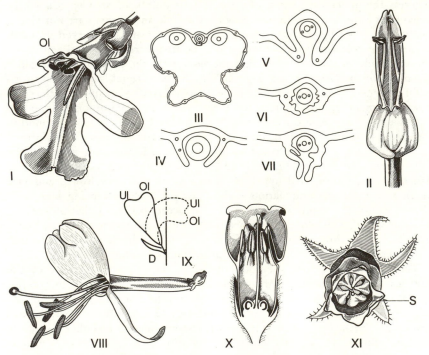

Fig. 42 I *Ajuga reptans*, flower. II–V *Jacobinia magnifica*, II distal part of the corolla seen from below, III cross-section through the limb of the corolla in bud, the rugula with the style on the lower side of the upper lip; IV cross-section through the rugula ("Griffelhalter") below the stigma. V–VII *Beloperone violacea*, cross-section through the rugula in the lower (V) and upper parts of the corolla. VIII *Lonicera periclymenum*, flower in anthesis. IX *Strobilanthes isophyllus*, diagram illustrating the resumination, Ol upper lip, Ul lower lip, D subtending bract. X, XI *Symphytum officinale*, vertical section of the flower (X), and diagonal view from the front, S coronal scales. II–VII after Troll, IX after Goebel, X after Baillon.

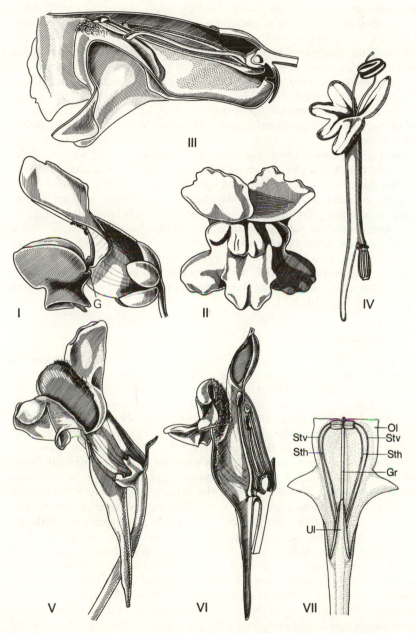

Fig. 43 I-III *Antirrhinum majus*, lateral view of flower (I), front view (II), and median cross-section (III), G articulated zone. IV *Centranthus ruber*, flower. V, VI *Linaria vulgaris*, lateral view of flower (V) and medial cross-section (VI). VII *Columnea picta*, distal part of flower seen from below. Ol upper lip, Ul lower lip, Gr style, Stv anterior stamens, Sth posterior stamens. I and II after Troll, VII after Trapp.

in the downward direction, are placed along the upper side of the corolla tube, and are rubbed on the back of nectar-seeking insects, so that nototribal pollination follows. Sympetalous corollas which are saccate or spurred at the base are also found in the Valerianaceae (Fig. 43 IV) and in the Lentibulariaceae, and also elsewhere. In *Valeriana* a nectary is found at the base of the saccate section, as well as in *Centranthus ruber*, whose long-spurred flowers are constructed in a totally asymmetric fashion (Fig. 43 IV). In the latter case, in contrast to *Linaria*, the nectar is secreted within the tip of the long spur. In Butterwort (*Pinguicula*, Lentibulariaceae) mucilage-secreting glandular hairs ("feeding hairs") are found at this position.

Another phenomenon involving localised impressions in the surface of the corolla is the **coronal scales**, which are found in the radially symmetric flowers of the Boraginaceae projecting into the corolla tube below the limb and narrowing the entrance (see Schaefer 1942). Seen from in front or from within they appear as scales growing from the base of the corolla lobes (Fig. 42 X). A section taken in the median of a corolla lobe shows, however, that there is a deep invagination in the corolla tube in this position (Fig. 42 X). There also exist, of course (as in many species of Gentian), coronal scales which could be described as "ligulate outgrowths" from the petals, and which can be compared with the scales which occur on the clawed petals of many of the Silenoideae (see Fig. 11 IV; Fig. 36 II, III, XII).

1.4.3.9   Floral dimorphism

The so-called ray florets of the capitula of the Compositae may be considered as an extreme modification of labiate flowers. As is well-known, the flowers of the sub-family Asteroideae (Tubuliflorae) are often conspicuously dimorphic: in the centre of the capitulum the small so-called disc florets occur, with a radially symmetric 5-merous tubular-campanulate corolla (Fig. 44 VI), whereas in the marginal "ligulate florets" the corolla is conspicuous and provided with a long "petal" with a three-toothed tip ("ray florets", Fig. 44 VII). It can readily be seen how these ray florets came about by considering the transitional forms which occur in the so-called "compound flowers", as for example in *Pyrethrum roseum* (Fig. 42 VIII). These show that the three outwardly-directed corolla lobes are transformed into a lower lip by the elongation of their common base, on the opposite side from the pair of lobes which are directed towards the centre and which form the upper lip. The latter are usually completely suppressed in ray florets, while the lower lip is elongate-spathulate, although its three-toothed tip still shows its derivation from three corolla lobes. In the tribe Mutisieae, which are chiefly found in the Andes (and partly bird-pollinated), the ray florets are usually still pronouncedly two-lipped (Fig. 44 IV). On the other hand, the corolla in all flowers in the capitula of the Cichorioideae (Liguliflorae) is unilaterally elongated with a five-toothed tip (Fig. 44 V). In this case obviously all five corolla lobes are included in the lower lip. The dimorphism in the flowers of the Asteroideae is also often expressed in the sex characters, and even to such an extent that the ray florets can become entirely female by reduction of the stamens, or sterile by loss of both the stamens and the ovary. The latter is also the case in the ray florets of the Cornflower

(*Centaurea*, Fig. 42 III) and other Cynareae in which the heads consist of tubular florets only (I, II). In these capitula the corollas of the marginal tubular florets are conspicuously enlarged in their entirety, but often become simultaneously zygomorphic and multi-lobed.

A dimorphism (the so-called "heteranthy" of Goebel) between sterile, marginal "display flowers" and central, fertile flowers is found in the structure of many inflorescences. As an example the inflorescence of Hydrangea (section 1.4.2.6) has been mentioned already, in which the display effect of individual

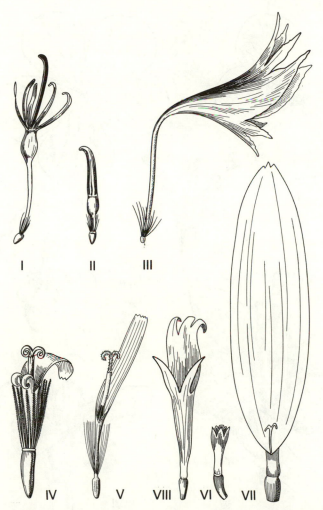

Fig. 44 Types of flower in the Compositae. I-III *Centaurea cyanus*. I, II disc floret, open (I) and in bud (II), III ray floret. IV *Nassauvia spicata* (Mutisieae), labiate flower. V *Taraxacum officinale*. VI-VIII *Pyrethrum roseum* (*Chrysanthemum coccineum*). VI disc floret, VII ray floret, VIII transitional form between the two from a "compound" head. After Troll (I-III, VI-VIII), Wettstein (IV) and Berg and Schmidt.

peripheral flowers comes about by a corolloid development of the complete calyx. A counter example to this is provided by many species of *Viburnum*, among which is our native Guelder Rose (*Viburnum opulus*, Fig. 45 I to III). Here the conspicuousness of the peripheral flowers of the inflorescence (I, II) depends on the fact that the corolla is greatly enlarged by comparison with the central flowers (III), while the calyx remains inconspicuous. Here too the corolla becomes enlarged at the same time as the flower becomes sterile.

We are already familiar with the example of sexual dimorphism or polymorphism in the different forms of the corolla which are provided by the male,

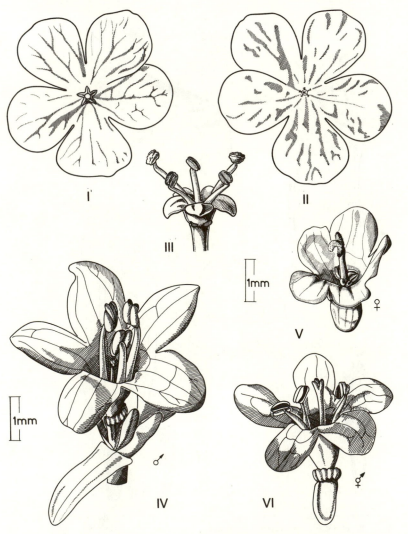

Fig. 45 I-III *Viburnum opulus*, sterile flower seen from below (I) and above (II), III fertile flower. IV-VI *Valeriana celtica* ssp. *pancicii*, IV male, V female, VI hermaphrodite flower. I-III after Troll.

female and hermaphrodite flowers of the Caricaceae (section 1.4.3.5). In not a few cases the flowers of differing sex are distinguished by having a corolla of a different size. This is true of many species of Valerian, such as *Valeriana montana*, *V. tripteris* or *V. celtica* (Fig. 43 IV to VI), in which the corollas of the male flowers are larger than those of the female flowers, whilst the hermaphrodite flowers which sometimes appear are intermediate. In other cases the male and female flowers are distinguished by the number and form of their perianth segments, as for example in *Begonia*, where the perianth of the male flowers consists of two differing pairs of decussate segments, while the female perianth is 5-merous. The most extreme example of sexual dimorphism is probably that of the genera *Cycnoches* and *Catasetum* in the Orchidaceae, in which the male and female flowers scarcely seem to belong to the same species or genus.

1.4.3.10 Nectar leaves

Apart from their function as protective and display organs, the petals of chori-petalous (e.g. *Berberis*) and sympetalous (e.g. Valerianaceae) flowers can also serve as a support for a great variety of nectaries. (The same is correspondingly true for sepals, stamens and carpels, see section 1.7.3.) In general the form of these organs is not fundamentally altered by the formation of nectaries. At times, however, the form of the petals is more strongly modified in relation to their function as nectary bearers, and they are then referred to as **nectar leaves** or **honey leaves**.

The great range of different types of nectar leaves is especially well-known amongst the flowers of the Ranunculaceae. A small selection of these examples is considered here, beginning with Love-in-a-mist (*Nigella damascena*, Fig. 46 I to III). Before anthesis its flowers are surrounded by an involucre of a variable number of delicately pinnate stem leaves, which are approached close to the flower and which greatly exceed the floral organs. It is this feature which has given rise to the plant's vernacular name. Five pale blue or white, shortly-clawed, elliptic perianth segments, which are arranged in a quincunx, follow the involu-cral leaves (the latter have been removed in Fig. 46 I), These perianth segments have evidently taken over the display function of the flower, for the nectar leaves which follow are relatively small. They appear constantly eight at a time, which in this context means that they are arranged in a phyllotaxy of 3:8.

Their form is evident in Fig. 46 II, III. The short pedicel bears a dish-shaped nectary pit, covered by a scale which is attached by an articulated base to the lower margin of the pit. This scale is seen folded over in Fig. 46 III. The nectar leaf is completed by a petaloid distal section which is deeply bifid, and which bears a tuberculate swelling at the base of each lobe. The honey leaf bears some resemblance to a stamen with a filament and its bisymmetric anther; this will be discussed further below. From the form of the nectar leaf with its petiolate, crateriform nectar pit and its scale-like appendage inserted on the ventral side, the general conclusion is that it is a peltate leaf organ (see section 1.5.6). This interpretation was confirmed by Troll (1928a p. 96) through ontogenetical studies. In other species of *Nigella*, and indeed in members of the section *Garidella*, the terminal part of the organ is much longer and more obviously petaloid than it is in *Nigella damascena*.

The differentiation into petiole, nectar-bearing and petaloid sections reappears in the long-stalked honey leaves of Monk's-hood (*Aconitum napellus*, Fig. 46 VI), but with the difference, amongst others, that the nectar-bearing portion is saccate in form (Fig. 47 VI). This plant also possesses eight honey leaves, as shown in Fig. 47 IV, which are arranged in the 3:8 ratio (III). In fact, of these, only those two posterior ones which are enclosed in the helmet-shaped, median perianth segment are well-developed. The rest are very rudimentary, and all the more so when they are inserted on the abaxial side of the flower. The formation of the honey leaves therefore expresses the same developmental favouritism as is recognisable in the large, brightly-coloured outer perianth segments, i.e. that the adaxial, posterior, helmet-shaped segment is the best developed, next to that the two upper, lateral segments, and particularly on the side next to the helmet, and lastly, and in contrast, both the adaxial perianth segments are very small (I).

In the flower of the Columbine (*Aquilegia vulgaris*, Fig. 46 IV, V) it is the floral organs with the upward-directed spurs which alternate with the lanceolate perianth segments (sepals) which will be referred to as the nectar leaves. The nectary here is found in the knob-like, thickened, upper end of the spur. Hence the region which corresponds to the nectar pit in *Nigella* and to the saccate nectar-bearing terminal section of the nectar leaf in *Aconitum* is particularly well-developed here; the petaloid distal part is also very prominent. In contrast the stalk-like section is completely suppressed. The entire organ is brightly coloured (or white), and moreover of the same colour as the lanceolate perianth segments, or else in vivid contrast to them. On the other hand the nectar leaves which appear in considerable numbers (8, 13, 21) in the Christmas Rose (*Helleborus niger*, Fig. 46 VII) and other *Helleborus* species are relatively small. They take the form of a stalked cup and the petaloid terminal section is practically absent.

In the nectar leaves in Meadow Buttercup (*Ranunculus acris*, Fig. 46 VIII, IX) and other Buttercup species the proportions are quite different. The nectar leaves are large, plane and petaloid in appearance, and are often simply referred to as "petals", which seems to be all the more justified by the fact that they are preceded by five small sepals in a quincuncial arrangement (Fig. 46 VIII). On closer inspection, however, a nectary, which is covered by a small "nectary scale", can be noticed at the base of the ventral side of these "petals" (IX). Therefore the "petals" of *Ranunculus* are nectar leaves whose form nevertheless seems to be mainly related to the function of display. According to the researches of Leinfellner (1958b, 1959a, 1964a) this is fundamentally, once again, an instance of a peltate (or peltate-diplophyllous) leaf organ. The same is true for the similarly petaloid nectar leaves of *Berberis* (Leinfellner 1955c).

It is not the case in all *Ranunculus* species that the petaloid "lamina" is larger than the nectar pit: in the South-American *Ranunculus apiifolius* the nectar leaves are smaller than the petaloid sepals, and are clearly peltate and broadly dish-shaped. The example of *Ranunculus acris* does show, however, that the nectar leaves, in spite of all their various forms, must be considered as petals whose shape may sometimes be more strongly adapted to the production of nectar, or at other times to the display function, or to both. Since there are also flowers in the Ranunculaceae which are totally without nectaries, but which prove nevertheless to be equivalent to those with nectaries, Hiepko (1965) prefers to distinguish between "nectar-fertile" and "nectar-sterile" petals.

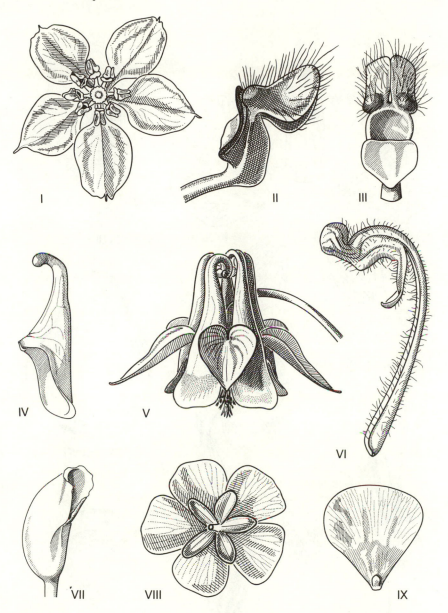

Fig. 46 Honey leaves. I-III *Nigella damascena*, flower seen from above, involucre, stamens and carpels removed, II, III nectar leaf in side (II) and top view (III, with the scale which covers the nectar pit turned down). IV, V *Aquilegia vulgaris*, single nectar leaf (IV) and flower (V). VI, VII nectar leaves of *Aconitum napellus* and *Helleborus niger*, VIII, IX *Ranunculus acris*, flower seen from below (VIII) and a single, petaloid nectar leaf. I-VI, VIII, IX after Troll, I altered.

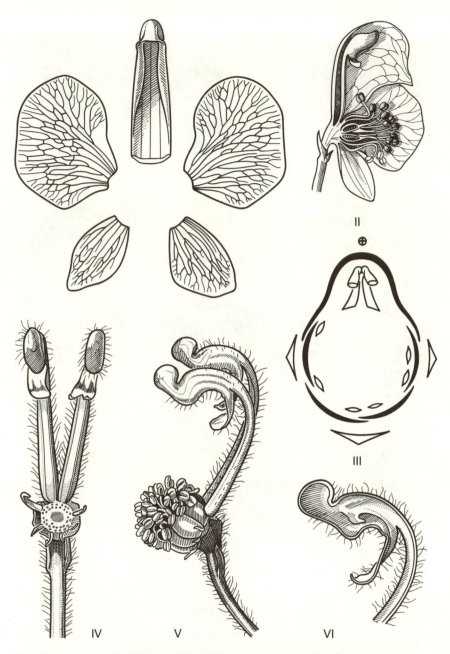

Fig. 47 Aconitum napellus. I Perianth, dissected into its components, II median cross-section through the flower, III floral diagram, including the perianth and the nectar leaves, IV nectar leaves in their natural arrangement, V lateral view of flower, after removal of the perianth, VI distal part of a nectar leaf in median section. After Troll (I, III–VI) and Warming (II).

1.4.4   The undifferentiated perianth (perigon)

1.4.4.1   Definition

A perianth which is not differentiated into a calyx and a corolla is usually called a perigon. According to Troll (1957, p. 12), this term refers to a perianth in which "the segments are essentially similar". For this reason it is also called a **homochlamydeous** perianth. We have already discussed in section 1.1 the perigon–like, and especially the corolloid structure of the perianth using the example of the Tulip as a member of the Liliaceae. In fact the perigon can be considered as a characteristic feature not only of the Liliaceae in particular but also of the Monocotyledons in general.

As already mentioned, the perigon occurs in two distinct forms; namely **corolloid**, as in the tulip flower, and **"prophylloid"** ("**sepaloid**", Engler 1926), as in the Wood-rushes (*Luzula*). The latter expression may be translated as "bract-like" (or "calyx-like"), and refers to the inconspicuous colours (green, bluish-green, yellowish or brownish) and to the generally diminutive size.

1.4.4.2   Forms of tepals

The form of the perigon segments, or tepals, is usually long- or broad-elliptic to ovate or triangular-ovate. Apart from certain prominent exceptions, such as for example when a labellum is constructed of perigon members, they are entire; nevertheless emarginate tepals are occasionally observed, as was remarked above for petals. Examples are found in the perianth of the Arrowhead (*Sagittaria*, Alismataceae, Fig. 48 I) and in the inner tepals of the Snowdrop (*Galanthus nivalis*, Fig. 48 III). As already mentioned, they are usually inserted on the axis of the shoot with a relatively broad base, and are often provided with three or more leaf traces, as was established by Kaussmann (1941). However, there are also exceptions to this, as for example the clearly clawed inner tepals of many Commelinaceae, as in *Tradescantia* or *Commelina* (Fig. 49 III), which are very narrow at the point of insertion, and which also have but a single leaf trace. Similar remarks apply to some of the Liliaceae, as the elliptic or lanceolate tepals of *Camassia* (Fig. 49 I), or the tepals of *Tricyrtis* (Fig. 49 II) with their abruptly contracted base; also in *Sagittaria* (Fig. 48 I) and in *Iris* (Fig. 48 VIII) the tepals are very much narrowed at the base.

Leinfellner (1960a, c, d, 1961a, b, 1963a, b, d, e) pointed out that the tepals of many Liliaceae "have a peltate structure, or at least can be interpreted as reduced forms of peltate leaves", in which "the more or less rudimentary ventral leaf blade..." is "...often not plane but developed as a nectary" (Leinfellner 1963e, p. 448). The appearance and the very varied form of these nectaries (Fig. 49 IV to XI) play an important role in the characterisation and classification of tribes within the *Liliaceae*.

1.4.4.3   Heterotepalous perigon

If at the outset the perigon was characterised by the statement that its "segments are essentially similar" (Troll), this does not mean to say that the two 3-merous

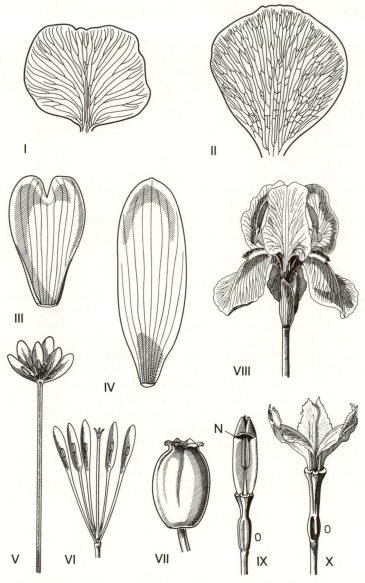

Fig. 48 Perigon. I–IV Tepals of I *Sagittaria engelmanniana,* II *Butomus umbellatus,* III, IV *Galanthus nivalis,* III inner, IV outer tepal. V *Colchicum autumnale,* syntepalous perigon with androecium, perigon tube split open on one side and spread out with the perigon limb. VI *Bulbocodium vernum,* flower, showing the stamens united with the tepals, spread out with one removed, VII *Muscari comosum,* side view of flower. VIII–X *Iris germanica,* flower in side view (VIII), and in bud with the tepals removed (IX) and with flower open (X). N stigma, O inferior ovary. After Kaussmann (I, II), Troll (V–X) and original (III, IV).

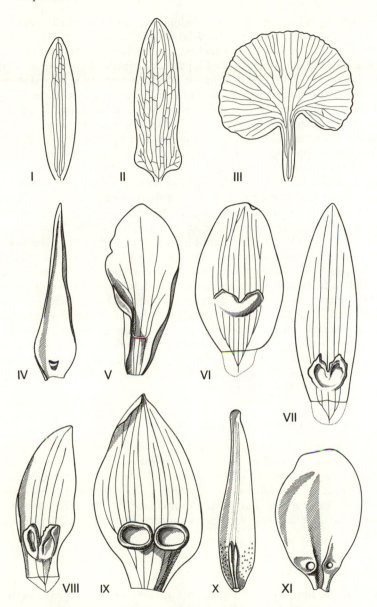

Fig. 49 Structure of the perigon. I-III Tepals of *Camassia cusickii* (I), *Tricyrtis hirta* (II) and *Commelina benghalensis* (III). IV-XI Nectary-bearing (peltate or diplophyllous) tepals of *Ornithoglossum viride* (IV), *Lloydia serotina* (V), *Anticlea chlorantha* (VI), *A. sibirica* (VII, VIII), *Zygadenus glaberrimus* (IX), *Lilium tigrinum* var. *flore-pleno* (X) and *Dipidax triquetra* (XI). After Kaussmann (I-III) and Leinfellner (IV-XI, partly altered).

whorls, by which the monocotyledonous perigon is usually represented, are always formed identically. The Wild Tulip (*Tulipa sylvestris*) is a case in point, but this can be taken further and even as far as the establishment of a sepaloid structure for the outer row of tepals, as is found for example in the genus *Calochortus*, or in the Commelinaceae, Bromeliaceae and in several other families, in which the outer perigon row is often referred to as a calyx. In the case of the Liliaceae the genera *Trillium* and *Paris* can be quoted as examples of perigons with a markedly heterotepalous structure, but it is important to point out that these are linked nevertheless to a homotepalous structure by a whole series of other forms. The Snowdrop (Fig. 48 III, IV) also has to be mentioned in this context.

The diversity of the tepal whorls in the genus *Iris* (Fig. 48 VIII) is of particular biological significance. The tepals of both whorls are corolloid in shape and are differentiated into a claw and a broadened blade. However, the inner ones ("standards") are erect and arch over together to form a kind of dome. The blades of the outer segments are spreading–deflexed (the "falls") and are often bearded with a brush-like band of hairs on the median line. The falls, together with the petal-like, coloured style branches ("crests") which stand up above them, form a labiate "flower". The crest is curved over the stamen which stands in front of it and is appressed to its lower side. Hence the crest protects the stamen in the manner of an upper lip (Fig. 48 IX). The anthers of the stamen are directed outwards. On the underside of the style branch or crest close to its outcurved tip a broad lobe with a receptive stigmatic surface is inserted, so that pollen brought by insect visitors entering the flower is rubbed off against it. Then, as the insect creeps out again, the fertile surface of the crest presses down on it. Hence the flower of the *Iris* can be described in biological terms as a combination of three labiate "flowers".

1.4.4.4   *Zygomorphic forms of the perigon*

Examples have already been quoted (*Gladiolus byzanthinus* and *Antholyza aethiopica*, section 1.2.4) of perigons with a zygomorphic structure, in which differences in the size, orientation or patterning of the perigon segments are observed as a result of the influence of gravity. The Scitamineae, and above all the Orchidaceae present additional examples of zygomorphic flower structure which are characteristic of these monocotyledonous families (see also section 1.4.3.4).

1.4.4.5   *Syntepaly*

Congenital fusion of the tepals, to a greater or lesser extent, occurs among various monocotyledonous families. The tepals of both successive whorls can be united to such an extent that they can appear as a single whorl. A perigon tube and a perigon limb can be distinguished, as in sympetalous corollas. It is evident, however, that the congenital fusion of monocotyledonous tepals does not approach the same systematic value as does sympetaly in the dicotyledons. In other words, free and syntepalous perigons can be found in closely related genera or even in different species of the same genus. For instance, the tepals in the genus *Ornithogalum* of the tribe Scilleae of the Liliaceae are free, as well as in

Scilla bifolia, S. amoena and other *Scilla* species, whereas in *Scilla non-scripta* they are joined at the base, and almost completely united in the urceolate perigon of *Muscari.* The degree of syntepaly and the relative sizes of the perigon tube and limb can be completely different in otherwise closely related species.

There is further scope for variation in the form of the perigon tube, which may be campanulate or urceolate (Fig. 48 VII), or elongated, as occurs in many species of *Aloe*, and in Meadow Saffron (*Colchicum autumnale*, Fig. 48 V). As in the sympetalous dicotyledons, the stamens can here also be united to a greater or lesser extent with a syntepalous perigon. This serial gamophylly, however, is not necessarily coupled with lateral gamophylly, as is shown in the example of *Bulbocodium vernum* as researched by Troll (1957, p. 47 et seq.). In contrast to the flower of *Colchicum*, which superficially resembles the flower of *Bulbocodium*, the tepals are only united for a short distance at their bases, and for the rest are long-clawed and free. They are "only held together by the closely enfolding foliage and scale leaves, as becomes clear when they are dissected out" (Troll, see Fig. 48 VI). It is therefore all the more remarkable that the filaments of the stamens are inserted above on the broadened distal section of the perigon segments.

The labiate flower of the banana (*Musa*) comes about from the congenital fusion of five tepals into a labellum, placed opposite the median, adaxial tepal which serves as an upper lip.

Amongst the dicotyledonous families there occur perianths whose segments are spirally arranged or in whorls of three (or occasionally in whorls of two) and which are not differentiated into a calyx and corolla. These occur in the families of the Magnoliales (e.g. Magnoliaceae, Illiciaceae, Schisandraceae, Lauraceae, etc.) and also in the Cactaceae, where the perianth is sometimes referred to as a "perigon of bracts". These are all considered to be genuine perigons. However, when further examples of perigon formation are sought among the dicotyledons, the question immediately arises as to whether a simple perianth represents a genuinely homochlamydeous perigon, or whether it has been derived secondarily, through the loss of the corolla (i.e. apopetaly).

1.4.5 Achlamydeous flowers, primary and secondary apetaly

The flowers of European Ash (*Fraxinus excelsior*) provide an example of secondary reduction. The flowers possess neither calyx nor corolla, i.e. they are achlamydeous, and are either hermaphrodite or unisexual. The hermaphrodite flowers (Fig. 50 IV) consist of a slender pistil, with a two-lobed stigma, and two stamens on short filaments which are placed opposite one another and decussate to the branches of the style. The male flowers consist of just two stamens (I), and occasionally a rudimentary pistil as well (II). The female flowers consist of a two-lobed pistil alone (VIII). Hermaphrodite and female flowers are connected, however, by a series of intermediate forms, which show a progressive reduction of the stamens (V to VII). These flowers represent the ultimate members of a reduction series, which is understood as a transition towards anemophily. The perfect flower of the Oleaceae consists in fact of a 4-merous, gamosepalous calyx, a similarly 4-merous sympetalous corolla, two stamens and a pistil composed of two carpels. This basic plan applies to the majority of the genera of the

Oleaceae, as in the Lilac (*Syringa*), the Olive (*Olea*) and in *Forsythia*. A loss of the corolla already occurs in many species of *Olea* in the section *Gymnelaea* and, on the other hand, male ash flowers occasionally appear which still have a calyx. Connecting links to the floral structure of other members of the Oleaceae are provided by the Manna Ash (*Fraxinus ornus*, Fig. 50 III) and some related species. Like the European Ash, the Manna Ash is also polygamous, and it also possesses both hermaphrodite and unisexual flowers, and even on the same tree. In contrast to the common ash, however, all these kinds of flower are usually provided with a calyx and corolla, although even here there are occasionally male flowers without a corolla.

For the ash it therefore seems obvious that the naked and unisexual flowers derive from hermaphrodite flowers with a complete perianth through the loss of the calyx and corolla as well as the stamens, or of the corolla only.

Caution is also necessary in the interpretation of the simple perianths from other families as to whether these are genuine perigons. This is particularly true of many of the families which were formerly grouped under the title of the Monochlamydeae. If the interpretation is accepted, that the Fagales is derived from the Rosales via the Hamamelidaceae, then its simple perianth (whenever

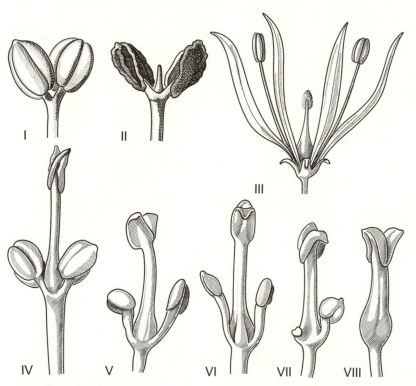

Fig. 50 I, II, IV-VIII *Fraxinus excelsior*, I, II male flowers, II with rudimentary ovary, IV hermaphrodite flowers, V-VII hermaphrodite flowers with more or less rudimentary stamens, VIII female flower. III *Fraxinus ornus*, flower. III original, the rest after Troll.

present) cannot represent a genuine perigon, but rather simplification by loss of the corolla, i.e. it is **apopetalous** (N.B. not **apetalous**, which denotes a primary absence of the corolla, which of course has not yet been conclusively established). Similar remarks must apply to the naked and **achlamydeous** flowers of the Salicaceae, if their derivation from the Tamaricaceae is to be accepted, since the latter possess a differentiated perianth. The lack of the perianth in the flowers of the Salicaceae is therefore not a primitive but a derived character. Questions of this kind may often be resolved via clarification of the systematic position of the families concerned on the basis of comparative investigations of other character complexes (on this subject see also Wagenitz 1975).

1.5   The androecium

1.5.1   The organisation of the stamens

The differentiation of the angiosperm stamen into a **filament** and an **anther** has already been discussed. The anther generally shows a longitudinal division into two (usually) equivalent **thecae**, which are joined to each other and to the filament by what is called the **connective**. Each of the thecae consists of two pollen sacs, which are usually elongate and united together. These pollen sacs have already been referred to as microsporangia. The thecae can therefore be compared with the synangia of an eusporangiate fern.

When considering the examples commonly quoted in an explanation of the differentiation of a stamen it may not seem to be justifiable to denote the unifying zone which appears between the two thecae, the **connective**, by what is admittedly nothing more than a descriptive and not a morphological term. This region can, however, be much more prominently developed and contributes considerably to the form of the anther. The many and varied outgrowths and appendages of the connective deserve special mention, since their occurrence is highly characteristic for many families, as for example the Melastomataceae. The special form of such appendages of the connective is not only an important diagnostic character for many plant groups, but is often also of biological significance. Furthermore the forms of the thecae or pollen sacs and their topological properties are extremely varied, as is made clear by way of introduction in the example with just one variable in the anther cross-sections reproduced in Fig. 52 II to V. In the cross-section shown in diagram III, both thecae are placed side by side on the lateral faces of the anther (the latrorse condition), whereas in IV they face inwards on the ventral side (introrse), and in V they are directed dorsally (extrorse).

1.5.2   Histogenesis and anatomical structure

The manner in which the stamens develop has already been mentioned above. According to the analysis of the histogenetical processes due to Kaussmann (1941), the mode of inception of the stamens is equivalent to that of the leaves. This will be explained by means of the example of *Cleome gigantea* which was investigated by Kaussmann. Just as in the foliage leaves and in the perianth segments, the production of the stamens often begins with the elongation and periclinal division of subepidermal cells (Fig. 51 I a_{1-6}); most frequently the divisions begin below the subdermatogen. After further divisions a subapical initial cell becomes recognisable (section 1.4.1.2), from which the longitudinal development of the primordium proceeds for a short time (II, III). After preliminary all-round growth of the primordium by periclinal divisions of the subepidermal cells the differentiation of rows of submarginal initial cells follows on both lateral faces of the primordium, and these take over its lateral growth by a series of alternating divisions periclinal and anticlinal. Furthermore a thickening of the median part of the primordium can take place as a result of periclinal divisions in the subepidermal cell layer. By this means the initially hemispherical primordium develops into a somewhat elongated organ which is usually elliptical in cross-section, and which tapers off towards its place of insertion. This section remains at this thickness and represents the beginning of the filament. This remains short while the internal and external development of the anther is completed, and does not lengthen again until the anther is ready. The filament often does not reach its full length until after the flower has opened. This is well-known in the flowers of grasses, where it is particularly impressive since it can take place within the space of only a few minutes (on this point, see Richter 1929, p. 61).

The beginning of tissue differentiation is often indicated by the fact that the anther takes on a four-cornered shape, due to cell divisions in the subepidermal layer by walls parallel to and below the surface of the longitudinal corner zones, which leads to the formation of four pollen sacs (Fig. 51 IV). From the first of these cell divisions one row of inner, archespore cells and one row of outer cells arise. Whilst the cells of the archespore continue to grow and multiply by repeated division, the outer cells divide, on the one hand, by their anticlinal walls, corresponding to the growth of the archespore and, on the other hand, by periclinal walls, and usually so that three layers of cells are created (V, VI, VII to IX). The innermost of the layers of the wall becomes the **tapetum**, whose cells serve for the nourishment of the archespore cells, from which in turn the pollen mother cells and finally the pollen grains arise. The tapetum can act as a **secretory tapetum**, which is a layer of glandular tissue covering the inner surface of the pollen sac. The cell walls of this layer remain intact during the complete development of the pollen grains and the cells secrete metabolic products into the pollen sac. Occasionally this tapetum consists of two or (more rarely) more layers. However, the tapetum cells can also penetrate between the pollen mother cells as an **amoeboidal tapetum** or even form a **plasmodial tapetum** by the disintegration of their walls (see further the researches of Carniel 1963 and Lersten 1971).

In some cases the inward part of the tapetum on the side of the connective is swollen owing to multiple layers of cells or due to cell enlargement ("inner tapetum", Carniel 1963). This should not be confused with the so-called **placentoid** (Chatin 1870), which is a longitudinally-directed "ridge of tissue which projects into the pollen sac on the side of the partition wall" (Hartl 1963; see Fig. 52 VI). It is probable that the placentoid could have value as a systematic character, as it is met with on the one hand in almost all families of the Tubiflorae, and on the other hand in the Zingiberaceae, Cannaceae and Orchidaceae.

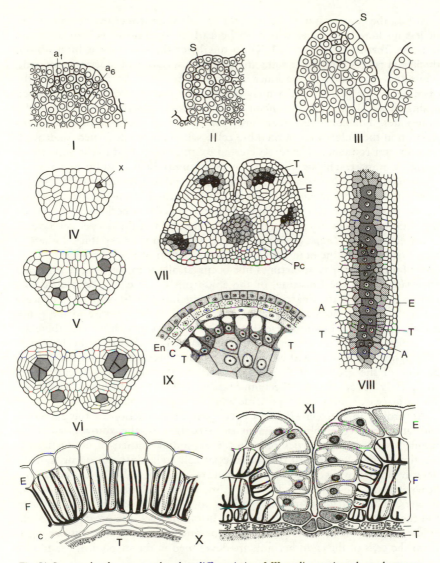

Fig. 51 Stamen development and anther differentiation. I-III median sections through stamen primordia of different ages in *Cleome gigantea*, I beginning of stamen formation, II, III, construction of the subapical initial cell S with periclinal-anticlinal mode of division. IV-VI cross-sections through anthers of different ages in *Chrysanthemum*, in IV at × the first periclinal division of a subepidermal cell, V archespore cells (shaded) in all four corners of the young anther, VI further construction of multi-layered anther walls. VII, VII *Vinca rosea*, cross-section through an anther (VII) and longitudinal section through part of an anther (VIII) at the time of pollen sac development, E epidermis, T tapetum, A archespore, Pc procambial strand. IX *Hemerocallis fulva*, cross-section in the region of a pollen sac, En endothecium, C disappearing layer which later disintegrates. X *Lilium pyrenaicum*, cross-section through the wall of a ripe pollen sac, F fibrous layer (= endothecium), XI *Tulipa gesneriana*, cross-section through the stomium region of a theca. I-III, XI after Kaussmann, IV-VI after Warming, VII, VIII after Boke, IX, X after Strasburger or Firbas.

Whilst the epidermis of the anther, the so-called **exothecium**, remains more or less unaltered, the layer which lies beneath it, the **endothecium**, develops into the **fibrous layer** (Fig. 51 X). In this layer the cell walls exhibit fibrous thickened ridges running at right angles to the surface and tapering outwards, while they are united at the inner cell wall. These cells later bring about the opening of the pollen sacs, namely when they wither away after the pollen grains ripen. By then they contain only a watery substance which gradually evaporates. The cohesive forces caused by this bring about considerable contraction in the outer walls of the fibre cells, while the strengthened inner walls offer greater resistance. The forces created by this finally lead to the anther wall being torn apart. This usually happens as a longitudinal splitting in a zone of preformed tissue, the **stomium**, which lies in the furrow between the pollen sacs (Fig. 51 XI). The endothecium as a rule consists of one cell layer; sometimes, however, it is also multi-layered (due to preceding periclinal divisions of the endothecial cells or corresponding differentiation of the deeper cell layers), as for example in *Tulipa gesnerana* (Fig. 51 XI), or in the very thick anthers of many orchids. On the other hand the differentiation of the subepidermal layer as a fibrous layer may sometimes not occur at all, as in the genera *Erica* and *Rhododendron* of the Ericaceae, or the epidermis can itself be developed as a fibrous layer, as in the Epacridaceae or in the genus *Loiseleuria* of the Ericaceae (see below). The latter also appears at first sight to be the case in the genus *Struthanthus* in the Loranthaceae (Fig. 52 XIV). In reality, however, there are still isolated cells of the epidermis situated between the endothecial cells which, although they are initialised in the usual manner, become pulled apart later on because of the growth of the cells beneath, so that the epidermis no longer completely covers the endothecium. Staedler (1923) found similar cases in the Casuarinaceae, Proteaceae, Piperaceae and the Euphorbiaceae.

According to Staedler (1923), a corresponding reduction of the epidermis and other reduction phenomena in the structure of the anther wall occur frequently among the Urticiflorae. In the exploding anthers of the Urticaceae the fibrous development of the endothecial cell walls does not occur, and the delicate anther walls rupture shortly before anthesis. Also in cleistogamous flowers the fibrous layer is reduced by comparison with that in chasmogamous flowers, and is usually completely lacking in plants which flower underwater. On the other hand, in *Phthirusa pyrifolia* (Loranthaceae), the greater part of the epidermis and the remaining otherwise parenchymatous anther tissues, apart from the cells of the endothecium, are developed in the manner of the fibrous cells.

The one or more intermediate cell layers between the tapetum and the endothecium usually become greatly stretched in the tangential direction because of the growth of the pollen sacs and are often destroyed during the further development of the anthers (**disappearing layers**).

The **longicidal** mode just described, in which the anthers dehisce by longitudinal slits as a result of forces of cohesion caused by drying out of the fibrous layer is not the only type which occurs in the Angiosperms. The dehiscence can also come about by means of **transverse slits, pores** or **valves**. Anthers which open by valves are found, for example, in the Lauraceae (Fig. 52 VII, VIII), Hamamelidaceae and the Berberidaceae. The development of the fibrous layer

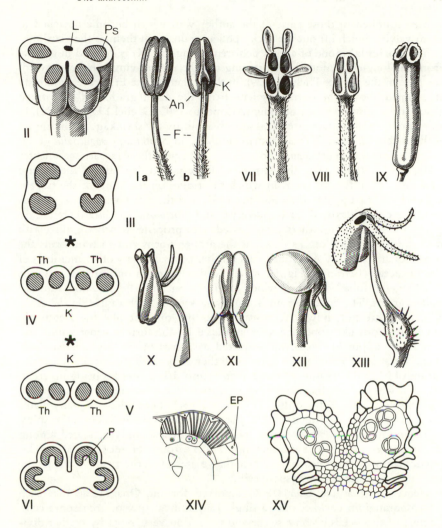

Fig. 52 I Stamen of *Hyoscyamus niger* seen from in front (a) and from behind (b), F filament, An anther, K connective. II anther in transverse section, schematic, Ps pollen sac, L vascular bundle. III cross-section of an equifacial anther of *Papaver rhoeas*. IV, V schematic cross-sections of an introrse (IV) and an extrorse (V) anther, Th theca, the * shows where the floral centre lies, VI schematic cross-section of an anther in which placentoids (P) are developed. VII-IX Stamens of: *Cinnamomum zeylanicum*, anther dehiscing by valves (VII), *Persea americana*, ditto, but valves still closed (VIII), *Ochna multiflora*, poricidal anther (IX). X *Vaccinium uliginosum*, poricidal anther inverted, with horn-shaped appendages. XI, XII young anthers of *Arctostaphylos alpinus* before (XI) and during (XII) the inversion. XIII *Arctostaphylos uva-ursi*, stamen with open pores. XIV *Struthanthus calobotrys*, transverse section of the anther wall, Ep separated epidermal cells. XV *Erica carnea*, transverse section of a theca; an endothecium is absent although the pollen tetrads have already developed. After Schimper (I), Schaeppi (III), Troll (IV, V), Hartl (VI), Goebel (IX, XIV, XV), Amberg (XI, XII, altered) and Warming (XIII), the rest original.

is here confined to those parts of the anther wall which become detached to form valves, which lift out the sticky pollen grains with them as they curl up.

The **poricidal** mode of anther dehiscence (Fig. 52 IX) is derived in general from the longicidal mode by shortening the slit and moving it either towards the base or the tip (see Richter 1929). In the anthers of the Ericaceae, as already mentioned above, the pores originate from the disintegration of preformed tissue at the position of the opening (Artopoeus 1903, Goebel 1933, p. 1982). In this case the pollen grains are often squeezed out by the shrinkage of the anther walls as they dry out.* This happens similarly in *Spathicarpa sagittifolia* (Troll 1928c), *Zantedeschia aethiopica* and other members of the Araceae (see Richter 1929, Goebel 1933). Here the extrusion of the pollen depends on the inverse structure of the fibrous layer, in which the ridges on the walls of the fibrous cells are reversed so that they fuse outwards and the outer wall of the fibrous cells is greatly thickened. In *Cryptocoryne* and *Microcasia pygmaea* there are no pores, but instead the anthers are provided with processes which are filled with delicate tissue which breaks down as the anther ripens and so transforms the processes into regular "exit tubes". The anther pores of many other members of the Ericaceae (*Vaccinium*, Fig. 52 X) are similarly drawn out into processes or "shaking-out tubes" (Goebel) which can function with the dry, non-adhesive pollen grains. The endothecium is usually lacking in such cases (Fig. 52 XV). The tubes are not, however, positioned at the tip, but like the frequently appearing horn-like appendages (as indicated by the former order name "*Bicornes*") on a spur-like, backward-directed basal part of the "hypopeltate" anthers. These are turned to face upwards, either early on during development by means of inward curvature of the stamen primordia or later by being inverted, as is shown in both the late stages of anther development in *Arctostaphylos alpinus* (Fig. 52 XI, XII, see also Leinfellner 1963c). The anthers of many plants do not burst open, but open gradually as the tissue of the walls decays or splits.

The number of microsporangia (pollen sacs) is seemingly increased among the members of not a few larger or smaller groups of related plants. This phenomenon is explained by the fact that plates of sterile tissue occur in the sporangia between the archespore cells.

Hence for example Beer (1906) reported for the Onagraceae that each microsporangium encloses only a single longitudinal row of archespore cells. Certain of these cells become sterile and form transverse septa by further divisions. In *Annona cherimola*, according to Samuelson (1914), each pollen mother cell is isolated by a sterile septum. In many instances the sterile cells apparently form tapetum cells, unless septa or individual cross fibres are formed already by the tapetum ("trabecular tapetum"), as is said to be the case for the Balsaminaceae. The most familiar examples of the formation of septa in the microsporangia must be the Loranthaceae, the genus *Rhizophora* (Rhizophoraceae) and the properties of the Mimosaceae which are described in detail by Engler (1876). In these also each archesporium consists of only one cell row which is interrupted by sterile cells. In this manner two or more mutually isolated archespore cells are sometimes formed in the four corners of the anther. Each of the archespore

---

* In *Rhododendron* the pollen is neither squeezed nor shaken out, but is removed from the open anthers as the visiting insects brush past by means of the sticky threads of viscin which bind the pollen tetrads together (Richter 1929).

cells then divides into four or sometimes sixteen daughter cells, all lying in one plane. Each of these daughter cells behaves like a pollen mother cell by undergoing a subsequent meiosis and creating four pollen grains. All the cells which originated from one archespore cell remain joined together even after the pollen has ripened, so that groups of sixteen, thirty two or sixty four pollen cells remain united together in what is called a massula. After the reduction of the sterile tissue between the individual massulae these are then left in groups of two or more in a pollen sac. The current state of knowledge in the existing literature about the structure and distribution of septate microsporangia was succinctly summarised by Lersten (1971), and Endress and Voser (1975) give further results (Flacourtiaceae, Myrsinaceae, Sterculiaceae).

1.5.3 Reductions of thecae and sporangia

The morphological aspects of sporangial reduction are of considerable importance. There are two different possibilities to be distinguished (Goebel, 1933, p. 1892; see also Trapp 1956a, b). That is to say, either one of both thecae may be reduced, which is called **lateral** reduction, or it is either only the two anterior or the two posterior sporangia which degenerate, and this is termed **facial** reduction.

Goebel quotes an example of facial reduction which was investigated in more detail by Demeter (1922) for the Asclepiadaceae-Asclepiadoideae. In this case there are only two pollen sacs to begin with, namely both the anterior ones, in contrast to the Periplocoideae and the Asclepiadoideae-Secamoneae. On the other hand, certain genera of the Lauraceae develop only the two posterior sporangia, whereas others have only the anterior pair (see Mez 1889). On the whole facial reduction is a relatively rare phenomenon.

The anthers of *Arceuthobium* (Loranthaceae) are "unilocular from the very beginning", and when mature a "central column of sterile tissue" is surrounded by the cylindrical archespore (Cohen 1968).

Examples of **lateral** reduction, i.e. loss of one theca, are quite numerous. The best-known example is that of Indian Shot (*Canna indica*, Fig. 74 I, II and also III) in which there is only one fertile stamen per flower and this bears only one theca, so that only one half is fertile, while the other half develops like a petal.

Thecal reduction can also occur in such a way that the form of the stamen is unchanged, apart from the actual halving of the anther. This is the case in more than a few genera of the Acanthaceae, in which the regression of one theca during anther development can readily be observed (Fig. 53 XIII to XV). The structure and development of such anthers has been studied very thoroughly by Trapp (1956a, b), whose research will often be referred to in what follows.

The characters of *Salvia* species are of particular interest, because the sterility of one theca is linked with the development of a lever mechanism which is important for pollination. In the labiate flowers of Salvias only the anterior or lower pair of stamens is functional, whereas the posterior pair of stamens is rudimentary (Fig. 53 IX). The filaments of the functional stamens remain short. In Meadow Clary (*Salvia pratensis*) only one of the two thecae is fertile; it is placed at the end of a greatly extended branch of the connective. The other theca atrophies completely, but the corresponding section of its connective expands considerably and forms a small plate (Fig. 53 VI). The distinctive

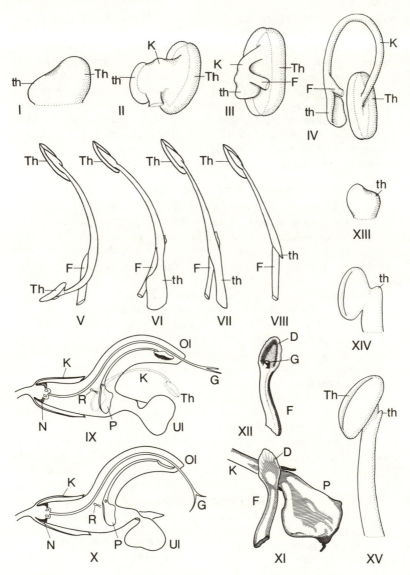

Fig. 53 I-IV Stamen development in *Salvia pratensis*, th sterile, Th fertile thecae, F filament, Ko connective. V-VIII Reduction of thecae within the genus *Salvia*, V *S. officinalis*, VI *S. pratensis*, VII *S. splendens*, VIII *S. verticillata*, IX-XII *S. pratensis*, IX, X flower in median section, schematic, showing the lever mechanism which effects pollination, K calyx, Ol, upper and Ul, lower lip of the corolla, N nectary, G style, P sterile part of half-anther (plate) of the fertile stamen, R rudimentary stamen. XI Operation of the anther joint: following pressure from the left the plate has moved up to the right and displaced the connective arm of the fertile theca (K) downwards, XII filament of the fertile stamen, where the connective is flexible and attached to the joint G, and the broadened "tip" of the filament forms the socket D. XIII-XV Stamen development with reduction of the theca in the genus *Crossandra* (Acanthaceae). After Trapp (I-IV, XIII-XV), Hruby (V-VIII, from Trapp), Knoll (IX, X) and Troll (XI, XII).

structure of the two thecae is clearly discernable from the very beginning in the ontogeny of the stamen (Fig. 53 I to IV). The filament is narrowed towards the insertion of the anther. Thus the elongated lever with the fertile theca at the end of the long arm and the plate ending the short arm is flexibly connected to the tip of the filament, which according to Troll (1957) widens to a shell-like projection below this point and functions like the socket of an articulated joint (Fig. 53 XI, XII). The plates of the two stamens are loosely attached to each other and close the throat of the corolla. As a nectar-seeking insect pushes against this barrier, so the long lever arms of the connectives swing down and apply the fertile thecae to its back (nototribal pollination, see section 1.4.3.8). The flowers are protandrous in this case and the receptive stigmas of older flowers are in the same position as the ripe thecae were previously, hence cross-pollination occurs.

The highly modified stamens of *Salvia pratensis* could be arranged as part of a sequence of forms which demonstrates the progressive reduction of a theca. To begin with, there is *Salvia officinalis*, in which the short arm of the connective still bears a fertile stamen (Fig. 53 V). After this would come *Salvia pratensis* (Fig. 53 VI), *S. sclarea*, and various others. Next there would immediately follow the species of the Section *Calosphace*, e.g. *S. splendens* (Fig. 53 VII), and then further species with yet further reduction until finally there is *S. verticillata* (Fig. 53 VIII) where one theca is almost entirely lacking. The final stage of reduction, as represented by *Salvia verticillata* in Fig. 53 VIII, also corresponds to the stamen type of *Rosmarinus officinalis*, although there, in contrast to the *Salvia* species, it is only the two anterior stamens which remain fertile.

Reduced thecae are also quite frequent in the Scrophulariaceae, and here too are occasionally linked with considerable modifications of the stamens. Hence for example in species of *Calceolaria* the two thecae can be separated some distance from each other by elongated, stalk-like connectives, and combined with this there is a greater or lesser degree of reduction of the pollen sacs in the diminished thecae. The lower stamens of *Pseudosopubia obtusifolia* (Fig. 58 XVII) even resemble the highly modified stamens of *Salvia*. In other members of the family the sterile thecae are transformed into longer spur-like appendages.

1.5.4 Divergence and dislocation of thecae, and transverse anthers

According to Trapp (1956b), other types of transformation in the form of the stamen occur quite frequently, namely **thecal divergence**, **thecal dislocation** and **transverse insertion** of the anther. In all cases it is a matter of displacement by modified and unequal growth in the connective region.

In **thecal divergence** what happens is that in the course of their development the thecae become more or less widely separated at their bases. According to Trapp this comes about by means of "an active movement of unfolding at a late stage of development, for which the intercalary meristem at the base of the connective is mainly responsible". Divergent thecae are not uncommon in the Labiatae, Scrophulariaceae (Fig. 54 I to IV) and the Bignoniaceae, and also in other families. The final angle of divergence reached in different species is rather different. It is not uncommon for the thecae to diverge by more than 180° so that they point obliquely upwards (Fig. 54 V).

Asymmetrical growth of the connective region can also bring about a vertical displacement of the thecae, a so-called **thecal dislocation**, which in detail may be very different. This can be observed in many members of the Acanthaceae (Fig. 54 XI to XIII, XIV to XVII). Asymmetrical growth is also the cause of the spiral twisting of the long-exserted anther tips of *Nerium oleander*, according to the researches of Kunze (1979).

Examples of **transverse insertion** of the anthers occur among the Labiatae, Scrophulariaceae, Gesneriaceae and the Bignoniaceae, and elsewhere. This comes about through asymmetrical growth in the connective region occurring immediately before anthesis (Fig. 54 VI to X). In many cases, however, a corresponding orientation of the anther is caused by distortions of the filaments.

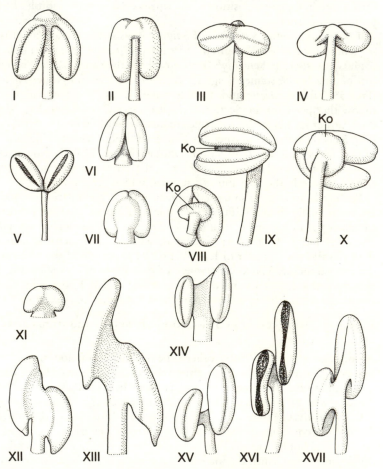

Fig. 54 I-IV Divergence of thecae in *Russelia juncea*. I, II Anterior and posterior surfaces of the anther before and III, IV after divergence. V Divergence of more than 180° in *Erythrococca aculeata* (Euphorbiaceae). VI-X *Gratiola officinalis*, development of the stamens with the anther later becoming transverse because of unequal growth in the connective region Ko. XI-XIII Stamen development of *Beloperone guttata* with dislocation of thecae, XIV-XVII, similarly in *Isoglossa lactea*. After Trapp.

1.5.5   Fusion of thecae and of sporangia

The properties of the stamens of *Scrophularia, Verbascum* and a series of other members of the Scrophulariaceae must be distinguished from the transverse insertion caused by asymmetrical growth of the connective. In the former the pollen sacs are curved and form two kidney-shaped swellings which run across the apex of the stamen (Fig. 58 XII). A median (Fig. 55 V) and a tranverse (IV) longitudinal section through a ripe anther of *Scrophularia nodosa* show that there are really only two pollen locules (labelled Loc) present here. They dehisce by means of a single slit which serves for the entire anther, after the septum which separates the internal "locules" has previously disintegrated. In other words, what we find here is the same opening mechanism as would otherwise occur in both the pollen sacs belonging to a theca. This has therefore been interpreted as an anther halved by lateral reduction, which has later been "set upright" and has taken up a terminal position.

The ontogeny of the stamens shows however that the later orientation of the anthers is already determined at their inception. At the beginning of their development (Fig. 55 I) all the stamen primordia are equivalent in form, even including the median primordium, which later develops into a petaloid, glandular staminode. All the primordia are orientated with their broader faces towards the flower centre and its periphery. At a somewhat later stage of development (Fig. 55 II), the distinctive development of the median stamen primordium and the transverse dividing line between what later become the two microsporangia over the apices of the other primordia (III), are already recognisable. The differentiation of the archespore begins in the familiar way by cell divisions leading to the growth of four edges which are recognisable in a cross-section (Fig. 55 IX). A series of transverse lengthwise sections through anther primordia of different ages (VI to VIII) shows, however, that the differentiation of the archespore very soon progresses from the margins towards the apex, so that a completely uniform pollen sac is created in the inner and outer halves of the anther respectively. According to Trapp (1956a, b), what we have here is therefore a **primary fusion of the thecae** occurring at an early stage, which leads to the formation of a **synthecous anther**, which is principally distinguished from a dithecous anther by the fact that in the latter "the apical part of the primordium is not utilised for the formation of sporogenous tissue and remains sterile". In anthers which are termed **secondarily synthecous** by Trapp the fusion of the thecae does not take place until the pollen ripens, or else occurs immediately before anthesis, as a result of the decay of tissue of the dividing wall. This phenomenon is found in *Digitalis purpurea* and other members of the Scrophulariaceae, and in the Bignoniaceae, Gesneriaceae (*Monophyllea,* A. Weber 1976) and many related families, and in polypetalous families also.

The same type of archespore formation as in *Scrophularia nodosa* is also observed in *Selago spuria* and *Hebenstreitia dentata* in the Selaginaceae (Fig. 55 X, XI). In this case the anthers in their initial stages also look like those of *Scrophularia nodosa*, but later on become laterally placed on the filament as a result of asymmetric growth processes, and therefore have often been interpreted as monothecous anthers derived by lateral reduction. The species of the genus

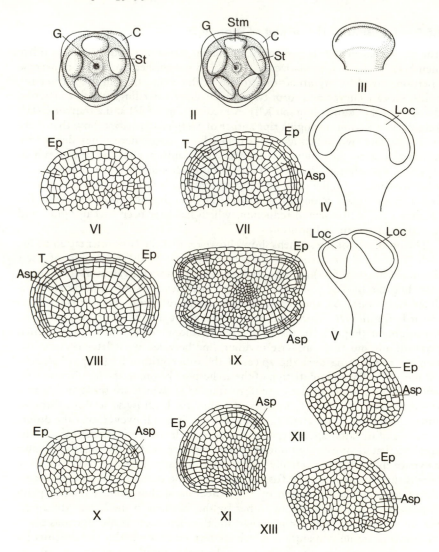

Fig. 55 I-IX *Scrophularia nodosa.* I, II Flower primordium shortly after differentiation of the stamens (I) and somewhat later (II), the corolla C removed, St stamen primordia, Stm primordium of the staminode, G primordium of the gynoecium. III Stamen of a somewhat later age; IV, V transverse and median lengthways sections through a ripe anther, Loc pollen locules. VI-VIII transverse lengthways sections through anthers in different stages of development, differentiation of the microsporangia, IX anther corresponding to stage VII in cross-section, Ep epidermis, Asp archespore, T tapetum. X, XI *Hebenstreitia dentata*, transverse lengthwise sections through anthers of different ages, differentiation of the microsporangia, XII, XIII the same in monothecous anthers of *Rosmarinus officinalis*. After Trapp.

Verbascum, section *Thapsus*, are especially interesting since they possess anterior stamens which are constructed like those of *Hebenstreitia* with "laterally decurrent anthers" (Fig. 58 XI), whereas the anthers of the posterior stamens (Fig. 58 XII) have the form corresponding to *Scrophularia nodosa*.

By way of contrast, the anthers of *Rosmarinus officinalis* develop in quite a different way, since they really have become monothecous by lateral reduction. This is confirmed by observation of the tissue development (Fig. 55 XII, XIII), where the early reduction of one theca is clearly visible. Moreover, it is clear that the process of construction of the archespore begins from only two places in the unreduced theca (hence in the transverse lengthwise section shown in the diagram from one place only).

1.5.6 Stamens as diplophyllous organs

The results established by Trapp concerning the structure of synthecous anthers, as well as a series of other stamen characters, are in good agreement with the ideas developed by Baum (1949c, 1952a, 1953a) and Leinfellner (1956b, d, 1957a, 1958a, 1960a; Baum and Leinfellner 1953a) about the fundamentals of stamen morphology in the angiosperms. According to these ideas we arrive at an understanding of the structure of stamens and their various modifications if we consider at least the majority of stamens to be peltate or diplophyllous leaf organs, like the carpels. Also stamens "which in their mature state, show no obvious similarity to peltate leaves" are, according to the researches of Baum (1949c), "*quite unmistakably peltate during early ontogenesis*".

The peltate leaves of the garden Nasturtium (*Tropaeolum majus*, Fig. 56 X) will serve as a well-known example of peltate leaf structure. A hypothesis which explains the formation of peltate leaves is that of the unifacial structure of the petiole, i.e. the suppression of the upper surface of the leaf in the region of the petiole. In bifacial petioles the upper leaf surface is always clearly developed (Fig. 56 I). Correspondingly the leaf margin here shows continuous progress from the point of origin of the leaf, along the margins of the petiole and as far as the leaf tip. In many petioles, however, the upper surface is more or less furrowed and relatively narrow so that the expansion of the lower surface is comparatively more pronounced (Fig. 56 II). This tendency can even lead to the complete suppression of the upper surface, in which the vascular bundles which originally ran along the outermost edges can be united into a single strand which lies opposite the middle nerve (median) as a ventral median on the adaxial side of the leaf (IIIa to d). The unifacial structure of the petiole produces an important effect on the direction taken by the leaf margin: in the transition zone between the petiole base and the petiole the two leaf margins run together in a sharp angle (III) or transversely across the petiole (IV). The same happens at the base of the lamina (compare also V). The marginal growth of the blade now overlaps this "cross zone", creating a peltate leaf blade, which appears to be "stalked on the back", so to speak (VI). In the region of the leaf basis, a corresponding activity in the cross zone leads to the growth of a ligule (or so-called median stipel), which is more or less cucullate in shape. If the expansion of the margin of the blade of a peltate leaf lags behind the development of the surface, then cone-shaped or even tubular (ascidiate) blades are produced in extreme cases (diagrams VII to IX); a phenomenon which is largely responsible for determining the form of the carpel (see Figs 77, 78).

The cross zones are already clearly evident at an early stage of the ontogeny of peltate leaves. According to Baum (1949c etc.) this is also true for the early stages of development of the stamens. This is shown in the median longitudinal section of a very young peltate stamen given in Fig. 57f; this is "completely in agreement with the corresponding view of a young carpel". According to the

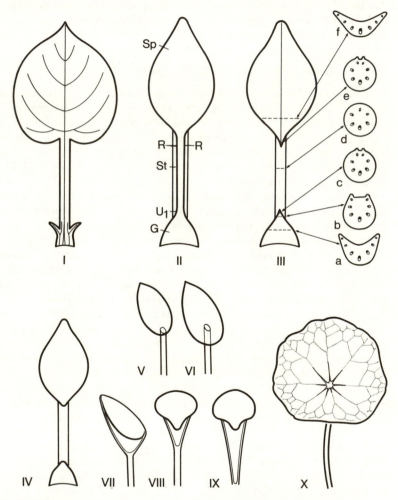

Fig. 56 Unifacial, peltate and tubular formations in leaf organs. I, II Leaves with bifacial petiole, base with (I) and without (II) stipels; the cross-section through the petiole of leaf II corresponds to diagram IIIb. III Leaf with unifacial petiole, a–f are cross-sections taken at different levels through leaf III; the leaf margins which run together and separate again at an acute angle in III are united transversely across the petiole in IV. V subpeltate, VI peltate leaf blades, VII–IX funnel-shaped and tubular blades. X peltate leaf of *Tropaeolum majus* (II–X after Troll, partly modified).

comparative researches of Baum, the further development of a stamen is "entirely a question of the intensity of growth in the cross zone of the peltate organ, so whether the peltate shape of the stamens remains clearly visible throughout, or whether it comes to resemble the normal four-edged shape to a greater or lesser degree during the course of ontogenesis, depends on the varied degrees to which the margins fold during their ontogeny". This is illustrated in the diagrams a–e of Fig. 57. It must be observed that in contrast to the tubular carpels (h) no central cavity is reserved, but that the tissue produced in the median area of the cross zone is congenitally fused with the rear faces of the anthers (i–l). The microsporangia then develop on the margins, which have remained free (marked with dots in Fig. 57). According to Baum and Leinfellner (1953a), it is perfectly clear from the ontogenesis that "the places where the pollen sacs are formed correspond with the morphological margin of the leaf". When the growth of the anther primordia takes place uniformly on all sides (Fig. 57 b, d) the peltate form of the anther is also still recognisable later on. *Alchemilla* and *Monophyllea* (Fig. 58 VI, VIa, VII; XIV, XV) are good examples

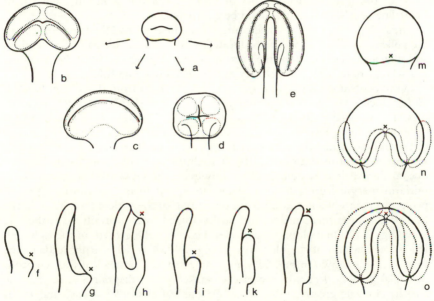

Fig. 57 Three schematic representations of the development of a ripe stamen from a young peltate leaf. a–e. Very young peltate stamen (a) which can develop ontogenetically into various different types of ripe stamen (b, c, d, e). f–l. Cross zone growth represented by median longitudinal sections (f–l): f a very young carpel or stamen primordium with a clearly developed cross zone, from which are derived, by various kinds of intensive growth, a latently (g) or a manifestly (h) peltate carpel or a stamen with a slightly (i), moderately (k) or strongly (l) developed cross zone. m–o. Folding of the 'leaf', represented in ventral view; m very young stamen with a (leaf blade) margin completely closed by the development of a cross zone, which is shown somewhat folded in n and highly folded in o. The highest point of the cross zone is marked with x, and in b–c, n and o the sporangia are outlined with dots. In n and o the secondary elongation of the two halves of the anthers is not taken into consideration (from Baum).

of this. The anthers of *Coleus thyrsoideus* (Fig. 58 VIII) are constructed like those of *Alchemilla*. These anthers were thoroughly investigated by Trapp (1956a, b), who cited them as an example of "superimposition of the sporangia". If the margins of the anther primordia are only slightly curved, then it is possible (as in *Verbascum* and *Scrophularia*), that two curved, parallel sporangia are formed, one behind the other, running over the apex of the anther (Fig. 57c, 58 XII). If the growth of the lateral parts lags behind, then as a consequence the margin becomes strongly curved in the shape of a noose (Fig. 57m–o, *Veronica, Gloxinia*, Fig. 58 XVIII). During their ontogeny the parts of the margin of each anther half which run roughly in the longitudinal direction of the stamen primordium can come to lie more or less parallel and side by side (Fig. 58 XIX), and the relation between them can become completely obscured as growth proceeds (Fig. 58 XX to XXIa). This results especially from the fact that on the one hand, in the vertexes of the curved margins the sporogenous tissue is usually interspersed with sterile tissue, and that on the other hand the development of the pollen sacs can be favoured during the further growth of the anthers, and hence determine the form of the ripe anthers to a great degree. This mode of development is evidently the cause of the form of many types of stamen. On this point the research of Leinfellner on *Viola* (1957a) should be compared.

The strongly curved or arched course of the margins and the later isolation of the pollen sacs which appear to be located on the ventral margins of the anthers and which are derived from the cross zone, agree generally with the phenomena which are termed **diplophylly** in the blades of some foliage leaves.

In this context the leaves of certain species of *Caltha* from the section *Psychrophila*, which were thoroughly investigated by Troll (1937), are suitable examples. The unusual property of these leaves (Fig. 59 VIII, IX) is that they are "double-bladed". That is to say, two additional leafy wings arise on the surface of the lamina, so that the whole blade appears to be four-winged, which is also shown by the cross-sections (II, III). The diplophylly of these *Caltha* leaves is due to their peltate structure. Here, however, the cross zone does not show uniform marginal growth, as in *Tropaeolum*, but forms two auricles, which project to the right and the left from the initials of the petiole (Fig. 59 I). In the bud primordium such auricles are usually folded over ventrally, as is the case here (Fig. 59 V). In diplophyllous leaves, however, the growth of the right and left lower margins of the blade (yy in Fig. 59 IV) which usually leads to the form of blade shown in Fig. 59 V is greatly reduced. Instead of this, vigorous marginal growth occurs in the places marked xx in diagram IV, so that the auricles elongate greatly, as VI shows for the leaf of *Caltha sagittata* and as VII shows for the leaf of *C. obtusa* shown in VIII. If in the course of further leaf blade expansion, the longitudinal growth occurs particularly in the region of the insertion of the auricles, they are later positioned as wings over the upper side of the blade, as is the case for *Caltha dionaeifolia* (Fig. 59 IX). The similarity of a leaf such as this to the anthers described above appears to be obvious.

According to the numerous studies of Baum and Leinfellner many, if not all, forms of stamen can be explained as a peltate or diplophyllous structure. This is also true for the stamens which Schaeppi (1939) termed **epeltate** or **impeltate**, according to Baum and Leinfellner (1953a). These (Fig. 60 III-VI, XI) were previously contrasted with the **peltate** stamens, in which "the tip of the filament is not inserted at the base of the connective, but on its surface", and in which the basal elongation of the anther can "originate towards the morphological upper face as well as towards the lower face" (Schaeppi 1939, p. 402), i.e. with adaxial or abaxial orientation.

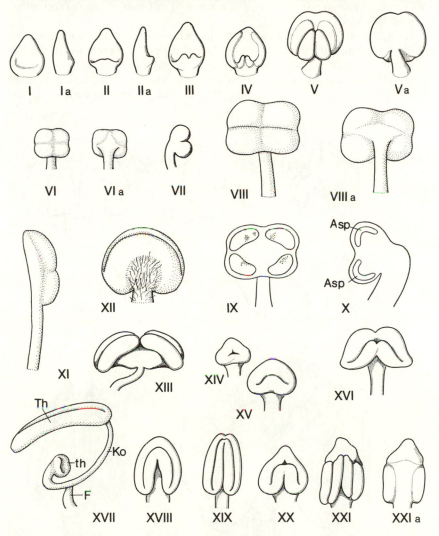

Fig. 58 I-V. Anther development in *Streptocarpus kirkii*. Ia, IIa lateral views of stages I and II, Va the reverse side of V. VI, VIa anthers of *Alchemilla mollis* in ventral (VI) and dorsal (VIa) view; VII *A. speciosa*, side view; VIII-X *Coleus thyrsoideus*, VIII ventral, VIIIa dorsal view, IX transversal, X sagittal sections; XI, XII *Verbascum olympicum*, anterior (XI) and posterior (XII) stamen; XIII *Lamium orvala*; XIV, XV young and older anthers of *Monophyllea horsfieldii*; XVI, XVII ripe anthers of *Streptocarpus gardenii* (XVI) and *Pseudosopubia obtusifolia* (XVII), Th fertile, th sterile thecae, Ko connective. XVIII, XIX young and ripe anthers of *Gloxinia hybrida*, XX, XXI of *Viola obliqua*, XXIa dorsal view of XXI. VI, VIa, VIII-X, XVII after Trapp, the rest after Baum.

Goebel (1931) named these two forms as **epipeltate** (Fig. 60 I) and **hypopeltate** (II). Instead of these terms the descriptive terminology uses the terms **dorsifix** and **ventrifix**; in particular moreover the **adnate** and **semiadnate** anthers are distinguished from the dorsifix, distinctions which finally are due to differences of proportion in the rate of growth of different regions of the anther (Fig. 60 VII-X).

According to Baum and Leinfellner the peltate form of the young stamen which they regularly observed is usually "obscured in the course of further development, since the cross zone of the stamen folds upwards in the manner of a diplophyllous leaf and hence forms a ventral blade section, which is fused medially with the dorsal blade section. This ventral section will be more or less

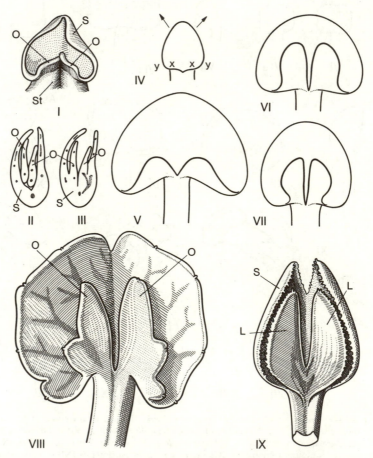

Fig. 59 Diplophylly. I-III *Caltha obtusa*, I blade of a young leaf, II, III cross-sections through a more or less completely developed leaf still enclosed in the bud, II taken in the centre and III at the base of the blade. IV-VII schematic diagrams for the explanation of leaf development in the genus *Caltha*, see text. VIII, IX unfolded leaves of *Caltha obtusa* (VIII) and *C. dionaeifolia* (IX). After Troll.

well developed, depending on the intensity of cross zone growth, and thus will essentially also determine the length and position of the pollen sacs which differentiate from the leaf margin" (1953a, p. 92). Correspondingly "the various forms of the mature angiosperm stamen..." "...are merely quantitative variants of the diplophyllous primordium. In undistorted and isodiametric further development this becomes a stamen with an introrse, basifix anther, which for us therefore represents the type of the angiosperm stamen" (p. 134). In these stamens which later exhibit a peltate form the "growth of the anther basis exceeding the attachment point of the filament takes place in a stamen which has already become diplophyllous..." and this development is "only a *secondary* growth process" (p. 96). On this point the development of the stamens of *Hemerocallis fulva* (Fig. 61 IV to VI) is to be compared with that of the ventrifix anthers (IX, IXa) with which the stamens of *Cobaea scandens* are provided (X to XII). Also in the case of the ventrifix stamen, which is illustrated here for *Cobaea scandens* (Fig. 61 IX, IXa), a normal (ventral) peltation and a subsequent diplophyllous folding can be recognised (X to XII, XIIa). The differentiation of that part of the anther section, which is here dorsal and downward-directed, first takes place at a late stage and has nothing to do with the peltation. The difficulties which arose in the interpretation of the ventrifix stamens as "hypopeltate leaf organs" (see Troll 1932b, p. 295) are hence resolved.

It is also secondary growth processes which bring about the form of sagittate (Fig. 60 V, Fig. 61 VII), x-shaped (Fig. 60 VI), H-shaped (Fig. 61 VIII, VIIIa) and **centrifix** anthers (Baum and Leinfellner 1953a). The latter are encountered among species of the Liliaceae and Crassulaceae (*Echeveria multicaulis*), and elsewhere. They are particularly unusual in that the filament is surrounded by the anther which in its basal part forms a tube, into which the filament may project as much as half the length of the anther before it is first attached there; in other cases (*Kniphofia aloides*) the filament tip is placed in a pit- or pocket-like hollow of the anther. Here also, however, the young stamens have a basifix structure, and only later the base of the anther begins to swell up around the point of attachment of the filament. After all one can also assume that the so-called **"undifferentiated"** stamens (Fig. 60 III, XI) are representing a secondary modification of the diplophyllous anther form. As already assumed by Troll and others, the equifacial, introrse and extrorse orientations of the thecae (Fig. 52 III to V) are due to later changes in proportion. Nevertheless, according to Baum and Leinfellner, the basifix and introrse anther corresponds to the original structure and is the type for the Angiosperm stamen, while the equifacial and extrorse anther structure is to be ascribed to a greater or lesser development of the "ventral blade".

If we have here compared the peltate or diplophyllous structure of the blade of a foliage leaf with the structures which occur in the anther, this does not imply any equivalence in the morphological sense. The observation that the stamens usually differentiate into a stalk-like filament and a broadened terminal section, the anther, does not prove that these are homologous with the petiole and blade of a leaf. This is all the more the case since the sporophylls appear to be modified in many respects by comparison with foliage leaves. Baum (1949c) clearly states that for this reason (in the context of the above quotation), we can only speak of a "blade-shaped leaf apex". Beyond this we can already establish

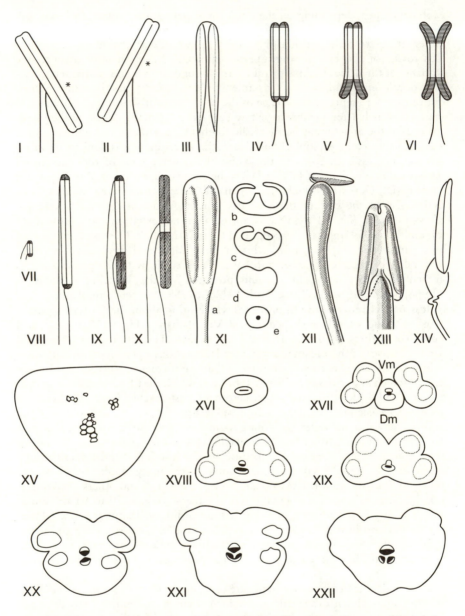

Fig. 60 I, II Diagrams of a hypopeltate (I) and an epipeltate (II) stamen in the sense of Goebel; the morphological upper surface is marked ★ (from Schaeppi). III–VI Diagrams of impeltate stamens in the sense of Schaeppi: III undifferentiated, IV non-sagittate, V sagittate stamens and VI stamen with X-shaped anther (after Schaeppi). VII–X diagrams for the explanation of the derivation of adnate (VIII), semiadnate (IX) and dorsifix (X) anthers from an undifferentiated primordium (VII), after Troll. XIa stamen of *Sparganium erectum* and b–e, a series of cross-sections through a stamen of *Smilax aspera* of similar structure. XII, XIII *Yucca filamentosa*, mature stamen in side view (XII) and young stamen from behind (XIII). XIV *Dianella caerulea*, stamen in side

for the foliage leaves, that the formation of the unifacial structures with the consequence of "peltation" of the associated laminar region is not confined to the petiole. In a leaf, unifacial and peltate leaf sections can alternate with one another several times, as is especially noticeable in the leaves of *Codiaeum variegatum* var. *pictum* f. *appendiculatum* (Fig. 86 VIII).

The interpretation proposed by Baum and Leinfellner, that the stamens of Angiosperms are a case of diplophyllous leaf organs, has certainly not remained unchallenged. The transverse furrows frequently visible during the development of the anther (see e.g. Fig. 57a, c) and interpreted by Baum (1949c) as incisions between the "dorsal blade" and the cross zone, is according to Kunze (1979) not a sign of peltation. He stresses, in contrast to the interpretation of Baum, that these are "by no means established so early in morphogenesis as is there maintained. Also, there is no outgrowth of this supposed cross zone on the dorsal blade. In any event diplophylly could only be understood as primarily and completely congenital." Nevertheless, exactly this widespread congenital formation of the anther structures was repeatedly emphasized by Baum and Leinfellner, as Kunze himself mentions. According to the view of Kunze the stamen is "to be regarded *as an originally bifacial organ*", where the transverse furrows which appear in the young anther primordia originate in his view by means of "the marginal growth which is initially uniform being then directed into two separate swellings, forming the transverse furrow between them". What is being discussed is then "merely a partial division of the periplastic marginal growth into a dorsal and a ventral swelling, by which an apparently peltate form of an early stage of the stamen primordium is simulated". The confirmation of this interpretation by means of histological illustrations (which indeed could also perhaps admit of other explanations) is not in fact to be found in the work of Kunze; on the other hand the drawings and photographs of anther development which are appended are often extensively equivalent to the illustrations in Baum and Leinfellner. Until this stage the controversy remained just a matter of differing interpretations (which is expressly confirmed by Kunze in his discussion referring to the anthers of the Violaceae, see p. 278). Meanwhile (i.e. after the appearance of the German edition of this book) valid arguments against the theory of Baum and Leinfellner have been made by Leins and Boecker (1982), who by sectioning young stamen primordia of *Cobaea scandens* and *Hemerocallis fulva* could never find a cross zone: "there is no 'Querzone' in any stage of early stamen development. The most important argument in favour of the peltation theory of the stamen is therefore untenable".

1.5.7   The filament

The filament is usually slender, as its name already suggests, but it is rarely really threadlike. There are, however, filaments which are much wider than the

Caption for fig. 60 (*cont.*) view. XV *Cobaea scandens*, cross-section through a filament with concentric vascular bundles, showing the 4 xylem groups. XVI-XIX *Pittosporum tobira*, divergence of the concentric filament bundle (XVI) in a dorsal median (Dm) and a ventral median (Vm) in the region of the anther. XX-XXII *Magnolia* × *soulangiana*, series of cross-sections through a stamen, XX through the middle, XXI through the base of an anther, XXII through the upper part of a filament. XI-XIV after Schaeppi, XV-XXII after Leinfellner.

anther. An example is the capitately thickened filaments in the genus *Dianella* of the Liliaceae (Fig. 60 XIV). In *Yucca filamentosa* (Fig. 60 XII) the filament is so greatly inflated that the anther comes to lie horizontally. This type of structure contrasts with the stamens which have what is called **versatile** anthers, that is to say, anthers which are mobile and are pivoted on the filament. A prerequisite for this arrangement, which is certainly advantageous for pollination biology, is that the tip of the filament should be much tapered down. Such attenuation of

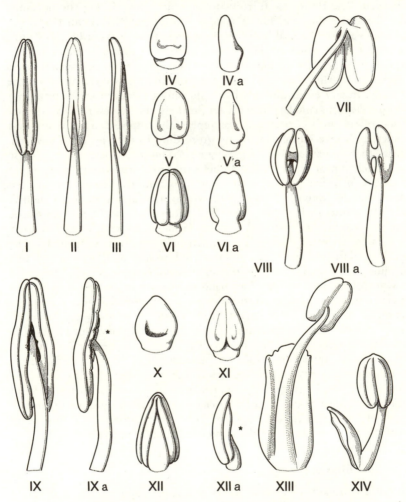

Fig. 61 I–VI *Hemerocallis fulva.* I–III Ventral view of mature stamen (I), dorsal view (II) and from the side (III); IV–VI stages in the development of the stamen, IVa, Va, lateral views of IV and V, VIa rear view of VI. VII *Rehmannia angulata*, stamen, dorsal view. VIII, VIIIa *Fagopyrum esculentum*, stamen, in ventral (VIII) and dorsal view (VIIIa). IX–XI *Cobaea scandens*, IX, IXa stamen in ventral (IX) and lateral view (IXa), X–XII stages from the stamen development, X–XII ventral views, XIIa side view, ★ marks the morphological upper surface. XIII, XIV stamens of *Zygophyllum album* (XIII, dorsal view) and *Alyssum armenum* (XIV, side view). After Baum and Leinfellner.

the filament is, however, also observable in *Yucca filamentosa* (Fig. 60 XIII), but it does not come into effect on account of the shortness of the tip section, and the pronounced clavate thickening of the remainder.

The Baum and Leinfellner theory of the peltate or diplophyllous development of the stamens assumes a unifacial structure, in the distal region of the filament at least. The question may therefore be asked, as to whether this unifacial property is also reflected in the anatomical structure of the filament. Nevertheless the ring-shaped arrangement of the vascular bundles which is so characteristic of the unifacial petiole is scarcely to be expected here, because the overwhelming majority of stamens are only provided with a single bundle. This vascular bundle is, however, often found to develop in a radially symmetric manner with central xylem. In some cases, as in *Pittosporum tobira* (Fig. 60 XVI to XIX), a division occurs in the anthers between a smaller ventral and a larger dorsal bundle, with the xylem portions facing each other, while there is one concentric bundle in the filament. These two bundles may be interpreted as a ventral median and a dorsal median. Also in *Magnolia soulangiana* "the anther possesses two bundles lying in the median, with their xylem portion turned to face each other, which indicates a rudimentary bundle ring, of which only the dorsal median and the ventral median are developed" (Leinfellner 1956d, p. 383, 1956b). The ventral median here shows itself to be the fusion product of two lateral nerves, which have separated from the single bundle of the filament (Fig. 60 XX to XXII). In *Cobaea scandens* one even finds four clearly separated xylem groups in the concentric filament bundle (Fig. 60 XV). The results of other investigations on the occurrence of vascular bundles in the stamens of *Lilium henryi*, *Tricyrtis pilosa* and *Trapa natans* enabled Leinfellner (1956c, 1957b) to support his theory that what are presumed to be the cross zones of anthers which appear peltate externally are formed by secondary growth processes.

It is not unusual to find filaments which show a leaf-like broadening, but then with the exclusion of at least the furthermost distal portion below the anther (Fig. 62 I to IV, VI). In spite of the greater likeness to a leaf which this suggests, these stamens may be just as little referable to the "primitive forms" as the flattened leaf-like stamens in general (Leinfellner 1956b), a point which we will need to consider again later. The sometimes very pronounced broadening of the filaments which we are currently discussing can lead to the formation of lobe-like outgrowths which project to the right and left of the anther (Fig. 62 V) or even beyond it (VII). Pointed filament appendages of this kind, which also occur directly on the right and the left at the insertion of "broadened or filiform" filaments in other stamens, are reminiscent in appearance to the stipules of foliage leaves, and are often so described (see for example Glück 1919). This identification does, however, assume that we can regard the filament of the stamen as homologous with a petiole and the anther as homologous with the leaf blade. However, the available basis for comparison is not adequate to justify this, as has already been stressed elsewhere. The same applies also of course to what Leinfellner (1956a) investigated more closely and interpreted as "median stipules". These are scales which are placed ventrally at, or slightly above, the base of the filament, as can be found in *Zygophyllum album* or *Alyssum armenum* and in not a few other species (Fig. 61 XIII, XIV). These basal scales also lack the characteristic features of stipules during ontogenesis, namely, their early initiation as well as the precocious growth compared with the rest of the leaf and their rapid differentiation. The latter phenomenon (known as prolepsis) is lacking in the supposed stipules of stamens in all cases which have been investigated so far. In *Allium rotundatum* (Fig. 62 VIII, IX) the anthers are perfectly formed in the stamen primordia when the appendages which are interpreted as "stipules" are only visible as minute points. Although this situation might be due to the change of the rates of growth within the floral region, one of the helpful criteria for determining whether or not these are stipules is inapplicable. We would therefore be well advised to speak of "lateral" or "basal stamen appendages".

Furthermore, basal appendages of the stamens can function as nectaries, as for example in the Polygalaceae and the Caryophyllaceae (see Glück 1919). It was

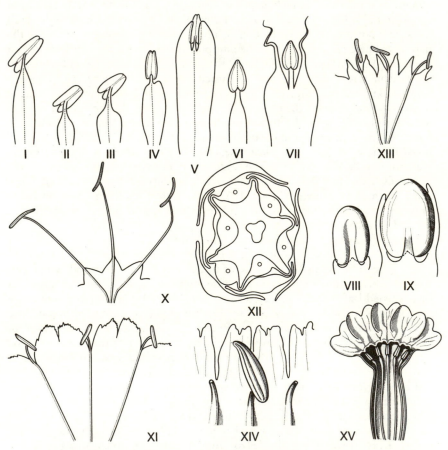

Fig. 62 I–VII Stamens of *Ornithogalum divergens* (I), *O. caudatum* (II, III), *O. nutans* (IV, V), *Allium porrum* (VI, VII), I, III, V, VII stamens of the inner ring. VIII, IX young stamens of *Allium rotundatum*. X, XI parts of the corona of *Hymenocallis illyricus* (X) and *H. lacera* spread out. XII, XIII *Pancratium maritimum*, cross-section of the flower (XII) and part of the corona spread out (XIII). XIV *Puschkinia* sp., part of the corona opened out, with the anthers removed from the outermost stamens. XV *Primula veris*, sympetalous corolla with stamens included within, cut open and spread out flat. XV after Troll, the others after Schaeppi.

already mentioned (section 1.4.3.4) that in *Viola* the anthers may bear nectar glands, which form spur-like appendages to the anthers.

1.5.8   Lateral fusion of the filaments, and coronas

These lateral stamen appendages which are often taken to be stipules have a special role to play in the formation of the so-called coronas in the Liliaceae and the Amaryllidaceae, whose structure results from the lateral fusion of flattened, leaf-like filaments.

Lateral fusions of filaments are by no means rare in occurrence. The best-known example is certainly the androecium of the Papilionaceae. In this family, either the filaments of all ten stamens are united into a tube which surrounds the ovary, or else (as for example in *Vicia*, Fig. 7 VII) only nine of them combine to form a channel which is open above and is covered by the tenth stamen left in isolation. Fusions of the stamens of this nature are described by the term **adelphy**. The androecium is termed **monadelphous, diadelphous** *or* **polyadelphous** according to whether the stamens are combined in one, two or several bundles.

The structure of the androecium is also monadelphous in *Melia azedarach* (Fig. 64 I, II). Here the product of fusion of the stamens is not unlike a corolla, and it is also spoken of as a staminal corolla. Similar structures are found in other members of the Meliaceae, and a further example is *Erythroxylon coca* (Fig. 64 III).

Very frequently the filaments of a diplostemonous or haplostemonous androecium are only fused for a short distance at their base. It is not uncommon for the monadelphous or polyadelphous arrangements to be linked with the development of polyandrous androecia (see section 1.5.9). This is especially true for the "Columniferae" families i.e. Malvaceae and Bombacaceae (and Sterculiaceae), in which usually all the stamens are united into a tube or column surrounding the style. Well-known examples for "bundling" of the stamens where the androecium has a polyadelphous structure are found in the Myrtaceae and the Hypericaceae, among others (Fig. 64 V). All these cases are not, however, comparable with those quoted above.

In the examples quoted so far we have been concerned exclusively with congenital fusion. There are also cases, however, in which the filaments fuse together postgenitally. In the Liliiflorae this occurs according to Schaeppi (1939) in *Iris unguicularis* (*I. stylosa*) and in several other Iridaceae (*Sisyrinchium, Tigridia*), whilst in *Ruscus aculeatus* it is a case of congenital fusion, which even extends to the anthers.

Among members of the Liliaceae and the Amaryllidaceae united filaments which are provided with lateral appendages can also occur. This leads to a situation in which either two petaloid lobes, or one broad and emarginate lobe, appear on the margin of the corolloid and coloured stamen tube in alternation with the stalked anthers (Fig. 62 XI, XIII). The latter can even exceed the anthers to a considerable extent. By this means the development of a **corona** (Fig. 63) of often considerable size is brought about, whose colouring may be the same as the perianth segments which surround it, or may contrast with it. Such coronas are particularly well-known among the Narcissi. Here, however, the anthers are not placed on the margin of the corona, but are inserted more or less deeply within it on the inner surface. In *Puschkinia* (Fig. 62 XIV) it can already be established that the corolloid lobes which occur between the anthers are fused over some distance to the back of the stalk-like sections of the filaments. It was already assumed by Schaeppi that the corona of the *Narcissus* species is only an extension of this phenomenon (definitely due to secondary growth processes). This interpretation can be confirmed by unpublished findings of Troll.

1.5.9   Fusion of the anthers

As already mentioned for *Ruscus aculeatus*, the fusion of the stamens can continue as far as the anthers. It can also affect the anthers alone. The best-known example of this must be the Compositae. The old order name *Synandrae* expresses the fact that the (introrse) anthers of the five stamens are here combined into a tube which surrounds the style. Nevertheless this fusion is actually very superficial, since it was found "that it is *exclusively the cuticles* of two neighbouring anthers which grow together over a short distance and stay permanently united" (Tschirch 1904, p. 52).* This appears also to a lesser degree in the related Campanulaceae and particularly in the Lobeliaceae.

Fig. 63 *Eucharis grandiflora (E. amazonica)*, flower. P corona.

* "The ligament, which also surrounds the complete anther tube in the mature state, is formed from the cuticle only, which detaches itself from the outer wall of the anther epidermis and forms a coherent band by means of the partial fusion of the neighbouring parts, as mentioned above." (Tschirch 1904, p. 52).

Postgenital anther fusion is also widespread in markedly zygomorphic flowers, where it is not without significance for pollination biology. Here the androecium is also influenced by the dorsiventral structure of the flower, which is apparent both in the variety of forms of the upper and lower stamens (didynamy), and in the variation in their position.

In such flowers a symmetrical arrangement of the stamens on both sides of

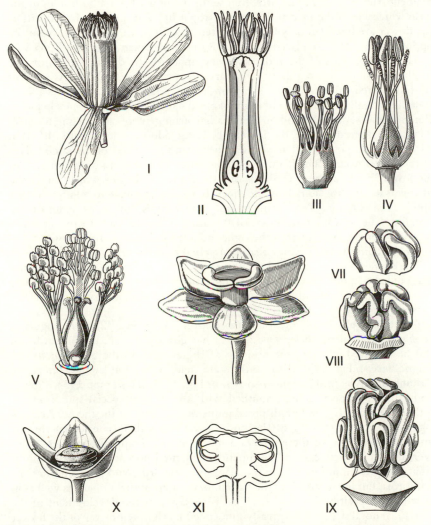

Fig. 64 I, II *Melia azedarach*, flower, II axial section through the central portion of the filament tube and pistil. III *Erythroxylon coca*, filament tube and pistil. IV *Linum usitatissimum*, do. V *Hypericum aegypticum*, triadelphous androecium and pistil, VI *Phyllanthus cyclanthera*, male flower. VII-IX *Cucurbita pepo*, development of the androecium. X *Cyclanthera explodens*, male flower. XI *C. pedata*, male flower in axial section. II after Harms, IV after Warming, III, VI-IX, XI after Engler, V after Baillon.

the style (Fig. 42 II), and in connection with this a convergence of the anthers one to another, can frequently be observed. This ensures at the same time that the anthers are fixed in a favourable position for pollination.

Columnea picta (Gesneriaceae), as investigated by Trapp (1956b), serves as a good example of this. We have already previously discussed the peculiar form of its corolla limb (Fig. 43 VII). That is, the two lateral lobes which originally belonged to the lower lip form a part of the upper lip in the opened flower. The filaments run in far-reaching arcs, following the upper lip, on both sides of the style which lies in the median below the upper lip. The anthers of each half of the flower are fused laterally and connected apically with the anthers from the opposite side. They are placed transversally to the median (Fig. 43 VII, 65 V).

The development of the androecium begins with the differentiation of five similar swellings in a regularly radial arrangement (Fig. 65 I, II). As growth continues a reduction of the median primordium sets in, which becomes a short style–like staminode that finally hardly appears at all. As the filaments begin to elongate the stamens abandon their radial arrangement and begin to push themselves forward along the lower lip, lying side by side in pairs (III, IV). Already in this position a lateral fusion between the anthers of the same halves of the flower occurs. The filaments still lie adjacent and more or less parallel to the lower lip. At anthesis vigorous one-sided growth begins which causes the filaments to separate laterally. Consequently the anthers have to relinquish their previous orientation. They come apart at their bases, but still remain united laterally. The upward movement of the anthers finally leads to the situation where the tips of the anthers of the right and left hand halves of the flower come into close contact, and become firmly attached to each other (V).

The characters just described occur widely in the androecia of the Gesneriaceae, and also in other related families. In some cases all the anthers coalesce into one (a **synanther**) and in others only the anthers of the anterior and posterior stamens are fused (Fig. 65 VI).

As the example already mentioned in section 1.4.3.8 (Fig. 42) of the rugula of *Jacobinia magnifica* and *Beloperone violacea* has already shown, the perianth can also be involved in fixing the position of the anthers. In the genus *Incarvillea* (Bignoniaceae, Fig. 66 I, II) the anthers are held in position by the connective clasping the style, in the manner shown in Fig. 66 III, IV (Trapp 1954, 1956b). Pollination by insects is accomplished with the help of special spines on the anthers (Fig. 66 IIID) and predisposed joints on the walls of the thecae. According to Trapp this "fixing of the position of the anthers is essential for the smooth functioning of the pollination mechanism". Trapp also stresses, however, that in all these cases a "dominating structural principle" comes to light, "that is expressed not merely in the diagrammatic order and disposition of the primordia, but is also visible in the varied formations of the stamens as well as in their diverse positional relations". That is to say, that the development of the individual floral structures is greatly influenced by the "symmetry of the flower in general", which "is the dominant reason why the stamens do not unfold independently, but develop according to a certain *order*" and "are integrated in a unity of a higher rank" (1956b, p. 60).

The process by which the fusion of anthers comes about is very well illustrated by the case of the African Violet, *Saintpaulia ionantha*, which was likewise investigated by Trapp. Here also the two individual stamens of the flower are similarly united at their apices here as well (Fig. 65 VII), and to such a degree that the two anthers cannot be pulled apart by a needle without damaging them. Both anthers are surrounded by a papillose epidermis, and the postgenital fusion depends on this property. If a median longitudinal section is made

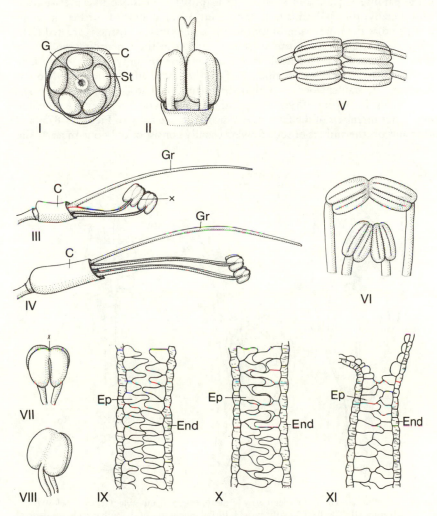

Fig. 65 I-V *Columnea picta*. Stamen development, I very young flower primordium seen from above; II lateral view of flower of medium age, corolla removed; III, IV stamen development in old flowers, most of corolla removed; V ripe coalesced (synandrous) stamens. VI Synandrous stamens of *Aeschynanthus pulcher*. VII-XI *Saintpaulia ionantha*, VII, VIII fused anthers from in front and from the side, x marks areas of fusion, IX-XI transverse longitudinal sections through the fusion zone, showing the successive stages of the dentonection of the epidermis, Ep epidermis, End endothecium, St stamen, G gynoecium primordium, otherwise as in Fig. 66. After Trapp.

through the tips of two young anthers (Fig. 65 IX), then it is seen that it is exactly here that the epidermis papillae are especially elongated. In older stages (X) the beginning of the fusion process is visible: the cells have elongated further and are beginning to grow into the spaces between the papillae belonging to the opposite anther wall. Hence a dentonection of the two epidermises comes about. The feature that makes it adhere so firmly is the fact that the heads of the papillae expand and so are held in position in 'slots' like press-button studs. Finally, the epidermis cells flatten out in the course of further development, so that the impression of uniform parenchymatous tissue is created (XI).

We encounter special forms of stamen fusion in the Cucurbitaceae, whose flowers are normally unisexual. Five free stamens are only rarely found here (*Fevillea*). In *Sechium* the filaments of all five stamens are united. Mostly, however, the stamens occur combined in pairs, so that a single stamen is left over. Whilst in Bryony (*Bryonia*) the anthers still remain free, they are united in most other members of the family. According to the view of Eichler (1875) and other authors the anthers of such flowers usually consist of only one theca.* The

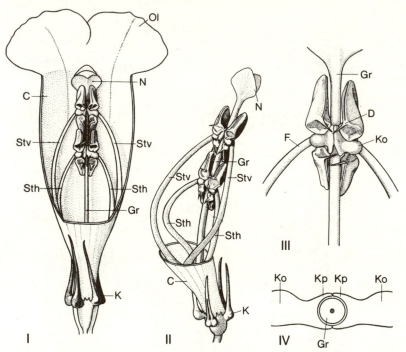

Fig. 66 Incarvillea variabilis. I, II General view of the androecium and style (Gr) from below (I) and from the side (II), corolla (C) partly excised. III Enlarged view of the anther region, anterior anther pair, IV schematic cross-section through the connective region of two neighbouring anthers. Ol Upper lip of the corolla, N stigma, Stv, Sth anterior and posterior stamens, K calyx, D thecal spine, F filament, Kp petaloid broadening of the connective (Ko). After Trapp.

---

* According to Bhattacharjya (1954) five stamens with two thecae are found in *Telfairia* and sometimes also in *Coccinia cordifolia*, and in *Benincasa hispida* the sole remaining stamen still has two thecae.

picture is made yet more complex by the fact that the thecae frequently undergo a pronounced and often S-shaped curvature during stamen development (Fig. 64 VII to IX). According to Kunze (1979), who studied the stamen development of *Bryonia dioica*, the stamen of the Cucurbitaceae does nevertheless correspond with "a normal one, but its form diverges on account of its specific marginal growth. It is *not* monothecous", "because the pollen sacs develop by means of lateral *and* apical meristem and only one side remains sterile". It is a case "of a primarily saddle-shaped anther, which in the end is reduced to a quarter of its length on one side".

The most extreme case of modification in this character is in *Cyclanthera*, whose name refers to the two pollen sacs which are placed in rings around a central column in the centre of the flower. Evidently what is happening here is that there is an entirely congenital fusion of presumably three stamens grouped around a rudimentary style (see Eichler 1875, p. 312, Chakravarty 1958, Leins and Galle 1971). The transversely inserted thecae which open by a horizontal slit are so thoroughly fused together that the pollen sacs merge into one another without any partitions (Fig. 64 X, XI). A contrast to this is provided by *Phyllanthus cyclanthera* in the Euphorbiaceae (Fig. 64 VI), in which, however, the pollen sacs of the stamens concerned remain separated. Synandries formed of numerous stamens occur frequently in the Clusiaceae.

Apart from the lateral, congenital or postgenital fusion of stamens one to another, there is the possibility, as already mentioned, of a **serial fusion** of stamens and petals. By this means the filaments of stamens which are inserted in front of petals may be united with the petals to some degree. In sympetalous corollas this fusion can go so far that the anthers can appear to arise directly from the inner wall of the corolla tube (Fig. 62 XV). The observation that fusions of this kind between petals and stamens often appear in very early stages of floral development (see also Fig. 18 V), has given rise to all kinds of false interpretations. Hence to begin with Pfeffer (1872) and later Mattfeld (1938), Vaughan (1955) and Roth (1959) believed that they could deduce that the petals originated on the outside of the stamen primordia from the observation that the stamens and petals in the Primulaceae proceed from a "common" primordium. Although this point of view had been disputed previously, it was recently disproved by Sattler (1962), who established that it is the point in time of the initiation of the individual organs (which varies within the family) which decides whether they remain separate or are fused together. Similar remarks apply to Markgraf (1936), who interpreted the tepals of *Potamogeton* as appendages of the connective (Sattler 1965, Posluszny and Sattler 1974).

1.5.10 Polyandrous androecia

We need to concern ourselves once more with the question of the **number and arrangement of the stamens** in the flower. We have already stated (section 1.2) that the spiral or **acyclic** arrangement, with numerous and most importantly, an indefinite number of floral organs in the different floral formations, which is thought to be "primitive", is a character of families which also show relatively primitive traits in other structures. Furthermore the spiral arrangement of the stamens (see Fig. 14) is in good agreement with those ideas by which the flower is thought of as a compressed sporophyll-bearing structure of restricted growth. The stamens of the polyandrous androecia of the Polycarpicae are in fact arranged in a markedly acropetal and often clearly spiral sequence. We can assume with reasonable justification that the **diplostemonous androecium** of many families has arisen by the reduction of the number of primordia from this spiral-polymerous androecium, and by making it cyclic, and from this in many cases the **haplostemonous androecium** by reduction to a single ring, which in other cases could be derived directly from the spiral-polymerous state.

According to Leins (1975), there are two further types of polyandrous androecia which must be distinguished from the (1) **spiral-polyandrous androecium** of the Polycarpicae, which are (2) the **androecium with secondarily inserted simple organs** and (3) the **compound androecium**.

For the **androecium with secondarily inserted simple organs** Leins puts forward the hypothesis that the initialisation of the numerous stamens requires a considerable enlargement of the receptacle. The initialisation of the stamens, which here are arranged in rings and never in spirals, usually proceeds even after the development of the gynoecium has begun. According to Kania (1973), the enlargement of the receptacle in the Rosaceae takes place between the stamen primordia which are already present and the gynoecium and the insertion of further rings of stamens in a centripetal sequence (Fig. 67 II). In the Helobiae (Alismatales) on the other hand the receptacle is enlarged in the region between the first stamen initials and the perianth. The expansion of the circumference which is linked with the centrifugal insertion of stamens (Fig. 67 III) also permits a collateral multiplication of the stamen primordia (Fig. 67 VI to VIII; Singh and Sattler 1972, 1973, 1974, Leins and Stadler 1973, Sattler and Singh 1973, 1977, and see also 1978). Also for several species of palm a centrifugal sequence of differentiation of numerous stamens can be observed in the context of a peripheral enlargement of the receptacle (Uhl and Moore 1977).

It remains to be seen whether or not the various cases of mere doubling of the stamen number (e.g. the inner, (*Monsonia*, Geraniaceae) or the outer, e.g. *Peganum*, *Nitraria*, Zygophyllaceae) through division of the primordia ("dedoublement", in the strict sense = "division") are to be classified here.

A **compound androecium** originates from primordia which arise from the receptacle as broad swellings arranged in one or two rings during the acropetal sequence of differentiation of the floral organs and which later divide into individual stamens ("dedoublement", in the broad sense = chorisis). The differentiation of the individual stamens from the so-called 'primary swellings' (compound primordia) may advance either towards the centre (**centripetal**, Fig. 67 IV) or towards the periphery (**centrifugal**, Fig. 67 V, 69 VI). The former is found in the Myrtaceae (Leins 1965, Mayr 1969), and in isolated cases in the Saxifragales (*Philadelphus*, Gelius 1967). A centrifugal sequence of differentiation is characteristic for many of the polyandrous families from the subclass Dilleniidae. Examples of this are the Paeoniaceae (Hiepko 1964), the Hypericaceae, Cistaceae, Capparidaceae, Resedaceae, but not, however, the Begoniaceae, Datiscaceae and Ochnaceae (Leins 1964b, c, Leins and Sobick 1977, Merxmüller and Leins 1971, Leins and Bonnery-Brachtendorf 1977, Pauze and Sattler 1978); also the Malvales (van Heel 1966, Fig. 69 VI), the

Loasaceae, however, only in part (Leins and Winhard 1973). From the order of the Centrospermae the Aizoaceae may be cited (Ihlenfeldt 1960), and amongst the Myrtales the androecia of the Lythraceae, Punicaceae and Lecythidaceae develop centrifugally, in contrast to the Myrtaceae (Mayr 1969, Leins 1972b). The varied mode of development of the polyandrous androecia therefore often presents a highly informative systematic character for families and taxa of higher orders, although the application of this as a means to divide the dicotyledons into their main evolutionary pathways (see e.g. Cronquist 1957, Leins

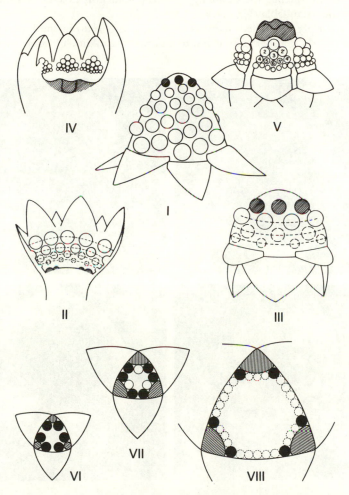

Fig. 67 Mode of differentiation of polyandrous androecia (gynoecium shaded in I-V, omitted in VI-VIII). I Spiral-polyandrous androecium with simple organs (Polycarpicae); II centripetal insertion of rings of stamens (Rosaceae); III centrifugal insertion of rings of stamens (*Hydrocleis*, Butomaceae); IV centripetal compound androecium (Myrtaceae); centrifugal compound androecium (various families of the Dilleniidae). Arranged from Leins 1975. VI-VIII Diagrams of flower buds of Alismataceae, the stamens which are initialised first in black, VI *Alisma*, VII *Echinodorus intermedius*, VIII *Echinodorus macrophyllus*. After Leins, adapted.

1964b) does not seem to us to be justified any more. There are meanwhile also families included which are known to have individual species where the development of the androecium is variable. Apart from the Loasaceae already referred to (*Mentzelia*; centripetal) the Hamamelidaceae (Endress 1976) must be mentioned. *Caloncoba echinata* (Flacourtiaceae) behaves "neither clearly centrifugally nor centripetally" (Endress and Voser 1975); see also Tucker (1974a).

In the compound polyandrous androecia the individual stamens produced from a compound primordium can be clearly united to a greater or lesser extent when mature. Hence the bundling of stamens in polyadelphous androecia is readily understandable (Fig. 64 V).

The multiply 'branched' stamens of the genus *Ricinus* in the Euphorbiaceae are particularly unusual in kind and therefore are also difficult to incorporate in

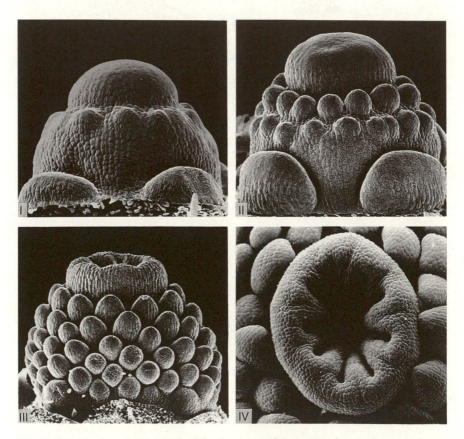

Fig. 68 Capparis spinosa var. *inermis* floral development. I Flower primordium with two petal primordia (below) and a ring-shaped swelling (above) with the uppermost ring of stamen primordia. II Appearance of further stamen primordia in basipetal sequence; the primordium of the gynophore arises in the centre of the flower primordium. III Androecium after the differentiation of all the stamen primordia; the primordium of the gynoecium is at the apex of the flower primordium. IV Gynoecium primordium seen from above. Magnification of I c. 260 ×. I, II from Leins and Metzenauer (1979), III, IV original, Leins and Metzenauer.

any comparative study. They were mistakenly employed by van der Pijl (1952) as evidence for the telome theory.

According to Leins (1975), primitive characters can be recognised in both the genuine spiral-polymerous androecium with its large and indeterminate number of organs and also in the compound androecium. This 'branching' has been repeatedly compared with the division of a pinnate leaf (see Leins 1964b, c, Hiepko 1964) and according to Leins could correspond to a primitive form of

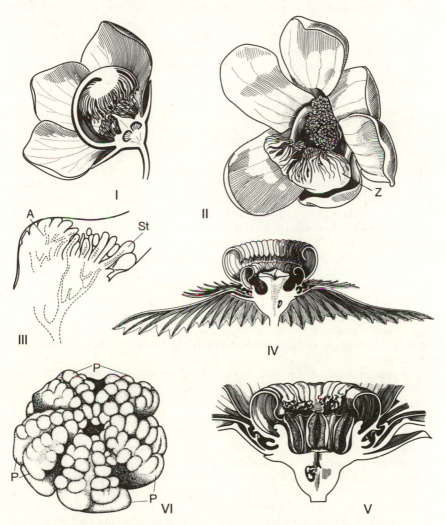

Fig. 69 I *Lecythis lanceolata*, flower in median section. II *Lecythis cf. urnigera* front view of flower, Z tongue-like organ. III *Couroupita guianensis*, median section through a young flower in the region of the developing tongue-like appendage A, St individual fertile stamens. IV, V *Napoleona imperialis*, IV flower in axial section, V do., central part more highly magnified. VI *Abutilon darwinii*, development of the androecium, P petal primordium. I, IV, V after Baillon, III after Leins, VI after van Heel.

stamen. This is all the more so when a spiral arrangement of the compound primordia is also recognisable as in the Paeoniaceae (Hiepko 1964), which is one of the families of the Dilleniales which is provided with various primitive characters.

In such circumstances it is reasonable to expect that androecia with one or two rings of stamens could not only be derived from spiral-polymerous, but also from compound androecia with centripetal (ventral) or with centrifugal (dorsal) differentiation.* In particular this possibility has been considered with reference to the development of obdiplostemonous flowers (Čelakovský 1894, Corner 1946, Huber 1963), that is to say, that in many families the obdiplostemonous androecium might be a special case of the polyandrous centrifugal one. So far, however, no clear evidence has been established on this point, and certainly not from the ontogeny of the relevant androecia (see Eckert 1966).

1.5.11   Obdiplostemony

The phenomenon of **obdiplostemony** has already been mentioned in section 1.2.1 as an "exception to the alternation rule" in the region of the stamens. What happens is that in flowers with two (isomerous) whorls of stamens of equal number, the "episepalous" stamens (inner row or calyx stamens) which alternate with the petals are placed further inwards than the "epipetalous" stamens (petal stamens) which are placed in front of the petals. The position of the carpels is also often affected by this anomalous arrangement of the stamens if it happens that the androecium and gynoecium are isomerous. In such cases the position of the carpels is epipetalous (Fig. 5 XIII), instead of alternating with the petals, i.e. both rings of stamens and the carpels appear to follow one another in regular alternation.

The best known examples of obdiplostemony appear in the Caryophyllaceae. It is also, however, a typical character of the Crassulaceae and the Saxifragaceae, and also appears in the Onagraceae, Combretaceae and Haloragaceae from the Myrtales as well as in the majority of the families in the Geraniales and the Rutales.

The interpretation of the special topological arrangement has long been in dispute. A comprehensive presentation of the various theories, which have already been mentioned above, is found in Eckert (1966), who has devoted thorough ontogenetical and anatomical research to a clarification of the problem.

The attempt made by Mattfeld (1938a, b) to explain the phenomenon through observations on the flowers of the Caryophyllaceae is interesting, although inappropriate. The position of the outer row of stamens directly in front of the petals comes about, in his view, because the supposed petals may really be homologous to stipules of the outer stamens, and that where they are united to the backs of the filaments, they correspond to dorsal median stipules. The starting point for his ideas was his observation that the pair of nectary glands at the base of the filament in *Stellaria filiformis* (*Drymaria filiformis*), which

---

*Leins (1979) considers that there are two possible processes by which a centrifugal compound androecium can be simplified: "by favouring the primary primordia and loss of their further division, *or*: reduction of the primordia and reduction in the number of stamens."

were interpreted by Glück (1919) as stipules of the stamens, were elongated in long petaloid lobes, "in such a way that these glands greatly resemble a deeply emarginate petal of a *Stellaria*" (Mattfeld 1938a, p. 89). In so far as they are united with the backs of the filaments, he interprets them as a kind of dorsal median stipule. According to Mattfeld, the emarginate to deeply bifid state which so often occurs in the petals of the Caryophyllaceae and other families, as well as the frequent and considerable degree of fusion between the petals and filaments in a "stamen-petal unity", is also explained by the origin of the petaloid organs from fused stipules. This latter basal fusion of the petals with the outer ring of stamens, which often appear to arise even from the same primordium, gave rise to an old theory about the Centrospermae as re-adopted by Friedrich (1956), according to which petal and outer stamen originate from one primordium through serial division. According to Eckert (1966 p. 591) the histogenesis reveals no sign of this.

In recent years various authors have undertaken thoroughgoing ontogenetical researches in families with obdiplostemonous androecia, such as Leins (1964a) on the Ericales, Eckert (1966) on families of the Geraniales and Rutales, Caryophyllaceae, Saxifragaceae and Crassulaceae, Gelius (1967) on the Saxifragales and Mayr (1969) on families of the Myrtales. All authors are in agreement on the result that obdiplostemony is a secondary phenomenon, which in each case is based on "primary differences in size and secondary displacements by growth" (Mayr, p. 268). This is expressed by the fact that the weaker petal stamens are thrust outwards, and especially so when the development of the petals is somewhat restricted. On the other hand a strengthening of the sepals leads to a slight elevation of the stamens which are positioned in front of them.

The anomalous positioning of the carpels is evidently explained by differences in the space available resulting from differential growth of the preceding rings of organs (Hofmeister's Rule). Hence if the upper margins of the primordia of the inner ring of stamens reach further up at the moment of initiation of the carpels than do those of the outer ring, then the initiation of the carpels takes place in an episepalous fashion, and if the upper margins of the (later initiated and weaker) inner row of stamens lie lower at this time, then epipetalous carpels arise. Hence this confirms the interpretations of obdiplostemony first put forward by Čelakovský (1875) ("Verschiebungstheorie" or displacement theory), and also in a certain sense the theory of Stroebl (1925), Goebel (1928), Troll (1928a) and Wassmer (1955) which speaks of a "preferential development of sepal sectors".

1.5.12   Staminodes

It has been mentioned repeatedly before now that individual stamens of a flower can be regularly transformed into nectaries, as in the case of the median stamen of *Scrophularia nodosa* (Fig. 55 I, II). They can also be reduced or can atrophy to a greater or lesser extent, as for example the two posterior stamens in *Salvia* (Fig. 53 IX, X), the median stamen in *Columnea picta* (see section 1.5.9), or the members of the inner ring of stamens as in *Linum* (Fig. 64 IV). Such sterile stamens, which show various stages of reduction, or which take over other functions, are generally termed **staminodes**. Staminodes often appear to be little altered in comparison with the fertile stamens (Fig. 73 X, XI), and also often take on a more or less flattened and petaloid form. In the latter case they can be brightly coloured, and then not uncommonly in a colour which contrasts with the perianth, as for example in *Loasa vulcanica*, in which the red and yellow staminodes stand out against the white petals, or in *Clematis alpina*, where the white staminodes provide a vivid contrast to the pale blue petals, and are linked with the stamens by a transition series of forms (Fig. 70 VIII, IX).

In the Malvaceae the compound polyandrous structure of the androecium determines to a great extent the form of the staminodes, so that for example branched staminodes are found (Fig. 70 X, XI). In particular the more or less petaloid staminodes of the double-flowered forms of *Hibiscus rosa-sinensis* and *Alcea rosea* (Fig. 70 XII to XVIII) are often divided into numerous lobes, which often bear rudimentary anthers on the margins.

There are quite a number of cases in which the occurrence of a large number of staminodes determines the appearance of the flower to a considerable degree, as for instance the flowers of the polyandrous Lecythidaceae. In *Napoleona* (named after Napoleon Bonaparte) the staminodes are united into a threefold brightly-coloured, cockade-like "corona ciliaris" (Fig. 69 IV, V), whilst the petals are lacking. In *Lecythis* (Fig. 69 I, II) or also in *Couroupita* they form a tongue- or helmet-shaped appendage on the anterior side of the flower. According to Leins (1972b and personal communication) in *Couroupita* this is formed as an appendage to the torus-shaped primordial zone of the androecium, which corresponds to the primary primordia.

The form of the ciliate staminodes of the Grass of Parnassus (*Parnassia palustris*), with their green, long-stalked, shining and minutely capitate "false nectaries" (see section 1.7.3) is quite bizarre. These alternate with the fertile stamens in their rings of five, and are placed in front of the large white petals, forming a striking contrast to them (Fig. 70 I to III). Their form and their subdivision into 11 cilia with capitate tips hardly relates at all at first sight to the form of a fertile stamen, and they have therefore been interpreted as "mere effusions* of the disc". However, in other species of *Parnassia* often scarcely more than 7, and often only 3 or fewer glands are found, and in *Parnassia tenella* the staminodes exhibit a clear division into a filament and an atrophied anther (Eichler 1878, p. 426). Finally, in *P. palustris* transitional forms are occasionally found (Fig. 70 IV to VII) which are intermediate between the fertile stamens and the ciliate staminodes (Wettstein 1890).

1.5.13   Petaloid transformations of the stamens

We have already encountered numerous examples of the "petaloid transformation" of stamens or the structure of intermediate forms between fertile stamens and petals, beginning with the petaloid organs of a "double" rose flower (see Fig. 22 VIII to XI). We have also already mentioned that the "nectary scales" with which the tepals of many of the Liliaceae are provided (Fig. 49 IV to XI) which have been the subject of many comparative studies have turned out to be "modified forms of peltate leaves" or diplophyllous

---

* outgrowths

leaves (see section 1.4.4.2). They clearly show a close morphological relationship with stamens; the same is true of the petaloid stamens of *Narcissus* (Leinfellner 1960a).

The line of reasoning taken by Leinfellner, (1963b, p. 369) that the underly-

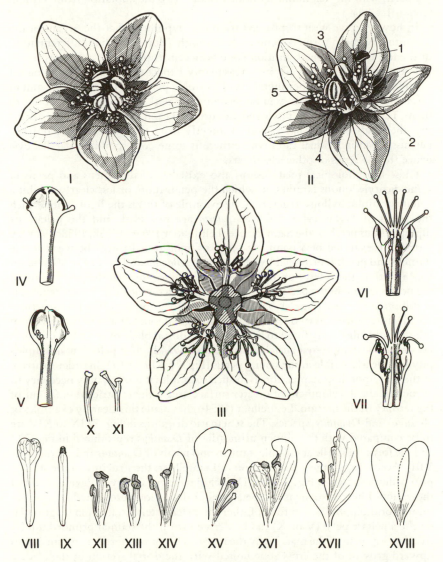

Fig. 70 I–VII *Parnassia palustris*. I, II Complete flower, from above, before anthesis (I) and in side view after the dehiscence of the first and the erection of the second stamen (the numbering gives the order of dehiscence of the anthers), III flower after removal of the gynoecium. IV–VII Intermediate forms between fertile stamens and ciliate staminodes. VIII, IX *Clematis alpina*, forms of staminodes. X–XVIII *Alcea rosea*, staminodes. IV–VII after von Wettstein, X–XVIII after van Heel.

ing plan of the extensive correlation between the perigon segments and the stamens of the Liliaceae "can only be understood by assuming that they are collectively sterile derivatives of the androecium" is nevertheless not at all convincing. In fact what is seen is "simply the morphological relationships of the petals with the neighbouring floral organs" (see the quotation from Troll in section 1.4.1.1).

If, however, we wish to proceed from the hypothesis that the organs of the perianth have been formed secondarily through a corresponding differentiation of the leaf organs in the transition zone between trophophylls and sporophylls (see section 1.4.1.1), then we may reasonably expect that close morphological relationships with the inner members of a double perianth, i.e. the petals, and in certain cases also the tepals, will be recognisable with the sporophylls closest to them, i.e. the stamens. A phylogenetic interpretation, however, which argues that the tepals or petals have evolved directly from the stamens, may only be permissible in individual cases where there is some corresponding parallel evidence from a series of other characters.

Close morphological relationships also exist between stamens and petals in some dicotyledonous families, in which the petals more or less clearly exhibit a peltate or diplophyllous structure. One example of this is the ligule with which the petals of *Erythroxylon* and *Koelreuteria* are provided, and the peltate or diplophyllous petals of the Sapindaceae (Leinfellner 1954c, 1955b, 1958c). A very instructive example of a complete series of transitional forms between fertile stamens and petals is provided by *Orthostemon sellowianus* (Fig. 71 I to VII, Baum 1951a), although no trace of the cross zone remains visible in the progression of the petal margin. In the petals of *Waldsteinia geoides* on the other hand the reverse is still often the case (Fig. 71 VIII to X); here it can be seen that the structures which either appear as two separate basal marginal lobes or as a uniform ventral lobe are quite distinctly fused to the base of the dorsal blade over a short distance, i.e. are not just appressed auricles (Leinfellner 1954d). The peltation and diplophylly is clearly visible in the young petals (Fig. 71 XI to XIV). In order to arrive at the corresponding stamen forms in such examples it is often only necessary to argue that the development of greater surface area is then restricted in favour of the fertility of the margin. Leinfellner (1954b) has made this clear by examples of *Aesculus*- and *Dianthus*- species. The schematic diagrams in Fig. 72 IX to XIV are to be compared with the stamen primordia of *Dianthus* reproduced in Fig. 72 I and the forms of sterile and fertile stamens and petals of *Dianthus* and *Aesculus* (II–VIII). According to the interpretation of Leinfellner, the structure of the stamen primordium shown in Fig. 72 I and IX includes the possibility of development of the various kinds of stamens and petals under discussion here, whether these are tubular or diplophyllous in form. Either the peltate sterile stamen in diagram IV or else a peltate petal (V or X) can be derived from the stamen primordium by free development of the margin; by the margins becoming fertile combined with upward growth of the cross zone united with the dorsal section of the blade a stamen is obtained, or in case of repression of fertility the result is a staminode in the form shown in III or XI (see also the staminode in VII). If the stalk-like section becomes broader "corresponding to the growth of the blade and if the slit which is formed by the deflexed margins is greatly increased in length by extension of the petiole of the primordium" (1954, p. 403), the form of the *Dianthus* petal

(VIII or XIII) is produced. The type of petal that is found in *Aesculus* (VI or XII) comes about as a result of vigorous surface growth with only insignificant upward growth of the curved ventral margins on the blade and with only a slight elongation of the stalk-like section.

A four-edged or four-winged form is sometimes visible in petals (Fig. 72 VIII), but is very often much more pronounced in stamens with petaloid structure. Čelakovský (1878) attempted to explain this with the aid of his "Überspreitungstheorie", or "super-lamination" theory. According to Leinfellner, however, it is a consequence of the fundamentally diplophyllous structure (compare Leinfellner's explanation of the *Dianthus* corolla with the critical remarks of Kunze (1979, p. 287 ff.).

The different forms of nectar leaves (see section 1.4.3.10, Figs. 46, 47) can

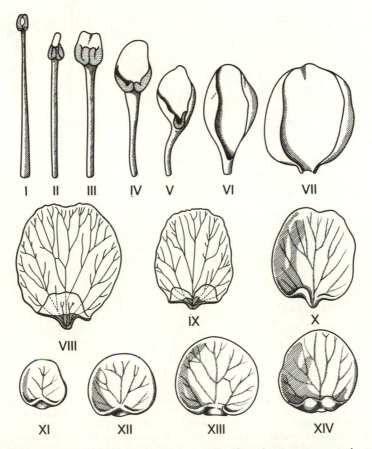

Fig. 71 I-VII *Orthostemon sellowianus.* Stamen (I), intermediate forms between stamens and petals (II-VI), VII petal. VIII-XIV *Waldsteinia geoides,* VII-X adult peltate and diplophyllous petals provided with a two-lobed ventral blade, made transparent in VIII and IX with the massive basal portion shaded. XI-XIV young petals in different stages of development. After Baum (I-VII) and Leinfellner (VIII-XIV).

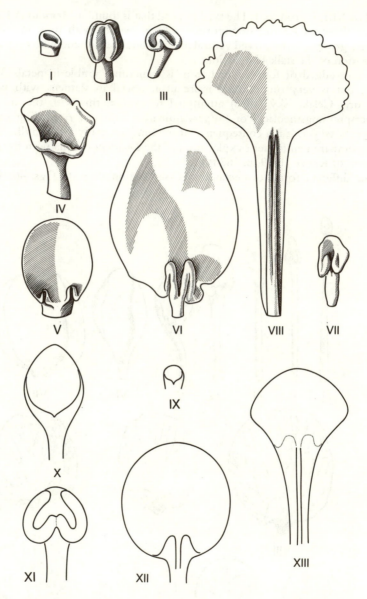

Fig. 72 I–IV *Dianthus caryophyllus.* I stamen primordium, II young stamen, III, IV deformed stamens. V *Aesculus hippocastanum*, young petal, VI *A. carnea*, older petal. VII *Dianthus semperflorens*, deformed stamen. VIII *D. barbatus*, petal. IX–XIII Diagrammatic view of the underlying structural scheme for a peltate stamen primordium of *Dianthus* (IX) and a peltate petal (X), a normal stamen (XI) and a petal of *Aesculus* (XII) and *Dianthus* (XIII). All after Leinfellner.

easily be placed among the range of different forms between fertile stamens and peltate or diplophyllous petals.

The leaf-like and flattened stamens of the Nymphaeaceae already mentioned which occur especially often among those families of the "primitive" Magnoliales are definitely diplophyllous, according to the researches of Leinfellner (1955b). They have however provided a "classic example" to prove that the stamens are leaf-like in nature (Fig. 73). Stamens of this form are still taken to be primitive even today, as for example in *Degeneria* (Takhtajan 1959). They also serve as an indication that during evolution "the broad, lingulate stamens of the primitive forms grade into highly specialised microsporophylls" (Takhtajan 1959, p. 85; see for example also Canright 1952). The presence of the three nerves which are frequently observed in these stamens is often interpreted in the same way (see for example Parkin 1951). According to Leinfellner (1955b), however, the occurrence of fringes or lamellae which cause a broadening of such stamens represents the result of secondary modifications (and likewise in the petals of the Bruniaceae; Leinfellner 1964c) In general these appear relatively late during ontogenesis (Fig. 73 I to VII). Also the "stamen primordia of *Nymphaea* are plainly diplophyllous"..."The leaf-like and flattened stamens are in no sense to be interpreted as simply constituted sporophylls, similar to bifacial leaves, but as derived structures" (Leinfellner 1955b, p. 288). The broad, leaf-like stamens of *Victoria amazonica* possess additional peripheral vascular bundles (partly inverted on the adaxial side), and thus likewise turn out to be secondarily compressed (Heinsbroek and van Heel 1969).

The petaloid transformation of stamens which is found in the families Zingiberaceae, Cannaceae and Marantaceae of the Zingiberales is quite definitely secondary in nature. It has continued here to such an extent that the display function of both the 3-merous perigon whorls, which are now scarcely conspicuous, is entirely taken over by the large, corolloid and coloured staminodes, which once again create the appearance of a labiate flower. This example demonstrates, how a distinct fashion – the "Gestalt" – of the flower (in this case the bilabiate form) can appear "independently from the basic organisation" and "by involving only some of its components while the remainder cease to have morphological meaning". In a case like this Troll (1928a, p. 89/90) speaks of a "secondary flower".

Curcuma australasica of the Zingiberaceae will serve as an example of such a two-lipped secondary flower (Fig. 74 V). As in all members of the Zingiberaceae only the median stamen of the inner ring is still fertile and the median of the outer ring is missing. In *Curcuma* and in all other Hedychieae (Fig. 74 VI) the other stamens are all developed to be large and petaloid, and the two innermost are united to form a lip-like organ, the labellum, which encloses the fertile stamen with its posterior margins and which functions as a "lower lip", as opposed to the "upper lip" (the median tepal of the inner perigon ring).

In *Kaempferia marginata* the upper lip is formed from the fertile stamen, whose connective region is expanded into two broad deflexed petaloid lobes; the theca is placed in the median and is curved inwards and also serves to hold the style (see Troll 1928a, p. 306 ff.). In *Roscoea purpurea* there is even a lifting mechanism, similar to that in *Salvia* species (see section 3.3.4.2b).

The petaloid modification of the stamens goes even further in the Marantaceae and Cannaceae than it does in the Zingiberaceae: here one half of the last stamen remaining in the Zingiberaceae becomes corolloid, and only one theca remains fertile. The most familiar examples of this are the *Canna* species (*C. indica, C. iridiflora*, Fig. 74 I to IV), in which the entire display function of the completely asymmetric flower is based on petaloid staminodes alone, and where one of them once again takes the form of a labellum (IV).

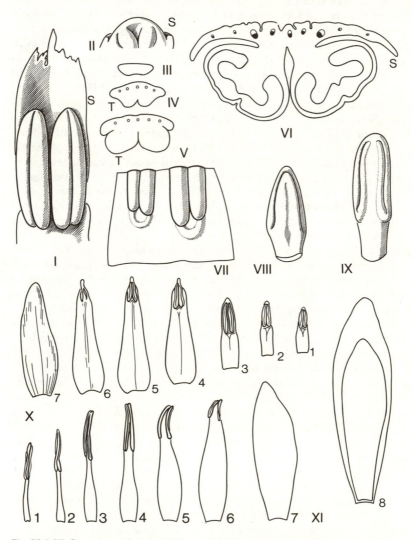

Fig. 73 I-VI *Costus igneus.* I stamen before anthesis, II stamen primordium, III-VI cross-sections through anthers in successive stages of development, S petaloid fringe, T protruding pollen sacs of the thecae. VII *Nymphaea stellata*, base of anther of an outer stamen, VIII, IX young stamens of *Nymphaea tetragona* in successive stages of development (IX somewhat less magnified than VIII). X, 1-7 and XI, 1-8 transition between stamens and tepals in *Nymphaea hybrida* and *N. alba.* I-IX after Leinfellner, X, XI after Troll.

1.5.14 Fusion between the stamens and the gynoecium

Since serial fusions between members of successive rings of floral organs occur so frequently, especially between stamens and petals, it is to be expected that serial connections between the stamens and the gynoecium may also take place. This is indeed occasionally the case, as for example in species of the Aristolo-

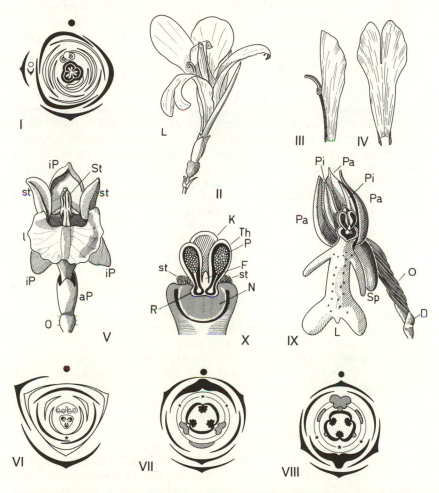

Fig. 74 I, II *Canna indica*, I floral diagram, II flower, L labellum. III, IV *Canna iridiflora*, III stamen and IV labellum. V, VI Zingiberaceae, V flower of *Curcuma australasica*, O inferior ovary, aP outer, iP inner perigon whorl, st staminodes, St the single fertile stamen, l staminodial labellum. VI *Kaempferia ovalifolia*, floral diagram. VII, VIII Floral diagram of *Cypripedium*, not yet resupinate, and *Orchis*, resupinate. IX, X *Orchis militaris*, IX flower with twisted inferior ovary (O), D bract, L labellum, Sp spur, N stigma, X the gynostemium further enlarged, Th theca, P pollinium and K connective of the fertile stamen, st staminodes, Pa outer, Pi inner perigon, R rostellum, F process of the rostellum, O inferior ovary (twisted), D pherophyll. After Eichler (I, VI-VIII), Schenk (II), Troll (III, IV), Hooker (V) and Berg and Schmidt (IX, X).

chiaceae, where the stamens and the style are united into a so-called **gynostem-ium** (see Fig. 189 III, IV); similarly for the Stylidiaceae.

In the Asclepiadaceae the fusion of the stamens with the gynoecium comes about either through mere cuticular dentonection (Periplocoideae), or through postgenital fusion (Cynanchoideae) (see Fig. 190). In the Apocynaceae the stamens are united with the gynoecium either by individual hairs (Plumerioideae), or else by tufts of hair and by lax proliferations (Echitoideae) (Baum 1948a).

In the Orchidaceae the fusion of the highly reduced androecium with the style and stigma is particularly intimate and of great significance for pollination biology. The structure of the gynostemium is a constant character and has given rise to the old name Gynandrae.

From among the two 3-merous whorls of stamens in the floral diagram of the Liliaceae it is only the three stamens which are originally orientated towards the subtending bract which reach maturity in the Orchidaceae. It is only exceptionally, however, that all three are fertile, as in the weakly zygomorphic flowers of the genus *Neuwiedia* (Apostasieae). In the rest of the Cypripedioideae, to which the familiar wild *Cypripedium calceolus* and the commonly cultivated *Paphiopedilum* (both under the same name, "Lady's-Slipper") each belong, the median stamen is transformed into a sterile scale (Fig. 74 VII). In the species of the Orchidoideae, which are much more varied and numerous than the other subfamily, only the median stamen is fertile (Fig. 74 VIII). The anthers only rarely release powdery pollen. Usually the pollen grains of a pollen sac or a theca are compacted into a "pollinium", which is removed in its entirety during pollination. The style and the fertile stamens are usually completely fused in what is called the **gynostemium**, which projects from the centre of the flower. One of the perigon segments, the one which is initially orientated towards the floral axis, is usually much modified and takes on a wide variety of different forms as a lip, the "labellum", which is often extended backwards into a nectar-bearing spur. During development the flower is often turned through an angle of 180° by rotation in the region of the inferior ovary or of the pedicel (VII), so that the labellum finally points downwards and can act as a landing platform for insects. The entrance to the spur then lies directly in front of the gynostemium (Fig. 74 IX). Normally no resupination takes place in the flowers of orchids which have pendent inflorescences.

The 3-merous ovary is usually unilocular with more or less projecting and often divided placentae, and is only rarely 3-locular with axile placentation. Only the two lateral lobes of the 3-lobed stigma are receptive to pollen; together they form a depressed or concave stigmatic surface. The third lobe of the stigma is fused with the fertile anthers as the "rostellum" (Fig. 74 X, R) and provides the viscidia and caudicles which are fused with each pollinium to form a "pollinarium". The pollinaria lie freely inside the thecae once these have opened, and the visiting pollinators, insects for example, push lightly with their heads against the viscidia of the pollinaria, pull these out of the anthers and carry them away to the next flower. In the meantime the caudicles bend over forwards by hygroscopic movement, so that the pollinium is pressed against the stigmatic surface as the insect pushes its way into the next flower. It is only by the transfer of an entire pollinium in a single pollination event that the fertilisation of the several thousand ovules often present in an ovary is possible.

1.6   The gynoecium

1.6.1   Structure of the carpel, apocarpous and syncarpous gynoecium, placentation types

The apex of the flower is occupied by the macrosporophylls (megasporophylls), which in the seed plants are known as the carpels. In many families there is generally more than one per flower, but they also occur singly, as for example in the legumes (Leguminales) or in the Rosaceae-Prunoideae. As already mentioned, the carpels of a flower are known collectively as its gynoecium.

By comparison with the stamens the leaf-like qualities of the carpels are a good deal more obvious. The principal reason for this is the intense green coloration caused by the high chlorophyll content. Furthermore the carpels in many families often undergo a considerable growth of their surface (Fig. 75 I), but even so, they do not unfold.

We have already briefly discussed (section 1.1) the fact that the ovules of the Angiosperms are enclosed by the carpels, contrasting to the Gymnosperms where the ovules are free and open to the air on the margins of the sporophylls, at pollination time at least. The example of the carpel of *Helleborus foetidus* (Fig. 75 III, IV) shows us what this means. Externally (II) there is little in this to suggest a macrosporophyll. It is only when such an organ is cross-sectioned (Fig. 75 V, 76 II), or is cut open lengthways (Fig. 75 III, IV) that the way the ovules (Sa) are housed inside the carpel is visible. The ovules arise from the margins of the carpel, which curve inwards ventrally and are fused together postgenitally along a suture.

The margins of the distinctly elongated tip of the carpel remain free at this point and diverge from each other in such a way that the upper surface is turned outwards (upwards) towards the open. This section of the carpel is the **stigma** (Fig. 75 III, IV, VII), which serves to collect the pollen brought to it by the wind or by animals. Its upper surface is therefore often densely covered with papillae which produce a slimy substance (stigmatic secretion) to which the pollen grains adhere and which can also stimulate the growth and development of the pollen tubes.

By comparison with the other nerves of the carpel, the middle nerve Fig. 75 (M) is especially conspicuous. There are likewise yet larger vascular bundles (V Plb) running through the margins which are united in the ventral suture over on the opposite side, and these obviously serve to supply the numerous ovules which are inserted here on the ventral side of the margins. The region in which the ovules are placed projects as an outgrowth of tissue shaped like a ridge, which is called a **placenta**. The vascular bundle which runs along the margin is known correspondingly as the placental vascular bundle (Plb).

The type of ovule arrangement being discussed is termed **marginal** (or submarginal), and predominates over other kinds of placentation. From this fundamental type, a second type known as **laminal** (or laminate) placentation can be distinguished, in which the ovules arise from the inner surface of the carpel, as for example in the Butomaceae (Fig. 78 XIV, XV), Hydrocharitaceae and the majority of the Nymphaeaceae (Fig. 14 I, 78 XVI). Usually the ovules are inserted between the margin and the median zone (laminal-lateral); only in a

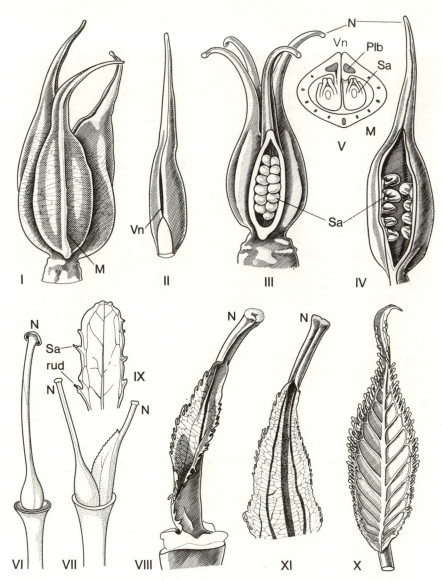

Fig. 75 I-IV *Helleborus foetidus*. I Gynoecium during development of the fruit, M median nerve of the carpel nearest to view, II single carpel during anthesis, ventral view, Vn ventral suture, III gynoecium during anthesis, rear wall of the nearest carpel cut away in order to show the ovules Sa placed along the ventral suture, IV single carpel during development of the fruit with the ventral suture pulled open. V *Delphinium elatum*, cross-section through a carpel, Plb placental vascular bundle, M median nerve, VI *Prunus avium*, flower with all the organs except the carpel removed. VII-X *Prunus paniculata*, VII gynoecium, here consisting of two carpels exposed as in VI, with one carpel of the leafy type, N stigma, VIII the same in a unicarpellate, foliose gynoecium, IX the same carpel spread out, X young ordinary leaf for comparison. XI *Barbarea vulgaris*, foliose carpel, Sa rud, rudimentary ovules. For further explanation see text. V, VI after Troll, VII after Rauh, XI after Rohweder.

few cases, which require careful verification, are the ovules found in the median of the carpel whilst the margins remain free. The latter is laminal-median placentation, as in *Brasenia* (Nymphaeaceae) and Ceratophyllaceae (see Troll 1934c). It may be possible "to derive the laminal mode of placentation from a marginal one" (Troll 1931, p. 5).

The foliar nature of the carpel comes to light particularly often in cases of abnormal floral development, as in the so-called phyllody or frondescence which is often observed in deformed flowers (Fig. 75 XI), and particularly in association with the "double" forms seen in cultivated plants. A striking example of this is provided by the Japanese Cherry (*Prunus paniculata* = *P. serrulata*), whose flowers are often doubled by the numerous stamens appearing in petaloid form. It is characteristic of *Prunus* species in general that the gynoecium has a single carpel only, consisting of an ovoid ovary, a long style and a dish-shaped stigma (Fig. 75 VI).

In Japanese Cherry, however, two carpels usually appear, of which at least one is generally foliose (Fig. 75 VII). Apart from the lack of a petiole the latter shows an astonishing similarity to the foliage leaves: the margins remain separate and show the same type of toothing as appears on the margins of the normal leaves (VIII). Even the venation is like that of the foliage leaf, except that two strong lateral veins are visible beside the midrib, where these may correspond to the placental vascular bundles (IX). The style and stigma, which continue the midrib, have obviously remained in a rudimentary form. The way in which the leaf is folded, as in aestivation, is likewise preserved.

In the case that several carpels are present in a flower, there are two situations which can be distinguished, namely the apocarpous (choricarpous) and the syncarpous (coenocarpous)* forms of gynoecium. The term apocarpous refers to a gynoecium which consists of free carpels, which are not united to each other. A good example of this is the gynoecium of *Delphinium elatum* (Ranunculaceae, Fig. 76 I). In this apocarpous type the gynoecium appears in what seems to be a particularly fundamental state, which has survived along with a variety of other primitive characters which are typical of the families of the Polycarpicae, the Dilleniales and many of the Helobiae, as well as certain other families (Sterculiaceae, Simaroubaceae, Apocynaceae, Asclepiadaceae, Palmae etc.).

The **syncarpous gynoecium** on the other hand is distinguished from the apocarpous gynoecium by the congenital or (at least partial) postgenital fusion of the carpels into a more or less uniform structure. Given the cross-section of the separate carpels of *Delphinium*, as illustrated in Fig. 76 II, it is only necessary to imagine these to be united laterally in order to arrive at the corresponding cross-section of a syncarpous ovary (Fig. 76 IV), of the kind which is characteristic of many members of the Liliaceae.

The structural properties and placentation of the individual carpels are not altered in any way by congenital fusion into a syncarpous ovary. The carpels have nevertheless given up their individuality in order to create an organ of

---

* We distinguish: I apocarpous (choricarpous) II coenocarpous – syncarpous (= syncarpous *sensu stricto*) and coenocarpous – paracarpous.

higher order, namely the **pistil**, named from the Latin for pestle, as in 'pestle and mortar'. The thickened basal portion which surrounds the ovules is known as the **ovary**.

The part of the pistil between the ovary and the stigma is often much elongated and attenuated and is called the **style**. In cases where no style is formed, as in the Tulip (Fig. 95 III, IV), the stigma is placed directly on top of the thickened, fertile section of the gynoecium, i.e. on the ovary. If, as in the above example, three completely closed carpels become fused, then an ovary with three inner cavities is the result, i.e. the pistil is provided with as many cavities as there were foliar organs which took part in its formation. The internal dividing walls which correspond to the fused lateral faces of the carpels are called **septa**, and the cavities are called **locules**. This arrangement of the placentae where the ovules are placed in the centre of the ovary in the angles between the septa is termed **axile placentation**. As already mentioned, the margins of the carpels remain sterile in the region of the style. In general, the margins separate from each other at this point and withdraw somewhat from the central axis of the ovary, so that a uniform cavity or **stylar canal** is formed. This canal is nevertheless often filled with a growth of secondary tissue, which can play some part in the guidance of the pollen tubes.

It is also advisable to distinguish the **paracarpous** gynoecium from the **syncarpous** form, *sensu stricto*. The latter is characterised by the fact that although the carpels are mutually united, the margins of the individual carpels do not meet and close on the ventral side, so that only a single cavity results, and the margins of the carpels which bear the placentae project into it a greater or lesser degree. In this way incomplete locules are created (Fig. 81 XII). If the margins of the carpels do not project into the ovarial cavity (paracarpous-aseptate) the placentae are placed on the inner wall of the ovary. This is termed **parietal** placentation (Fig. 76 VI). Paracarpy is often thought to be a secondary phenomenon, since it is presumed that once the carpels have united to form a syncarpous ovary, the ventral connection can be abandoned without putting the protection of the ovules at risk.

The withdrawal of the septa can also bring about what is called a **central placenta** (Fig. 76 VII, VIII). This can be achieved in later stages of development, i.e. postgenitally by disappearance of the septa, or occur congenitally as well, so that the carpel margins do not reach laterally as far as the centre of the ovary. In this way an ovary is created with a uniform cavity, in which a central ovule-bearing column is found. This column can be further reduced, as long as only the lower portion bears ovules, so that the ovules are positioned in the centre of the ovary at the base. This may give the impression that the ovules may have originated from the apex of the floral axis, instead of being produced by leaf-like organs.

In order to be able to really understand the form of central placentation just described, and even more in order to gain a better understanding of the structure of the gynoecium as a whole, it is necessary to turn once more to the morphology of its parts, i.e. to the structure of the individual carpels. The apocarpous gynoecia of the families already named offer various opportunities for this.

1.6.2   Carpels as peltate-ascidiate organs, U-shaped placentation

In the case of the carpel we are dealing with a leaf organ which usually differs from other leaves in more ways than just by being folded and having the margins joined. The fundamental form of the carpel is often really that of a peltate, more or less tubular leaf, which bears the placentae with the ovules along the margins of its morphological upper surface.

We have already discussed the origin and structure of a peltate foliage leaf in section 1.5.6 with the example of the leaf of *Tropaeolum*. The process by which a tubular leaf is derived from a plane, peltate leaf such as this is obvious. It is only necessary to imagine that the surface of the *Tropaeolum* leaf grows vigorously while its margin keeps the same circumference as before. What must happen is that the blade first becomes concave and that this then leads to a funnel-shape and finally, in extreme cases, to an elongated, tubular leaf.

Such elongated tubular or conical leaves are known from various insectivorous plants; *Sarracenia* or the pitcher-shaped leaf of *Nepenthes* come to mind. Correspondingly long and tubular carpels are found in the Potamogetonaceae or in the Zannichelliaceae (Fig. 77, I, II).

The peltate-ascidiate nature of such carpels appears at an early stage as a

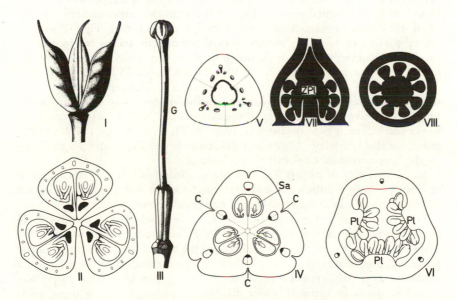

Fig. 76 Structure of the gynoecium. I, II Apocarpous gynoecium of *Delphinium elatum* after flowering, general view (I) and in cross-section (II). III–V Syncarpous gynoecium of *Lilium candidum*, III general view, G style, IV cross-section through the ovary, V through the style, Sa ovules, C commissures in which the carpels are united, VI cross-section through the paracarpous-aseptate gynoecium of *Passiflora*, Pl placentae. VII, VIII axial and cross-sections through the gynoecium of *Lysimachia*. ZPl central placenta. All after Troll.

swelling in the cross zone (Fig. 77 VI Q, VII). In this way a more or less obliquely inserted, cup-shaped primordium is formed, which expands into a cylindrical structure with a more or less oblique orifice by means of a completely encircling ring of marginal meristem (Fig. 77 III to V, 78 I, XVI). In such cases the stigma can be formed by the complete circular margin (Fig. 78 I) or even be expanded into a saucer-shape (Fig. 77 I, II).

Peltate-ascidiate carpels are found particularly frequently, or may even dominate, amongst the Magnoliales, although admittedly there are exceptions with an O-shaped placenta (see below). This appears to justify the assertion of Leinfellner (1969b, p. 124), that they "are the fundamental type of angiosperm carpel, not only in the morphological sense, but also phylogenetically" (see also Baum 1952c), or at least are one of the fundamental types.

It is a fact that clearly peltate carpels of uniformly tubular form are encountered rather rarely. Usually the ventral part of the carpel wall does not reach to the apex, but remains significantly lower than the rear wall; see for example the median section through a young carpel of *Thalictrum* (Fig. 77 X). In other words: the dorsal region of the carpel is often continued considerably further distally than the tubular part. To this extent therefore there is agreement with the scheme of a funnel-shaped leaf blade which is shown in Fig. 56 VII. Nevertheless the marginal growth of the primordium is not directed outwards in the dorsal zone, but upwards, and finally the margin curves inwards, so that the carpel cavity culminates in a slit-like opening, the ventral suture, which is later closed off by postgenital growth (Fig. 78 II). Thus the carpel orifice, which is often still directed obliquely upwards in the primordial stage (Fig. 77 VIII), is brought into a more or less vertical adaxial position by means of vigorous or even overwhelming growth of the rear wall of the carpel and elongated to form the ventral suture (see also Fig. 77 VI, VII, XI to XV and XVI to XVIII). This is also true for the ascidiate carpel of *Degeneria vitiensis* (Degeneriaceae), which is thought to be relatively primitive, in which the oblique margins of the "apical suture" are differentiated to form a "stigmatic crest" with the exception of the lower end which is bent downwards (Fig. 78 VI, VII; Swamy 1949). On this point, see also Leinfellner 1969b, and the researches of Payne and Seago (1968) on the "open conduplicate" carpel of *Akebia quinata*.

In the majority of carpels it is possible to distinguish, in descriptive terms at least, between a **tubular** (**ascidiate**) basal section and a non-tubular, but **plicate** section, whose dimensions may be very different.

Let us first consider the carpel of *Clematis cirrhosa*, which is shown in Fig. 78 II in a ventral view and in diagrams III to V in longitudinal sections. Since the lower end of the ventral slit reaches almost to the base of the carpel, the ascidiate portion is presumably very short here. This can be confirmed by means of a median section (III). Diagram IV shows the lower part of another carpel sectioned in the same manner, whilst diagram V provides a view from inside towards the ventral slit and the placenta with the dorsal part of the carpel removed. It is obvious here, as in almost all the Anemones, that only one ovule is developed completely, and that this is inserted on the cross zone of the carpel. The presence of more or less rudimentary ovules on the plicate margins of the carpel indicates, however, that it is fundamentally possible for ovules also to be formed on the two upwardly-directed margins, which are at first free and then later united, as well as on the cross zone. In this form the placenta of the angiosperm carpel often presents itself as a **U–shaped** curve, which encloses the lower end of the ventral slit (Fig. 78 XIII; see Leinfellner 1951a). In species of *Drimys* there are, however, tubular carpels where the elliptic and closed carpel

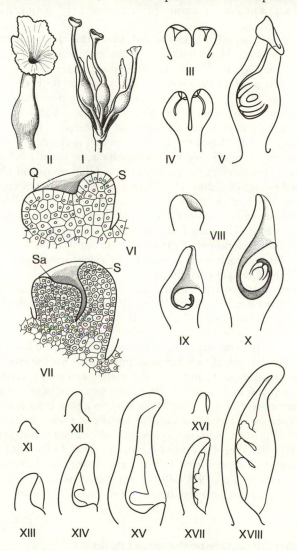

Fig. 77 I *Althenia filiformis*, gynoecium, II-V *Zannichellia palustris*, II young carpel, III-V sequential stages in the development of the gynoecium (III, IV) or of a single carpel (V). VI, VII *Potamogeton alpinus*, stages of carpel development, both in median section. VIII-X *Thalictrum flavum*, median sections showing stages in carpel development, XI-XVIII stages from the development of the peltate carpel of *Amorpha fruticosa* (XI-XV) and the epeltate carpel of *Astragalus galegiformis* (XVI-XVIII) in median sections. Q cross zone, S subepidermal apical initial cell, Sa ovules. I after Prillieux, II-V after Eber, VI, VII after Kaussmann, VIII-X after Troll, XI-XVIII after Leinfellner.

margin can be fertile along its entire extent (Fig. 78 I, VIII, IX). "The ventral slit is here therefore surrounded by a ring- or O-shaped placenta. Hence the U-shaped placenta turns out to be...an (excessively frequent) secondary form, which is determined by the differentiation of the normal angiosperm carpel into a fertile (ovary) and a sterile section (style and stigma)" (Leinfellner 1966a, p. 95).

The individual sections of the **U-shaped placenta** can be developed or inhibited in various ways. Hence either only the cross zone may bear ovules (often only one, see Fig. 77 X) or else only the postgenitally fused carpel margins, which then produce two separate lateral placentae (Fig. 75 III to V, Fig. 77 XVIII, Fig. 78 XII). This variability in proportion results from an unequal development of the ascidiate and plicate sections, which is often already recognisable at an early stage in the ontogeny of the carpel (Fig. 77 VIII to X, XI to XV, XVI to XVIII). When the ascidiate section is extremely suppressed, then the resulting carpel form is said to be **subpeltate**, or latently peltate. Some examples of this are provided in the carpels of *Eranthis hyemalis* and *Delphinium requenii* illustrated in Fig. 78 X and XI, XII. In a certain number of carpels, as in *Prunus* (Schaeppi 1951) or in most of the Leguminales (Leinfellner 1969d, 1970a), the ventral slit even continues as far as the beginning of the carpel stalk, which is often unifacial. The carpel of *Astragalus galegiformis* (Fig. 77 XVI to XVIII), in which no development of any cross zone can be seen, will serve as an example of this. Such carpels with a completely absent cross zone are consequently plicate only: **epeltate carpels**.

The relationship between epeltate and peltate carpels becomes clear in the example of *Helleborus foetidus* used by Baum (1952e): here "manifestly peltate, latently peltate and epeltate carpels" all appear side by side in one and the same gynoecium. These are distinct ontogenetically by the moment of initiation of their cross zones: "as the late initiation of the cross zone of the latently peltate carpel is yet further delayed, then it becomes impossible that the cross zone still might appear during ontogeny, and the result is then an epeltate carpel".

The question remains as to whether the peltate structure of the carpel is also reflected in its nervature. This is actually the case. In general the visibly peltate carpels show a median and inverted vascular bundle in their ascidiate sections, opposite the dorsal median, with the xylem portion turned to the abaxial side (Fig. 79 I, II, VII, VIII); that is, a ventral median appears. In the plicate section, on the other hand, this divides itself into the two marginal placental vascular bundles, as is shown in Fig. 79 VI-III, X for the densely veined carpel of *Thalictrum flavum* (see also amongst others Troll 1932a, Schaeppi and Frank 1962b).

1.6.3   Unifacial structure of the carpel stalks

We have already referred (section 1.5.6) to an important hypothesis concerning the origin of a peltate leaf zone, i.e. the unifacial structure of the petiole, or in more general terms; the unifacial structure of one of the segments which precedes the peltate zone. In general it is usual for the carpel "blades" to be inserted directly onto the receptacle, or at most to be quite shortly stalked. In the Zannichelliaceae (Fig. 77 I, II), the Ranunculaceae and in several other families, however, gynoecia are produced in which the carpel stalk (or petiolus) is quite considerably elongated (Fig. 79 IX). In such cases it is of interest to determine whether carpel stalks of this nature may show evidence of unifacial structure, not only in their more or less circular cross-section, but also in their anatomy.

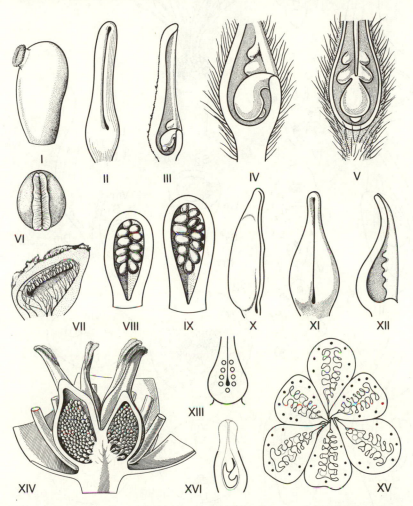

Fig. 78 I *Drimys winteri*, carpel. II–V *Clematis cirrhosa*, carpel in ventral view (II), in median cross-section (III), in IV the basal part of another carpel in section, further enlarged, V the same carpel opened from the dorsal side. VI, VII *Degeneria vitiensis*, carpel from above (VI) and in median section (VII). VIII, IX *Drimys winteri*, carpel after removal of the rear wall, showing the arrangement of the ovules around the ventral slit, in VIII the uppermost ovule lies presumably in the median, and in IX the lowest. X *Eranthis hyemalis*, subpeltate carpel. XI, XII *Delphinium requenii*, epeltate carpel in ventral view (XI) and median (XII) section. XIII Schematic diagram for the explanation of U-shaped placentation, XIV, XV *Butomus umbellatus*, flower in axial section (XIV), perigon and stamens removed, XV gynoecium in cross-section. XVI *Cabomba caroliniana*, tangential longitudinal cross-section through a young carpel. I, VIII, IX after Leinfellner, III–V, XI, XII after Payer, VI, VII after Swamy, X after Baum, XIII after Hartl, XVI after Troll, the rest original.

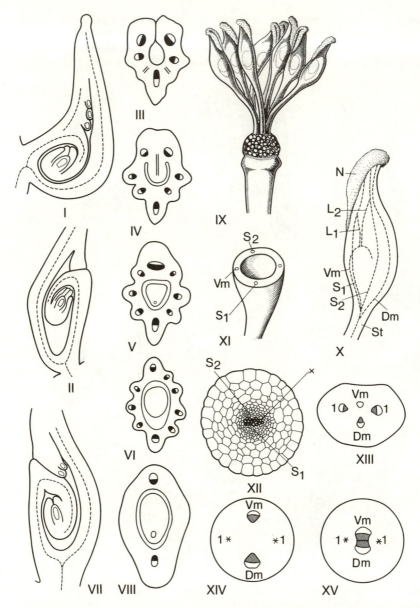

Fig. 79 I, II Carpels of *Adonis annua* (I) and *Thalictrum flavum* (II) in median cross-section. III-VI series of cross-sections through the carpel of *Thalictrum flavum*. VII, VIII *Clematis vitalba*, median section of the carpel (VII) and transverse section in the ascidiate portion (VIII). IX-XII *Thalictrum aquilegifolium*, IX gynoecium, with the perianth and androecium removed, X individual carpel, made to appear transparent, Vm ventral, and Dm dorsal median, S_1, S_2 and L_1, L_2 lateral branches of the ventral median, N stigma. XI Schematic cross-section of the ascidiate portion of the carpel, XII cross-section through the carpel stalk (St). XIII *Eranthis hyemalis*, cross-section through the carpel stalk. XIV, XV Schematic diagrams for the explanation of the unifacial structure of the carpel stalks of *Thalictrum* and *Eranthis*. In I and VII

According to Troll (1932b), the carpel stalk in *Thalictrum aquilegifolium* possesses a central vascular bundle, which divides into two branches as it enters the carpel proper (Fig. 79 X): the dorsal median (Dm) and the ventral median (Vm). The latter gives off from its base two lateral strands S_1 and S_2, and two more side branches (L_1 and L_2) before it enters the funiculus of the single ovule which is inserted on the median of the cross zone. The ascidiate section of the carpel is reproduced in schematic form in diagram XI.

When a cross-section of the carpel stalk (XII) is examined, a conspicuous compressed central xylem body (x) is found, which is bounded dorsally and ventrally by two small sieve elements (S_1 and S_2). It follows from this that two bundles are united into one strand, which separate higher up into a dorsal median and a ventral median. These have been made visible in the transparent carpel in diagram X.

If we refer to the carpel stalk of *Eranthis hyemalis* (Fig. 79 XIII), then we find here that three separate collateral vascular bundles are recognisable in cross-section, while a fourth median bundle (Vm) is only very weak and possesses no xylem. The remaining strands (Dm and 1–1) have their xylem portions turned towards the centre of the stalk, which is also typical of a unifacial structure.

The characteristics of *Thalictrum* correspond to the diagram of Fig. 79 XIV. Here both the lateral strands are reduced, which is to be expected when only one median ovule is formed. Both bundles have their xylem portions on the inward side. Hence here also there can be no doubt that the structure of this zone is unifacial, even if the two xylem bundles are fused together as in diagram XV.

1.6.4 Marginal or submarginal and laminal placentas

There has long been a controversy as to whether the ovules are really marginal, i.e. arise directly from the "involute" carpel margins which are rolled inwards towards the central cavity and which are fused postgenitally at the points of contact (Puri 1961, Guédès 1965, Rohweder 1967b, Leins and Galle 1971, Endress 1969), or whether they are submarginal and arise on the surface near the margin, and the carpel margins come into contact through simple "conduplicate" folding and fuse postgenitally. In recent years most authors have inclined to the latter view.

One of the most important arguments to be given for the first point of view is that the ventral vascular bundle can be rotated by up to 270° from the dorsal median (see Puri 1961, Rohweder 1967b), which can be explained if the inrolled margin of the "involute carpel" is assumed. In this case the regions which lie on either side of the line of fusion in the plicate region would not be the actual carpel margins. Furthermore, the ridges which often appear on either side of the ventral slit and project outwards to a considerable extent, as for example in the Ranunculaceae, would be interpreted as secondary structures, as ribbed processes on the underside of the carpel. According to Rohweder these are pro-

Caption for Fig. 79 (*cont.*) there are rudimentary ovules above the one which is fully developed. In XII–XV S_1, S_2 sieve elements, x xylem, 1–1 lateral strands, if reduced marked as *. I–VIII after Schaeppi and Frank, IX–XV after Troll.

duced by periclinal divisions in the subepidermal layer, and in this respect they resemble the wing-like ridges which follow the dorsal median and the lateral nerves in *Caltha, Thalictrum* and *Helleborus*. In extreme cases, during the development of these ridges even "a formation of pronounced marginal meristems can take place, which do not therefore in any way have to be confined to the primary phyllome margin" (1967b, p. 420). "Their cells form vacuoles at an early stage, whilst the placental regions which are apparently distant from the margins remain fully meristematic" (p. 419). According to Rohweder, corresponding processes are "also visible both beside and particularly below the point of insertion of the ovule" (p. 381) in the carpel of *Ranunculus sceleratus*, which has its single ovule placed on the cross zone. Hence these phenomena cannot refer to the primary carpel margin, since this "is of course already occupied by the ovule". Endress (1972a, 1973a, b, 1975a) shows that the same is true for the Lauraceae and the Monimiaceae.

On the other hand, Leinfellner (1973b) has shown by means of *Sedum hybridum* and other examples that the growth produced by continual bilateral segmentation of subepidermal marginal cells also extends to the region which later projects outwards as the margin. "The cells of the placental meristem ... are probably derived from marginal meristem, however the fact that the distance of this meristem from the margin and the direction of division of its cells, which is perpendicular to the rows of cells built from the marginal cells, clearly indicates that here no leaf margin is formed, but only an outgrowth (placentas and ovules) from the surface" (p. 299). Leinfellner found a corresponding situation in *Butomus umbellatus*, which is the textbook example of laminal placentation. Moreover Puri (1962) admits that in cases such as *Butomus, Nymphaea* etc. the placentation of the ovules is clearly superficial, and quotes (p. 196) a point of view expressed earlier with respect to the placentation of the Gentianaceae (Gopal and Puri 1962) "that the so-called 'superficial placentation can be looked upon as a function of growth in which one part of the carpel (the marginal region) grows more than the rest".

The forms of frondescent carpels, which are often lobed, have often been taken as clear indications of the marginal origin (and the homology) of the ovules. Hence for example Velenovský (1910, p. 978) writes: "on the margins of those carpels of frondescent flowers which have once again taken on the form of green leaves we find flat lobes which show all stages of transition to the normal ovules". Also Guédès (1965, 1966a, d) and Rohweder (1967b, see also 1959) refer back to the analysis and interpretation of teratological phenomena of this kind, which does not appear to be acceptable without criticism. However, the identification of the frondescent ovules with the lobes of a normal leaf which had already been made by Čelakovský (1884b) is neither by Puri (1961, 1963) nor by Rohweder. Moreover it does not have any bearing on the question of the position of the placentas. On the question of this proposed homology, which has been taken up again by Guédès (1965, 1972) and Guédès and Dupuy (1970), Leinfellner aptly remarks: "the ovules ... originate relatively late in the ontogenesis of recent carpels as tuberculate outgrowths (nucellus primordia). If the further development of the carpel is modified in the direction of a frondescent organ, how can these tiny tubercles of tissue grow out in order to form a flat, leaf-like structure? If, however, the conversion begins to take place after the initialisation of the integument, then various forms of these conical-tubular structures are produced, as have been repeatedly described. Their form is readily explained as the result of excessive growth of the ring-shaped integument primordium". Furthermore Leinfellner was able to show, in contradiction to the results illustrated by Guédès (1967) for *Trifolium repens*, that the more or less modified ovules of frondescent carpels of *Trifolium hybridum*, arise from near the margin of the inner surface of the carpel blade, just like the normal ovules of normal carpels, and that in the winged ridges of the young carpel the "characteristic cell arrangement is unmistakably the result of a bilateral segmentation of subepidermal marginal cells" (1973b, p. 295).

If we try to balance the remaining arguments advocated by Puri, Rohweder and others, for marginal placentation on an involute carpel, against the arguments for the point of view which has been represented here by the statements of Leinfellner, for submarginal placentation on a conduplicate carpel, then it seems to be quite possible that both forms of placentation and carpel margin fusion could exist, and that they are not so greatly different from one another as it might appear on the basis of the foregoing discussion. In this context, the comparison of the often greatly augmented thickness of the carpel margins with the normal leaves is of particular importance (as Puri has also pointed out).

1.6.5   The structure of the syncarpous gynoecium

Since we already know that the individual carpel does not abandon its characteristic structure when it is fused with several more of the same kind into a more or less compound gynoecium, we can say that the main features of the structure of a syncarpous gynoecium are determined by the proportions between the

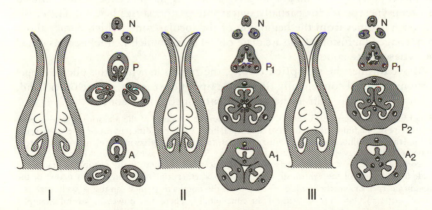

Fig. 80 Schematic longitudinal and cross-sections through an apocarpous (I), a partially fused (hemisyncarpous, partially syncarpous) (II) and a completely syncarpous (eusyncarpous) gynoecium (III). N Cross-section in the stigmatic zone, P in the plicate and A in the ascidiate (tubular) zones, P_1 in the partially symplicate-aseptate (hemisymplicate-aseptate) zone and below P_1 in II the partially symplicate, completely septate zone (hemisymplicate-holoseptate), P_2 in the symplicate, completely septate (symplicate-holoseptate), A_1 in the partially synascidiate (hemisynascidiate) and A_2 in the synascidiate zone. After Leinfellner.

individual carpel zones and the partial sterility of the U-shaped placenta, by the various numbers of carpels and the differing degree to which they are fused.*

As already mentioned, the distinguishing character between the syncarpous (coenocarpous) and the apocarpous (choricarpous) gynoecium is given by the lateral fusion of the carpels. The individual carpels may be combined equally well by congenital fusion as by postgenital growth. Commonly the opinion is held that whether the gynoecium is viewed as apocarpous or syncarpous (coenocarpous) is sufficient to decide whether there is congenital fusion or not.

However, the postgenital fusions which occur during development of the flower frequently imitate the characteristic type of congenital fusion in every detail, overlap the congenital fusion and often lead to the same results in the tissue structure, so that these processes can no longer be distinguished in the mature gynoecium (Baum 1948c, f, 1949a, Leinfellner 1950, 1951b). In order to elucidate the type of fusion it is therefore often necessary to observe the young stages of development of the gynoecium, in which the postgenital processes have not yet occurred, or where they are at least readily recognisable.

Beyond this the ascidiate and plicate zones of the singular carpel are to be distinguished in the structure of the syncarpous gynoecium as well. From this follows that there are various possibilities for the structure of the ovary, which are brought together into a scheme which was originated by Leinfellner (Fig. 80).

In Fig. 80 I an apocarpous gynoecium is illustrated. Cross-sections through the ascidiate and plicate zones of the carpels, with a cross section of the stigmatic region drawn in above are shown alongside a schematic longitudinal section.

The two other schematic longitudinal sections with the corresponding cross-sections show syncarpous (coenocarpous) gynoecia, in which only congenital fusion of the carpels is considered. The gynoecium illustrated in diagram III was termed "eusyncarpous" by Leinfellner, i.e. completely syncarpous. The term "holocoenocarpous" could also be used, as well as "hemicoenocarpous" instead of "hemisyncarpous" for partially syncarpous gynoecia (see below). The cross-section A_2 comes from the beginning of the locular zone, which comes about from congenital fusion of the ascidiate carpel bases, which Leinfellner has called the **synascidiate** zone.

Above this lies the **symplicate** zone, the region in which the plicate carpel portions are fused congenitally (P_2).† The tips of the carpels, on the other hand, usually remain free – this is the **asymplicate** or stigmatic zone.

* According to Saunders (1925–29 ff.), what underlies the structure of the majority of angiosperm gynoecia is not one fundamental form of the carpel, but a variety of different forms (the valvate, the solid and the semi-solid carpel) which must participate to differing degrees in the formation of different gynoecia. This view was already rejected by Eames (1931) as well as by Joshi and Rao (1933).
† According to the view of Hartl (personal communication), it is not usually possible in practice to decide how the loss of the ventral slit comes about: whether it is by the ascidiate structure of the individual carpel, or whether a plug of tissue from the floral axis (which can just as well serve as a 'plastic' material) closes up the ventral slit congenitally. There are, however, several examples where the latter is evidently the case (*Geranium*; several members of the Rutaceae, e.g. the navel orange *(Citrus)*; and the Nolanaceae), and where there is still a short conical section of axial material which projects above the "false cross zone". The observation that there is very often a double ventral row of ovules which extends a long way down in the locules of the primarily syncarpous zone, is similarly an argument against an ascidiate origin, and can be better explained as an indication of the secondary closure of a "falsely ascidiate" zone with axial material. It would therefore be better to only refer to a "synascidiate" zone in those few cases where its presence can be unequivocally established. In other cases only the term syncarpy should be used. In conclusion, therefore, the Leinfellner scheme is essentially correct, but it is still incomplete, and may not even apply to the majority of cases just without further comment.

The transition from the asymplicate to the symplicate zone does not in general take place suddenly, but the folded carpel blades gradually join more and more together, from the outside inwards, until they become completely fused. The slit which separates the carpel surfaces becomes gradually narrower below until it finally disappears. Since this transition zone is a structural feature which occurs constantly, Leinfellner proposed the special term **hemisymplicate** for it.

Therefore according to Leinfellner, a typical syncarpous (coenocarpous) gynoecium exhibits four vertical zones, which are determined by the extent of congenital fusion of the carpels and by their fundamental structure. These are as follows:

asymplicate
hemisymplicate
symplicate, and
synascidiate zones.

According to the principle of variable proportions, the individual zones can develop to a very different extent on different occasions. The development of the synascidiate zone can dominate, etc. In the case of a totally asymplicate structure we would have an apocarpous (choricarpous) gynoecium. The apocarpous gynoecium therefore fits into the Leinfellner scheme without difficulty.

On the other hand, the zone of incomplete fusion can also become very extensive. In this case, along the entire length of the carpel blade inclusive of the ascidiate section, the fusion of the carpels does not progress beyond the condition of the hemisymplicate zone of the syncarpous gynoecium, because only a part of the contiguous lateral carpel faces is congenitally fused. This case is illustrated in diagram II of the scheme; it is very often found in the Liliflorae. Such a gynoecium can be called partially syncarpous (hemisyncarpous, hemicoenocarpous), to contrast with that shown in diagram III, which is completely syncarpous (eusyncarpous, holocoenocarpous).

The structure of a partially syncarpous gynoecium may be made clear by a series of cross sections through the gynoecium of *Asphodelus albus* (Fig. 81 VIII to XI) which, like the later examples, was investigated and illustrated by Leinfellner (1950). The style (VIII) exhibits a continuous central hollow, into which the carpel margins project somewhat. The lateral faces of the carpels are united at this point by postgenital fusion exclusively, the only exception being the transition zone to the plicate region of the carpels. In this plicate zone the hemisyncarpous structure of the gynoecium is evident (IX): the lateral faces of the carpels are congenitally united to one another in the outermost region of the ovary, but otherwise remain separated from each other by distinct slits, the septal slits. These septal slits continue so far downwards as to enter into the ascidiate zone of the carpels (X, XI). Even when they subsequently become increasingly narrow, the hemisyncarpous structure of the ovary still remains in evidence even here. As far as the narrowing of the septal slits ensues from the centre of the gynoecium it probably results at least partially from the integration of the upward-projecting floral axis, the free tip of which is visible in the upper part of the hemisymplicate zone (IX A), in the centre between the three carpels (see moreover the researches of Sterling (1972, 1977a, b) on other members of the Liliaceae).

The incompletely fused ascidiate zone can be distinguished by means of the term **hemisynascidiate**.

As we already know, the central cavity of the gynoecium can take on different forms: the folded carpel "blades" may not project, as in the syncarpous gynoecium, as far as the centre of the ovary, i.e. the margins of each carpel do

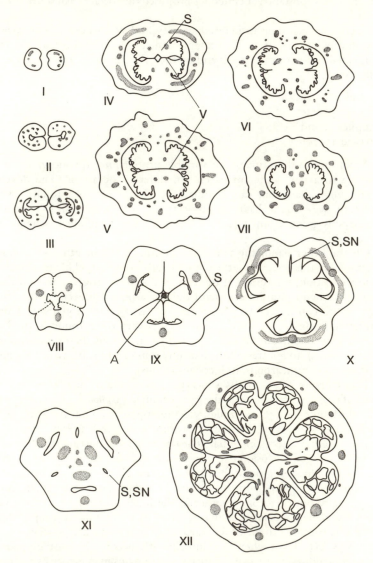

Fig. 81 Series of cross-sections through the completely syncarpous (eusyncarpous) gynoecium of *Boykinia tellimoides* (I-VII), with fertility in all four zones, and through the partially syncarpous (hemicoenocarpous) gynoecium of *Asphodelus albus* (VIII-XI), A apex of axis. XII Cross-section through the upper half of the ovary of *Philadelphus coronarius*, symplicate structure. S septal slit with SN, septal nectary, V ventral slit. Further explanations in text. All after Leinfellner.

not close together ventrally, but gape open, so that a more or less continuous inner cavity is formed. By this means the form known as paracarpous or syncarpous–paracarpous arises. There is only a difference of degree between the complete closure of the margins in the plicate portion and the structure which Troll termed coenocarpous–paracarpous.*

The uppermost point at which the margins of the individual carpel usually separate from one another is of course in the region of the stigma. However, a continuous transition can often be observed even in the fertile portion of the gynoecium extending from the completely closed carpel margins to the type of structure which has been termed paracarpous by Troll and others.

The forms with complete or incomplete closure of the carpel margins which correspond to a completely or incompletely locular ovary are not in any sense fundamentally distinct from one another, but represent variations on the form of structure of one and the same zone, the symplicate region.

Leinfellner distinguishes these forms by the terms **holoseptate** (completely locular) and **aseptate** or **hemiseptate**, respectively. It is much more usual nowadays to use the term paracarpous to apply to the symplicate region in general, whilst Troll initially used this term only for the description of an aseptate or hemiseptate structure.

Although it is possible for any of the four zones (the synascidiate, symplicate, hemisymplicate and the asymplicate) to be fertile, the formation of ovules is mostly confined to the synascidiate and symplicate zones. The completely syncarpous gynoecium of *Boykinia tellimoides* (Saxifragaceae) will serve as an example where all four zones are fertile. It consists of two carpels, which are united at their base with a receptacle, but this latter fact is not important for our purpose.

The series of cross-sections (Fig. 81 I to VII) from the tip to the base of the gynoecium shows us first of all the free carpel tips in the stigmatic region (I), then the two carpels in the asymplicate zone (II, III). This is already fertile in the lower part, i.e. the carpel margins which are folded inwards ventrally bear placentas with ovules on their inner surface (III). Below this begins the hemi-symplicate zone (IV), in which the contiguous lateral faces of the carpels become continuously fused to each other beginning from their outer parts. It is only in the region of the channel which is visible in the centre of the cross section of the ovary that there is no fusion. Further downwards this so-called septal slit also disappears, and only the suture which originates from the fusion of the ventral slits is still recognisable (V); hence, this is a holoseptate region. The symplicate zone is preceded by the synascidiate zone further below, shown by the sections VI and VII. Nevertheless the ventral slit in the centre of the section VI does reach somewhat further down. The synascidiate zone is still fertile.

In *Bergenia cordifolia* (Fig. 82 I to IV), on the other hand, we have an example of a gynoecium which is only fertile in the asymplicate and hemisymplicate zones, whilst the real symplicate and synascidiate zones remain sterile. The basal junction of the unusually large U-shaped placenta (I) remains free of ovules.

---

* The use of the terminology coined by Leinfellner should not be taken too literally, particularly in the context of the statement of systematic problems.

The gynoecium of *Digitalis lutea* (Fig. 82 V to VIII) will be used to demonstrate the continuous transition between the hemiseptate and the holoseptate zone of the symplicate zone. Not only the hemiseptate zone (VI) but also the holoseptate-symplicate zone (VII) bear ovules, as well as the synascidiate zone (VIII). We also have here a U-shaped placenta which is completely fertile in all its parts (V).

1.6.6   Central placentas

Let us suppose, on the other hand, that in a normal ovary the lateral faces of the carpels in the symplicate zone are suppressed, and that only the centrally fused

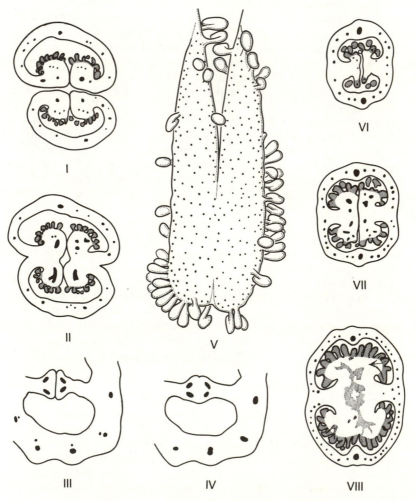

Fig. 82 Series of cross-sections through the gynoecium of *Bergenia cordifolia* (I-IV) and *Digitalis lutea* (V-VIII), V the substantial U-shaped placenta in side view. Further explanations in text. After Leinfellner.

cross zones are fertile, and that they form a massive common central placenta, then the result of this is a primarily aseptate gynoecium with a central placenta. We have already met an example of this in the Primulaceae (Fig. 83 I). A central placenta of this kind is totally free from the ovary walls from the moment of its inception, as a result of what might be called the "congenital loss" of its septa (Hartl 1956b). Free central placentas are also found in the Lentibulariaceae which are similar to those of the Primulaceae, and Hartl (1956a) was able to demonstrate this congenital disappearance of the septa through the structure of the ovary and especially the placentation characters via a series of clearly transitional forms in the closely related members of the Scrophulariaceae. The genus *Limosella* of the Scrophulariaceae (and probably some related genera), "bridges the last remaining gap between the Scrophulariaceae and the Lentibulariaceae". The central placenta of the Lentibulariaceae, just like that of the Primulaceae (Fig. 83 I), exhibits a sterile conical structure at its apex which projects into the stylar channel, and which evidently serves as an "obturator" for the guidance of the pollen tubes. It has been interpreted as the sterile end of the floral axis (e.g. Schaeppi 1937a), and also as a mere excrescence (Schlagorsky 1949). Pankow (1959) was even of the opinion that the entire central placenta of the Primulaceae must be understood as an organ of the central axis, on the basis of histological similarities which he observed in the development of the placenta and the initiation of axillary buds.

The creation of a uniform cavity in the ovary which results from the formation of the central placenta is, like paracarpy (see section 1.6.1), to be interpreted as "a sign of a special kind of internal synorganisation" (Hartl 1956b), and is found correspondingly in other derived families.

In the case of axile placentation the formation of a "free central placenta" (Pax and Hoffmann 1934) can be simulated by the later disappearance or the rupture of the septa which are present initially. This phenomenon, together with the initially basal position of single ovules in unilocular ovaries (basal placentation), as is well-known, has led to the naming of an entire order as the Centrospermae (Eichler 1878). It has likewise given rise to the theory that in this and in other similar cases (e.g. Piperaceae, Polygonaceae, Urticaceae) the "central placenta" is to be interpreted as an axial structure and that the ovules, and basal ovules in particular, are derived from axial material (Hagerup 1936). The distinction between stachyspory and phyllospory (Lam 1948 ff.) which we have already mentioned (section 1.1) is also based on this conception, and is contradicted by the results of comprehensive modern investigations, particularly those of Eckardt (1954, 1955, 1957b). It is a fact that within the Centrospermae, all intermediates are found between axile placentation with numerous ovules, through a reduction in the number of ovules and the later rupture or the early decomposition of the septa (Lister 1883) to simple gynoecia with just one basal ovule (Eckardt 1955, Rohweder 1965a, 1967a, 1970, Rohweder and König 1971, Rohweder and Huber 1974, Rohweder and Urmi-König 1975, Rohweder and Urmi 1978 etc.). The disappearance of the septa during the development of gynoecia which are still fully locular at the beginning of their development can be observed in detail in the Caryophyllaceae by investigations of the gynoeceal development in the Silenoideae, Alsinoideae and Polycarpeae. Cross-sections taken from a young gynoecium of *Agrostemma githago* (Fig. 83

III, IV), and again just before anthesis (V), may help to explain this. In the first two it can be seen that the ovary is divided into an upper holoseptate-paracarpous zone and a lower ascidiate zone. The numerous ovules are clearly arranged in two axile rows in each locule, which is still visible after the disappearance of

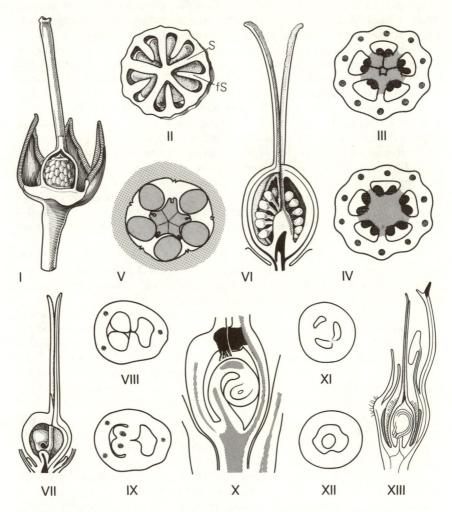

Fig. 83 I *Lysimachia vulgaris*, young fruit with the ovary wall partly removed in order to show the central placenta. II *Linum* species, ovary in cross-section, S normal septa, fS false septa. III-V *Agrostemma githago*, cross-section through a young ovary in symplicate (III) and synascidiate (IV) zones, and V through an ovary shortly before anthesis, with the outer wall of the ovary omitted. VI *Cucubalus baccifer*, median section of the gynoecium at anthesis. VII-IX *Drypis spinosa*, VII median section of the gynoecium at anthesis, VIII, IX cross-sections of young gynoecium in the zone of the insertion of the ovules (IX) and further up (VIII). X-XII *Uebelinia kiwuensis*, median section through gynoecium at anthesis (X) and transverse sections through a very young gynoecium (XI, XII), after photographs by Rohweder, schematised (E. Urmi), XIII *Paronychia przewalskii*, longitudinal section through the gynoecium. III-V after Rohweder, VI-IX after Rohweder and Urmi, XIII after Urmi-König.

the septa. The latter process begins shortly before anthesis – in *Vaccaria hispanica* (= *V. pyramidata*) during the development of the conducting tissue (see section 1.6.16), according to Rohweder (1967a, p. 144) – with the formation of large cells of low content in the mesophyll of the septa. "A little later on large gaps appear in the mesophyll at the relevant places, and finally the epidermis also ruptures, so that contact between the placental–column or the septum margins, respectively, and the ovary wall, is broken. It seems, at least partially, that the mesophyll is actively dissolved, whereas the epidermis evidently becomes torn mechanically by the resulting tension" (see also Hartl 1956b, p. 472). The abscission of the septa begins at the apex, below the upper wall of the ovary and continues down to the base of the carpels, while the space between the central column and the ovary wall is enlarged. The "free central placenta" so created can still remain fastened to the ovary wall only at the base and below the roof of the ovary, by means of narrow septum margins (the so-called "Leitfäden" = conducting threads, Fig. 83 VI), and the vestiges of where the partitions were attached can still be seen in cross-sections through the ovary (V). In some cases, e.g. *Telephium imperati*, the partition walls of the synascidiate zone survive, so that the lower half of the ovary still remains locular at flowering time (see Rohweder 1970). Hartl suggests that the "congenital loss" of the septa could be referred to as "congenital lysicarpy", as in the Primulaceae, Lentibulariaceae and in the genus *Limosella* (Scrophulariaceae), whereas what we are now dealing with is an instance of "postgenital lysicarpy".

In the case of *Drypis spinosa* (Fig. 83 VII to IX) we have an ovary which shows features of high reduction. The ovary consists of three carpels and possesses three style branches but only two septa and two ovules which must be attributed to the two adaxial carpels (VIII). Only the third and abaxial carpel (in diagrams VII to IX on the right) still exhibits a clearly ascidiate zone. Here too the septa vanish in the upper part of the ovary, into which the two ovules project, borne on the short placental column (VIII). The reduction in *Uebelinia kiwuensis* (Fig. 83 X to XII) is yet more pronounced. Here we find only a single axile ovule, which projects from the very short synascidiate basal portion of the ovary (XI), into the long symplicate portion, which only becomes partially locular at a later stage, so that the placentation appears basal (X). In this case too the septa, which are formed late and are only partial, become separated from the ovary wall by disintegration of the tissue and form the so-called "conducting strands" by the development of numerous long hairs (**conducting tissue**) for the pollen tubes (see section 1.6.16). Examples such as *Paronychia przewalskii* (Fig. 83 XIII) stand near the end of this reduction series. In this case the single ovule originates at the base of the ovary from the last vestiges of the synascidiate zone, whereas the symplicate zone is better developed and is aseptate, with the only exception of a slight, septum–like swelling, which is recognisable at the boundary between the median abaxial carpel and one of the pair of carpels which lie diagonally behind it (Rohweder and Urmi-König 1975).

In many other of those members of the Paronychioideae and the Chenopodiaceae for which basal ovules have been described, at least some trace of the septa of the synascidiate zone can be recognised at the base of the ovary. The endpoint of this reduction series is represented by cases like *Spinacia turkestanica*, in which there is no longer any sign of the synascidiate zone of the ovary in the form of septa (Urmi–König, personal communication), but this does not of course imply that the carpel bases are also lacking. To the contrary: the numerous examples detailed above in order to give evidence of the reduction series clearly show that even in extreme cases, an axial origin for the ovules is out of the question (for the anatomy of the vascular system, see Bocquet 1959). However, according to Rohweder (1967a), it is not impossible that there could be an element of axial tissue taking part in the structure of the placental column of the Silenoideae. It would not be unusual for the axis to take part in the formation of the ovary in this way. We have already come across this in *Asphodelus albus* (section 1.6.5). Klopfer (1971) was able to demonstrate that also in *Philadelphus*, the laterally fused carpels are united to one another in the peltate zone in the centre by a cone of axial material (see also Morf 1950).

Klopfer (1972a) records for *Francoa sonchifolia*, that the remains of the apex of the axis after the differentiation of the carpels expand between the enfolded carpel margins at the base of the gynoecium into a recognisable, few-celled cone of axial material. Troll (1931) already made a similar observation for *Parnassia palustris*, which was, however, not quoted by Klopfer (1972a).

1.6.7   False septa

It is also possible for the otherwise uniform interior of an aseptate gynoecium to be subdivided by partitions, the so-called **false septa**, which do not correspond to the lateral walls of the carpel. The best known example of this is the gynoecium of the Cruciferae (Fig. 99 VII). Contrary to previous interpretations, which explained the partition wall as "an outgrowth of the placenta", it is due to contact of the relevant placentas at a very early stage of ontogeny (Hartl, personal communication). The septum is not presumed to come about through "subsequent proliferation" of the placentas, but would be a genuine septum, which could be expected to show a ventral suture in its middle line. The fact that the ovules remain in a parietal position could be interpreted as a result of an intercalary elongation of the sides of the carpel between the outermost carpel margin and the submarginal placentas. In the Poppy (*Papaver*) and other Papaveraceae the ovary is unilocular, although often formed from many carpels, and is subdivided by placentas which project some way out from the walls.

The appearance of false septa in ovaries which are already completely or partially septate is also not a rarity. Hence for example the five locules of the ovary in *Linum* (Fig. 83 II) are subdivided by false septa, which project from the ovary wall in the region of the median nerve of each carpel to nearly as far as the halfway mark in each of the locules. The kernel of the drupe which is the fruit of the Walnut (*Juglans regia*) is also subdivided by true and false septa, which in this case even achieve a woody character. The lower half of the ovary, which consists of two carpels in median orientation, corresponds to the synascidiate zone and thus exhibits a complete, transverse and genuine septum (Fig. 84 III), whilst the upper half of the ovary has a partially septate structure. Here the carpel margins only protrude from the sides into the central cavity. The synascidiate basal part is further divided by false septa (IV), which run from the medians (!) of both carpels at right angles to the true septa, to which they are firmly attached. The lines of their insertion correspond to the "crack in the shell", which has swollen margins and is recognisable from the outside of the kernel, which is where the nut easily splits open. This does not, however, in any way correspond to the "suture lines" of the two carpels (see the floral diagram in Fig. 84 I), but runs at right angles to the plane in which they are united.

1.6.8　　Ovaries with lateral or basal styles (anacrostyly, basistyly)

There is another way in which false septa are formed, which is by no means rare. This underlies the phenomena of **anacrostyly** or even **basistyly** (basi-

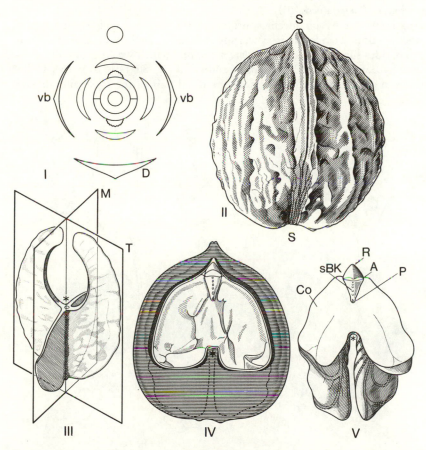

Fig. 84 Juglans regia. I Floral diagram of female flower, vb prophylls, D subtending leaf, II kernel after the exocarp has decayed, where the vascular bundle which runs in the suture S-S corresponds to the carpel median, III orientation of the system of septa in the fruit, M median, T transverse plane, ⋆ position of the hilum of the single seed, IV kernel split by its median plane (i.e. in the plane of the sutures in II), ⋆ position of the single seed, V seed (or the embryo enclosed in the scarious testa) in side view, after removal of the cotyledon which faces the observer (cotyledon insertion at A), Ru radicle, Co cotyledon, P plumule of the embryo, sBK series of accessory buds. II original, the rest after Troll.

gyny, or gynobasy), which will now be explained. By **anacrostyly** is under-stood that the style and stigma do not represent the apical continuation of the ovary (acrostyly), as in Fig. 75 I and VI or Fig. 83 I and VI, but that their insertion appears to be near the base of the ovary, as is shown in Fig. 85 I for *Alchemilla alpina*, whose superior ovary consists of only one carpel. We have already come across the same phenomenon in *Hirtella butayei* in the Chrysobala-naceae (Fig. 20 VI). This position is explained by arguing that the growth of the ventral side of the carpel lags behind that of the dorsal side, so that the style appears to grow out from the base of the carpel. This difference of proportions may be explained by the schematic diagrams IV to VI of Fig. 85, of which IV corresponds to the outline of the carpel of *Rubus* (II), and VI to that of the Wild Strawberry, *Fragaria vesca* (III).

Anacrostyly is by no means confined to the single carpels of apocarpous gynoecia, but is often found in syncarpous ovaries as well. As an example, let us now consider the gynoecium of *Ochna multiflora*, which was thoroughly inves-tigated by Baum (1951b). Diagrams III and IV of Fig. 86 show us an ovary after anthesis in general view (III), without the distal part of the style, and an axial

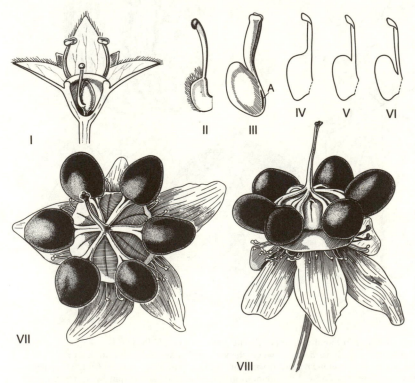

Fig. 85 I *Alchemilla alpina*, longitudinal section through the flower. II, III Individual carpels of a species of *Rubus* and of *Fragaria vesca* (III). IV-VI Schematic diagrams to explain the derivation of basistyly. VII, VIII *Ochna multiflora*, young fruit from above (VII) and in side view (VIII). I after Engler, II-VI after Troll.

section of a young ovary (IV). It is immediately obvious that the style which is common to all the carpels is inserted deeply into the centre of the ovary between the swollen carpels. This is therefore a case of "basistylous" carpels (or of a "gynobasic" style). The carpels appear to be separated by deep longitudinal grooves, so that they seem at first sight to be completely free from one another, i.e. apocarpous. The series of cross-sections in Fig. 86 V to VII teaches us, however, that further down, and indeed throughout the plicate and ascidiate zones, they are fused laterally (VI, VII), and in a section taken even further down at the level of the carpel bases, it is evident that they are even more completely united. The "free" zone in the upper region of the ovary in contrast represents nothing other than the inflated dorsal faces of the carpels (cocci). It can be seen in a cross-section (V) that the carpels appear to be completely closed, and that the dorsal median which passes through the rear wall is encountered twice, once on the outer side of the coccus and again on the inner side.

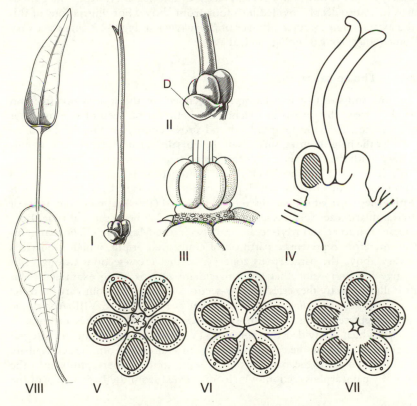

Fig. 86 I, II *Ajuga reptans*, gynoecium at anthesis, II with the ovary at higher magnification, D the disc structure which functions as a nectary. III-VII *Ochna multiflora*, III general view of an ovary after flowering (uppermost part of the style omitted) and IV in schematic longitudinal section; V-VII cross-sections through a young ovary at the level of the inflated dorsal faces of the carpels (V), the hemisymplicate zone (VI) and at the transition to the hemisynascidiate zone (VII). VIII *Codiaeum variegatum* var. *pictum* f. *appendiculatum*, leaf with intralaminar peltation. III-VII after Baum, VIII after Troll.

The impression that this is an apocarpous ovary is even stronger when the fruit ripens, because as a result of the expansion of the receptacle and multi-lateral growth of the fused, sterile carpel bases, the cocci which enclose the ovules become yet further separated from one another, and the style base is expanded into a stellate pattern (Fig. 85 VII, VIII). The region of insertion of the cocci is now so narrow that these now contrast even more clearly with the base of the ovary, which is now emphasised by strongly contrasting colours: the cocci are black when ripe, whereas the rest of the ovary is red – presumably an adaptation of the "ecologically apocarpous" fruit to dispersal by birds.

We come across a similar structure in the ovary of the Boraginaceae and the Labiatae, where there are two carpels. Here, however, there is the difference that each of the two carpels, which each enclose two ovules, is folded across the median in such a manner that each ovule is closely surrounded by the carpel wall and forms a so-called "mericarpic nutlet" on its own, which expands separately from the others and detaches itself as a nutlet (mericarp). The gynoecium is therefore deeply divided into four from above. For illustrations of this see the axial section through a flower of *Symphytum* in Fig. 42 X and the ovary of Bugle (*Ajuga reptans*) in Fig. 86 I, II.

1.6.9   The apical septum

Let us now suppose that in a basistylous gynoecium the style is not free and inserted between the carpels in a channel, whether it be deep or shallow, as is shown in the schematic longitudinal and cross-sections in Fig. 87 I and Ia. Instead, let the neighbouring surfaces of the dorsally inflated carpels be congenitally fused with the style and each other, as in diagrams II and IIa. Diagrams III and IIIa illustrate a further development of the process. Then the result is a septum of uniform structure which projects into the ovary from above. This is the **apical septum** which was discovered by Hartl (1962) in many members of the Scrophulariaceae, Solanaceae, Convolvulaceae, Verbenaceae, Labiatae, Boraginaceae, Ericaceae and Myrtaceae, and also in the Musaceae (*Heliconia*). This septum therefore represents a partition of congenital origin, which subdivides the ovary above the paracarpous zone (!), so that cross-sections through this zone give the impression that the synascidiate zone is being examined. This point is illustrated by the series of cross-sections through the four carpels in the ovary of *Orthostemon sellowianus* (Myrtaceae) in Fig. 87 V to VIII. Above the fertile holoseptate- and hemiseptate-symplicate zones (VI, VII) lies a section of the ovary which is divided by four septa (V), which apart from the integral stylar channel visible in the centre is scarcely any different from the synascidiate zone (VIII). Towards the base the apical septa grade imperceptibly into the septa of the paracarpous section, which are formed from the lateral faces of the carpels.

There can be no doubt that the apical septum represents a tissue structure of completely uniform and congenital nature. This is borne out by examination of the cross-section of the apical septum of the young ovary of *Kickxia elatine* (Scrophulariaceae) reproduced in Fig. 87 IV, where there is absolutely no trace of postgenital fusion that can be seen. Here also a cavity has been left in the centre as a continuation of the stylar channel.

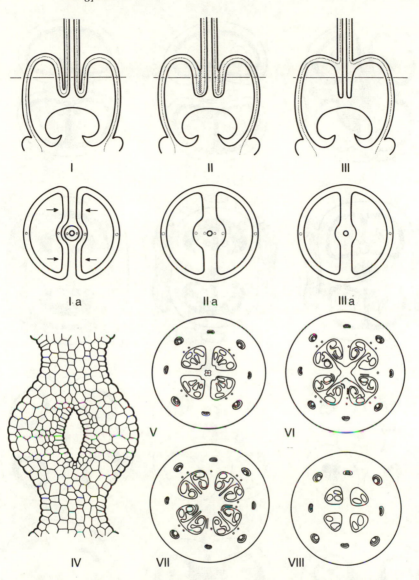

Fig. 87 I–III Scheme to explain the origin of an apical septum; axial sections (I–III) and corresponding cross-sections (Ia–IIIa) through the apical region (horizontal line) of various forms of bicarpellate syncarpous ovaries: I, Ia anacrostylous (basistylous) ovary, II, IIa basistylous ovary with congenital fusion of the style and carpel faces (arrows in Ia), III, IIIa ovary with apical septum. IV *Kickxia elatine*, cross-section through the apical septum of an ovary before flowering in the region of the pollen tube canal. V–VIII *Orthostemon sellowianus*, series of cross-sections through a gynoecium shortly before anthesis. V Zone of the apical septum, VI paracarpous zone, VII fertile and VIII sterile region of the syncarpous zone. All due to Hartl (V–VIII adapted).

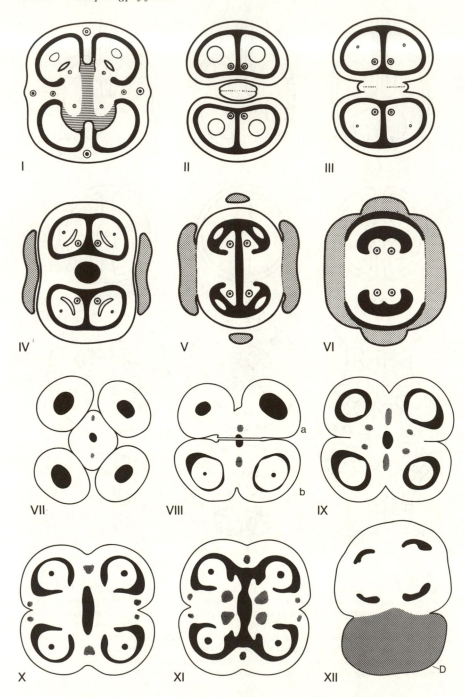

Fig. 88 I *Mina lobata*. Cross-section through the paracarpous zone of the ovary; in the diagram two different sections taken at different levels are shown combined: the upper half at the level of

The way in which the style is inserted in a channel or hollow between the carpels which is characteristic of basistylous ovaries is no longer to be expected in every case where an apical septum is formed; see the schematic diagrams of Fig. 87 III, IIIa. It is more usual in such cases for the style to be inserted in a shallow groove or depression between the carpels or at least to be sharply demarcated from the ovary. It is even possible in the formation of an apical septum for the ovary and style to grade continuously from one into the other, in spite of the fundamentally basistylous structure, e.g. in the Convolvulaceae. The opposite situation, i.e. the style placed in a deep channel between the highly inflated dorsal faces of the carpel is found in the Convolvulaceae-Dichondroi-deae. Here, however, an apical septum is also formed. Let us consider a series of cross-sections through the deeply bifid ovary of *Dichondra repens*. In Fig. 88 II in a section taken just before anthesis we see the region of the free, inflated, dorsal faces of the carpels, which occupy almost half of the overall height of the ovary. It can be seen that the dorsal parts of the two carpels and the style which lies between them are all free from one another. The two style branches which are free at the apex are postgenitally fused at this level. Further downwards (III) they are united in a syncarpous style and their dorsal faces are attached to the adjoining raised dorsal parts of the carpels, but otherwise the latter still remain separate from each other, and the lateral parts of the styles remain free. In a zone further down again (IV), the surfaces of the carpels which face each other and the style which they enclose are fused into a continuous septum, which has a stylar canal visible in the centre. Below this in the next section (V), the carpel margins become visible in the fertile hemiseptate-symplicate zone. The synascidiate zone which is adjacent below remains sterile (VI); apart from the absence of the stylar canal it exhibits the same arrangement of septa as does the previous ovary section which is divided by the apical septum (IV). Weber (1971) was able to demonstrate the presence of an apical septum in several members of the Gesneriaceae.

1.6.10   Formation of cells in the ovary

In certain species of the Convolvulaceae, as in the genera *Calonyction* and *Mina*, the structure of the ovary is complicated yet further by the combination of an apical septum with a false septum in the median of the carpel. In *Mina lobata* the complete upper part of the ovary, which is already partitioned by the apical septum, is divided into four chambers by this "arch of tissue of congenital origin between the rear wall of the ovary and the apical septum" (Hartl 1962, p. 282), so that these are practically indistinguishable from the locules of a tetramerous ovary. They also project some way further into the locules below the apical septum. This is shown in Fig. 88 I, in which a cross section through the ovary at the level of insertion of the ovules (in the upper half) is combined with another section taken somewhat lower down, into which the ovules project

Caption for Fig. 88 (*cont.*) insertion of the ovules, the lower somewhat further down; shaded: pollen tube conducting tissue which occupies the ventral slit and (at the lower level) which occupies the margins of the false septum. II–VI *Dichondra repens*, series of cross sections through the ovary, VII–XII similarly for *Ajuga reptans*, D ventral disc formation. I–VI after Hartl.

from above (lower half). The (ventral) slit, which would otherwise appear in this paracarpous zone between the divergent carpel margins, is closed by meristematic tissue for conducting the pollen tubes, which extends even above the slit and intimately surrounds the edge of the false septum (it also borders the micropyles of the ovules!). Since in the synascidiate zone of the ovary the edge of the false septum is congenitally fused with the cross zone of both carpels, the complete ovary is divided from top to bottom into four "cells", each of which includes an ovule and corresponds to the mericarpic nutlets of the Boraginaceae and the Labiatae already mentioned. In *Mina lobata* as well as in *Calonyction*, neither the false septum nor the apical septum (via the anacrostylous insertion of the style) can be detected externally, although in *Mina lobata* the style is more sharply demarcated from the ovary, as is otherwise usual in the Convolvulaceae-Convolvuloideae. This is therefore an important difference from the four-celled ovaries of the majority of the Boraginaceae and the Labiatae already referred to (section 1.6.8), whose differentiation is immediately and obviously visible from the outside, via the anacrostylous to basistylous insertion of the style and the median folding of the carpels causing the formation of a false septum. On the other hand, this means that the type of apical septum which is formed by the congenital fusion of the style and the dorsal faces of the carpels as in Fig. 87 I to III only occurs, if ever, as a very inconspicuous projection. This happens in the ovary of Bugle already illustrated in Fig. 86 I, II, of which we have prepared a sequence of cross-sections (Fig. 88 VII to XII). In diagram VII the only partially developed dorsal faces of the halves of the carpels were cut through at anthesis, and the cavities of the cells appear to be rather small. This is because the section was taken at a superficial level, and so the outer walls appear rather thick as a result. The four carpel halves and the anacrostylous style are completely separate here. In diagram VIII two sections taken at different levels are combined, and demonstrate the fusion of the style with the walls of the swollen halves of the carpels and the progressive combination of the respective halves of the carpels as well as the two carpels one with each other. Even in the lower of the two sections (lower half b), the existence of four parts is still visible in the general outline, just as in diagram IX: there is the median fold of the false septum and the deep transverse cleft which marks the boundary between the two carpels. The stylar canal becomes wider here to form the ventral slit which runs down the median, and more so further down between the more divergent carpel margins in the insertion zone of the ovules (X). The carpel margins remain congenitally (!) fused, however, with the "emarginate false septum" (Hartl), so that the ovary still has four locules. It is not until a lower zone that the paracarpous structure of the fertile ovary region becomes clearly visible (X), through the complete separation of the carpel margins and the retreat of the false septa. In the short and sterile synascidiate zone (XII), on the other hand, the two halves of the ovary which are separated by the cross zone are again divided in the median. Whether this should be regarded as a reappearance of the false septum or as a relatively disorganised outgrowth of median sections of the ovary wall which brings about a subdivision of the locules (which are in any case very narrow at this point) remains to be seen. Diagram XII also shows at the same time a horizontal section of the nectary (D) on the ventral side of the gynoecium (see also Junell 1934).

From a comparison of the ovary structures discussed above it becomes clear that the false septa with median orientation are to be explained correspondingly as the fusion of median folds of the carpel, just as the apical septum is the congenital fusion of the dorsal faces of the carpel which are turned towards one another whenever the basic structure is anacrostylous. It does seem surprising, however, that the possibility of postgenital fusion of the false septa with the carpel margins in the paracarpous zone, which by definition are free, is already anticipated by congenital processes. According to Hartl (1962) and Vogel (1959), this abbreviation of the expected processes of structural growth is to be understood by the fact that the "more additive type of development in the vegetative region" which is characteristic of plant growth reaches its end with the formation of the flower. In the flower the generally open form of a plant turns to the principles of development of the closed form of the animal body, by the nearly or entirely simultaneous formation of the primordia and a variety of congenitally developing processes of transformation.

1.6.11   Congenital and postgenital fusion of carpels

It has already become clear, in the context of the derivation of the syncarpous form of the gynoecium from the apocarpous (section 1.6.1), and the discussion of the vertical zonation of the syncarpous gynoecium (section 1.6.5), that both types of architecture are linked by intermediates. Both the apocarpous and syncarpous types as well as a very varied degree of fusion between the carpels can be observed side by side within different families. At the same time, the correspondence between the gynoecia which is shown by the homologous parts of the carpels appears again and again and with remarkable constancy, independent of the degree of congenital or postgenital fusion. This is all the more significant, because the congenital fusion of the organs takes place at the same time as their differentiation, and is not therefore a "process" which can be demonstrated as it occurs, but rather can only be deduced by a painstaking comparison of the relevant parts of the organs one with another, and in this particular case by comparing the individual zones of the carpels in various gynoecia.

Baum (1948f) has assembled an impressive list of examples of the various degrees of congenital and postgenital fusion of carpels within a closely related group, the subfamily Rosaceae-Spiraeoideae: in *Exochorda* the five fertile carpel sections are united by a central, conical receptacle (similarly in *Sibiraea*); the carpels of *Vauquelinia corymbosa* are originally united by their bases to the axis to only a very small extent but converge above the apex of the axis later on during ontogenesis and fuse completely and postgenitally, with the exception of a small recess above the apex, which still remains free and visible; in species of *Sorbaria* "the carpels are united peripherally at the base of the ovary from the very first, but in such a way that the appearance of the individual, slightly peltate carpels is not significantly altered, neither microscopically nor macroscopically!". In *Physocarpus opulifolius* the "base of the ovary, which is primarily syncarpous", reaches as far as half the height of the gynoecium. In the two- to four-merous gynoecium of *Neillia rubiflora* it is only the carpel stalks which are united. Similar series of examples can be found in the Saxifragaceae, even within the genus *Saxifraga* itself.

There can be many different ways in which postgenital fusion comes about.

At the beginning it is not uncommon for it to resemble the dentonection of epidermis cells, or the formation of interlocking papillae, as we have already encountered in the concrescence of sepals (section 1.4.2.4) and in the formation of falsely sympetalous corollas (section 1.4.3.6). According to Hartl (1956a), the postgenital fusion between symplicate carpel margins as, for example, in the Convolvulaceae, is established by means of a more or less homogeneous fusion tissue which originates via divisions in the subepidermal and epidermal layers. This begins shortly after the differentiation of the ovules in the upper part of the symplicate zone when the epidermal cells of the opposing carpel margins mesh together and then divide periclinally. *Verbascum thapsiforme* may serve as another example, in which in fact there is relatively little tissue involved (Fig. 89 I). Here the subepidermal cell layers also show periclinal divisions, "so that the epidermises are somewhat raised from their initial level and pressed against each other". Whilst this process continues downwards, a fusion tissue is formed at the tip by further divisions. Baum (1948b) describes similar processes for the fusion of carpel

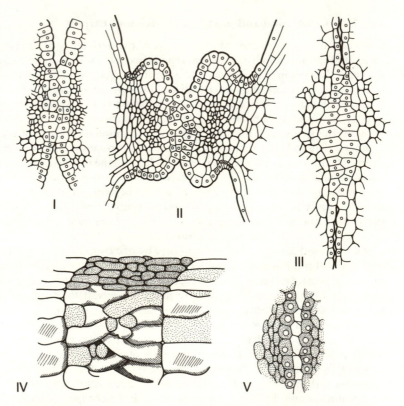

Fig. 89 Postgenital fusion of carpel margins. I *Verbascum thapsiforme*, symplicate zone, beginning of fusion; II, III *Rhinanthus alectorolophus*, beginning of the postgenital fusion in the sterile (II) and fertile (III) region of the symplicate zone; IV *Sutera coerulea*, block diagram from the region of postgenital fusion, formation of a "plectenchyma" from the development of tubular epidermal cells; V *Clandestina purpurea*, threads secreted between the epidermal cells in the symplicate zone. All due to Hartl.

margins in apocarpous gynoecia in many families, and they also occur in the postgenital fusion of the lateral faces of carpels in hemisymplicate gynoecia (Hartl 1956a, Baum 1948c, Leinfellner 1951b). The union of regions which have fused postgenitally can also be achieved by other means, and not just by dentonection of the epidermal cells, e.g. only by division of the epidermis in the zone of contact, in the course of which an alteration in the structure of the cell walls obviously takes place (repolymerisation of wall constituents etc., Walker 1975a, b, c).

In addition to fusion by **dentonection** of the epidermis cells, and by **division (and dedifferentiation) of the epidermis**, Hartl (1956a) describes the juncture of carpel margins by the **formation of papillae** within the Scrophulariaceae, e.g. in *Rhinanthus* (Fig. 89 II, III) and in *Sutera coerulea*, in which the tubular epidermal cells can intertwine as in plectenchyma (Fig. 89 IV).

In *Linaria vulgaris*, *Clandestina purpurea* and in other species the formation of papillae is suppressed, whereas the epidermis of the faces in contact has the character of a glandular epithelium which produces a viscous secretion, and it is this which causes the **adhesion** of the carpel margins (Fig. 89 V).

In the ripening fruiting stage, postgenital fusions sometimes break down again.

In the Asclepiadaceae, the union of the two carpels, which are free except at the level of the stigmatic head, is also brought about by postgenital fusion (Baum 1948a, Boke 1949, Walker 1975a). When the fruits ripen, the apical union of the carpels is dissolved again, because the whole stigmatic head is shed by means of an abscission layer (Baum 1948a).

According to Baum (1949b), it is the omission of postgenital fusion between the margins of the individual carpel in the apocarpous region of a gynoecium which explains the occurrence of the "open angiosperm gynoecia", of the kind which are found in the fruit of *Reseda* (see the series of cross-sections in Schaeppi 1937c), or still at flowering time in the dwarf palm *Chamaerops humilis* (Juhnke and Winkler 1938). In *Reseda* the separation of the carpel margins at fruiting time is that much more obvious, so that the seeds become visible inside the almost completely aseptate, syncarpous-paracarpous ovary, similarly as in *Houttuynia cordata* (Saururaceae) and *Datisca cannabina* (Datiscaceae).

1.6.12   The inferior ovary

Hartl (1962) has rightly stressed that the congenital fusion of the carpels of a syncarpous ovary is not to be explained by means of the "principle of variable proportions". On the contrary, the essential feature of the congenital fusion consists precisely in the fact that the relative proportions of the united organs in comparison with those which remain independent are not changed in any way (or only by a small amount). Hartl clarified his remark by saying that the congenital fusion affects zones which could also be combined by postgenital processes. In such cases the evidence for a similar underlying plan is the fact that all the parts of the congenitally fused organs can be unequivocally compared to others (in a related plant) which have remained free.

These remarks also apply to the inferior ovary! According to the view often advocated today, which we have already subscribed to in section 1.3.1, the inferior gynoecium is brought about in the following way. The floral axis is united congenitally with and around the dorsal faces of the carpels and is raised above the insertion zone of the carpel primordia so that it encloses the entire ovary in a cup-shaped receptacle. The latter bears the calyx, corolla and stamens on its upper margin (Fig. 10 III) and leaves only the styles or the tips of the carpels free.

It is also a fact that the congenital fusion between the floral axis and the dorsal faces of the carpels is not a "process" which can be demonstrated. It is also necessary to remind ourselves that in groups of related species in which there appears a sequence of forms from a superior ovary via various different kinds of semi-inferior ovary to an inferior ovary – e.g. the Liliineae – the characteristic carpel structure and zonation remains unchanged regardless of all the different degrees to which the ovary is set into the receptacle. This has been shown by series of cross-sections (see for example Leinfellner 1941, 1950). On the other hand, the question as to whether and to what extent the boundaries between axial and carpellary tissue can be detected in inferior ovaries by means of histological criteria, has not yet been clearly answered. Of course certain differ-ences can occasionally be recognised between the peripheral and the inner tissue zones. In the Onagraceae, for example, the cells in the outer parts of the tissue are more elongated and do not stain so readily (see Pankow 1966, Bunniger and Weberling 1968), and similarly in the Myrtaceae (Hartl 1962) and several mem-bers of the Saxifragaceae (see Klopfer 1968b, Fig. 9, b. *Tellima*; 1972a, Fig. 6, b. *Francoa*, transition to an inferior ovary). Klopfer does not interpret these illustrations in the same sense as above, but writes on the other hand of "a differentiation of outer cortex layers" in the inferior part of the ovary of *Tellima*, or of *Astilbe*, so that "the partially free carpels are coated with axial tissue". In contrast to the earlier examples, the wall of the carpel stands out "clearly from the receptacle by the fact that it does not stain so well" (1970b).

These differences can, however, have other causes, and cannot therefore be interpreted as clear evidence of the axial origin of the peripheral tissue layers.

In contrast to the interpretation given above, in which the inferior ovary is explained by the congenital fusion of the gynoecium and the floral axis, Leins (1972a) made an attempt to explain this phenomenon by means of the "princi-ple of variable proportions". He accepted a common intercalary growth of the floral axis and the carpel bases in the region of insertion of the carpels, according to Fig. 90. "In an inferior ovary, the dorsal side of the individual carpel is shortened", whilst correspondingly "the insertion (carpel base) is lengthened". That part of the wall which encloses the style and stigma in the inferior ovary, and only that part, should correspond to the total wall of a superior ovary. Intercalary growth processes of *expansion* of the zone of the carpel insertion and the displacement of the placentas from a primarily axial position to a parietal position in a semi-inferior or superior gynoecium are of course known from the Aizoaceae and the Cactaceae (Fig. 16 VI, 91, see Buxbaum 1937, 1948 etc.). It does seem quite possible that these ideas of Leins might prove correct in many cases. However, the fundamental considerations already discussed prevent the generalisation of these ideas. The presumed intercalary growth of the floral axis and the carpel bases in the insertion zone would have to lead to modifications in the structure of inferior ovaries as opposed to superior ones, which we do not observe, and which are not in agreement with the observed constancy of the form of the individual carpel in inferior and superior gynoecia. One only has to imagine how an inferior bicarpellate-paracarpous gynoecium would have to behave under these assumptions! The results of ontogenetic investigations show, moreover, that the carpel primordia in semi-inferior and inferior gynoecia usually originate on the inner surface of an apical depression of variable depth and that their dorsal faces are fused with the inner wall of this depression from the very beginning, which corresponds to the hypothesis of congenital fusion with the floral axis. Hence this ready-made "miniature" receptacle, so to speak, only needs to increase in size accordingly (see for example Leinfellner 1941, Bunniger 1972, Klopfer 1968, 1969, 1970 etc.). A good example of how an inferior ovary develops is provided by the flower of the Umbelliferae (Fig. 92 I), whose ontogeny has already been investigated by Leinfellner (1941), and more recently by Magin (1977). This also exhibits the type of carpel differentia-

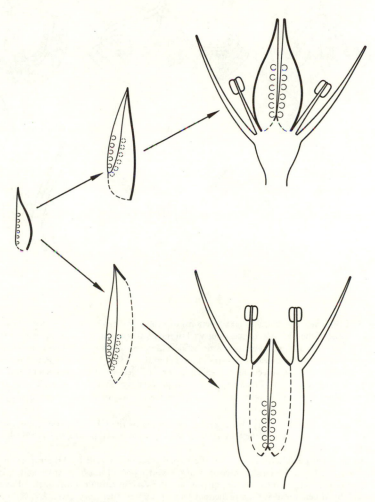

Fig. 90 Derivation of the inferior ovary by intercalary growth of the floral axis and the bases of the carpels in the zone of the carpel insertion (broken line). Dorsal faces of the carpels in bold. After Leins, highly modified.

tion just described, and the intercalary growth of the receptacle which is fused with the dorsal faces of both the carpels (Fig. 92 II to V), but not, however, intercalary growth in the region of insertion of the carpels.

In contrast to the interpretation which we have been following, which goes back to Čelakovský (1874) and Goebel (1884), that the inferior gynoecium is a product of congenital fusion between the carpels and the receptacle, another opinion has been advocated since de Candolle (1813) and van Tieghem (1868, 1875) – the "appendix theory"*. This supposes that the enclosure of the inferior

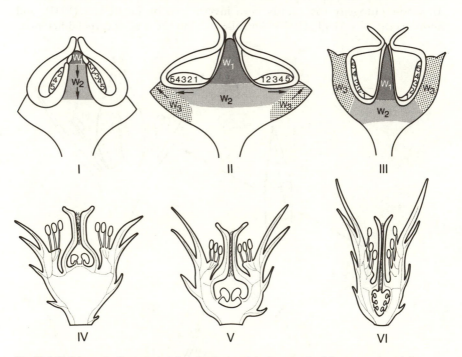

Fig. 91 I-III Schematic diagrams (axial sections) to explain the growth of the receptacle and the displacement of the placentas to the outer wall of the gynoecium in the Aizoaceae; W_1-W_3 growth zones of the receptacle; numbers 1–5: order of development of the ovules. I corresponds to the properties of the Aptenioideae, the forms illustrated in II and III appear in the Ruschioideae. IV-VI Schematic diagrams (axial sections) to explain the gradual descent of the gynoecium into the receptacle in the Cactaceae, IV corresponds to the properties of *Pereskia aculeata*, V to *Pereskia* subgenus *Rhodocactus*, VI to more highly developed Cactaceae. All after Buxbaum, partly redrawn.

* Sarkany and Kovacs (1971) also interpret the results of their histogenetic experiments on the development of the ovary of the Umbelliferae in the sense of the appendix theory (for discussion see also Leins and Erbar, 1985).

We do not need to discuss in detail the hypothesis put forward by Hagerup (1936 etc.) and Thomson (1934), in which the inferior ovary is said to be an axial structure and only the free portions of the styles and stigmas are carpels, nor the various combinations of the above ideas put forward by other authors. Comprehensive descriptions of them are given by Wilson and Just (1939), Douglas (1944, 1957) and Puri (1951, 1952b). The advocates of these theories often support their arguments with data from the anatomy of the vascular system (see the discussion at the end of section 1.1.).

gynoecium, and particularly the development of a receptacle which projects above it – a hypanthium – is brought about by the (congenital) fusion of the bases of the calyx, corolla and stamens, and that the wall of the inferior ovary is a fusion product of all the floral organs. Fundamentally, this possibility cannot be excluded. We have already mentioned (section 1.3.2), that the tubular portion of the flower, the so-called "perianth tube" which is typical of the Lythraceae was interpreted as a fusion product of the sepals and petals (Mayr 1969). Also the encasing of the receptacle with the bases of sepals can be taken as an accepted fact in several cases, as in *Rosa* (see section 1.3.2) and in *Eryngium* (Leinfellner 1941).

1.6.13   The pseudo-syncarpous ovary

We have already briefly discussed the fact that the floral axis can take part in the formation of the superior ovary with the examples of *Nelumbo* (Fig. 14 III) and *Nuphar* (section 1.3.1). According to Troll (1934c) the floral axis of *Nelumbo* modifies the formation of the apocarpous gynoecium in such a way that "it rises around the carpels, surrounds them and encloses them in the well-known obconical body which projects from the centre of the flower". Similarly, Troll states that the cyclically inserted free carpels of *Nuphar* are fused by the axis, and in such a manner that it "does not merely surround the carpels, but that the growth of the carpels and the axis takes place jointly"; even the "tissue of the

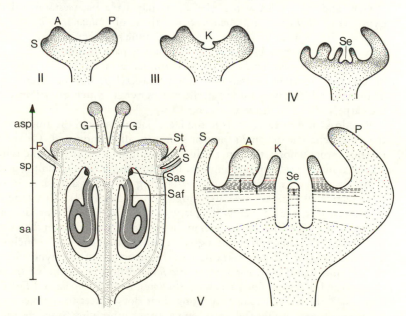

Fig. 92 I Schematic axial section through a flower of the Umbelliferae. II–V Ontogeny of a flower of the Umbelliferae, illustrated by means of partially schematised median sections. S sepals, A stamens, K carpels or their primordia; Sas sterile, Saf fertile ovules, Se septum of the ovary, G style branch, St stylopodium, asp asymplicate, sp symplicate and sa synascidiate zone of the gynoecium. All due to Magin.

stigmatic disc between the stigmatic rays" belongs to "the axis, in which the carpels are indeed completely embedded". He also states that the carpels in the inferior or semi-inferior gynoecia of *Nymphaea*, *Victoria* and *Euryale* are fused laterally and in the centre by axial tissue. The fundamentally apocarpous nature of the gynoecium is still recognisable here by the fact that, amongst other things, in *Nymphaea stellata* and related species, the lateral faces of the carpels remain separated from each other by "inter-carpellary slits" (Fig. 93 IV). There is also an indication of this character in *Victoria amazonica* and *Euryale ferox*. In all such cases therefore a "pseudo-syncarpy" is present, and moreover in the superior ovary of *Nuphar* this is abandoned in the ripe fruit, when the external coat of axial material as well as the tissue beside and between the carpels is shed completely (Fig. 93 V).

The statements made by Troll about the pseudo-syncarpous structure of the gynoecia of the Nymphoideae have not remained unchallenged, in particular where it concerns the lateral and central fusion of the carpels by axial tissue (Winkler 1941, Moseley 1972a, b). Further investigations may need to be made before a definitive conclusion can be reached.

According to Troll (1934b), a central and lateral fusion of the primordia of free carpels by axial tissue also determines the occurrence of the apparently syncarpous zones at the base of the gynoecia of *Helleborus* and *Aquilegia*, and the fusion of the carpels which reaches as far as the stigma in some species of *Nigella*. In the meantime, and particularly because of the researches of Rohweder (1967b) and also of Kaussman (1972), it may have been established that the gynoecia of *Helleborus* and *Aquilegia*, and those of *Caltha* and *Trollius* as well, possess a short syncarpous zone, and that genuine syncarpy is the rule in *Nigella*. Nevertheless, the possibility that the axis takes part in the formation of these gynoecia has not been excluded. Kaussmann has even established that in *Nigella* "a residual meristem of star-shaped appearance" persists between the carpel primordia, which gives rise to "sterile parenchyma, which grows upward with the carpels" (see also, however, Lang 1977).

Pseudo-syncarpy also occurs in the Helobiae, namely in the Butomaceae and the Hydrocharitaceae. In *Butomus* it is not at all pronounced. Of the two 3-merous carpel whorls, the inner originates somewhat later than the outer. The carpels remain free from one another, but they are, however, united at the base at the periphery by a ring of material arising from the axis, as is seen in the cross-section in Fig. 93 I (see also Singh and Sattler 1974). This zone remains short, however, and lies below the plane of the cross-section reproduced in Fig. 78 XV. Nothing of this can be seen in the bisected gynoecium in the view given by Fig. 78 XIV, although it could give a confirmation of Troll's conclusion that a central cone of the axis takes part in the structure of the gynoecium. Troll (1932c) also reports this for *Limnocharis flava*. The gynoecium of the Hydrocharitaceae is inferior. The carpels of *Stratiotes*, *Boottia*, *Ottelia* and *Enhalus* (Fig. 93 II, III)* are united exclusively along their dorsal faces to the receptacle, but they remain completely free from one another otherwise. Since moreover the

---

* In the cross-section of the ovary of *Ottelia* which is illustrated in Fig. 93 II, disc-shaped mucus glands which are inserted between the ovules are recognisable. These secrete a slimy substance with which the entire ovary cavity is filled. In other members of the Hydrocharitaceae this mucilage is secreted from the entirely gland-covered inner wall of the carpel.

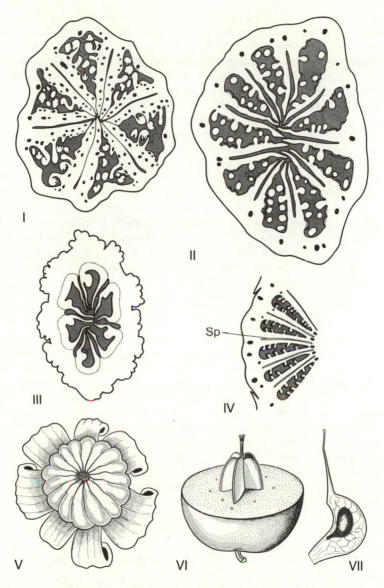

Fig. 93 "Pseudo-syncarpy". I *Butomus umbellatus*, cross-section through the basal region of the gynoecium, the 6 carpels are fused peripherally. II, III Hydrocharitaceae, cross-section through the inferior gynoecium of an *Ottelia* species (II) and *Enhalus acoroides* (young gynoecium III). IV *Nymphaea stellata*, segment of gynoecium in cross-section, Sp inter-carpellary slits. V *Nuphar lutea*, uncovering of the apocarpous carpels by shedding the coat of axial tissue when the fruit ripens. VI, VII Apple after the upper parts of the carpels have been uncovered, and an individual carpel dissected out (VII). III after Eber, I, II, IV, V, VII after Troll, VI after Rauh.

ventral margins of the carpels are not closed, the carpel "blades" project freely into the central cavity of the ovary. Because of this the adjacent halves of neighbouring carpels are arranged in pairs. Corresponding to the laminal placentation which predominates in the Hydrocharitaceae, these inward-projecting portions of the carpel blades bear ovules in greater (*Ottelia, Stratiotes, Boottia*) or lesser numbers (*Enhalus*, in a single row?) on their inner surfaces. It is not surprising therefore that these are taken to be "split placentas" of a paracarpous ovary! In *Vallisneria* and *Halophila* these carpel blades only project inward for a short distance, and in *Hydrocharis morsus-ranae* the adjacent carpel blades are fused together (see Troll 1931, Eber 1933, Kaul 1969).

By far the best-known example of pseudo-syncarpy is the semi-inferior gynoecium of the apple (*Malus sylvestris*), in which the carpels are free from each other to a considerable extent, but are nevertheless fused with the floral axis on the dorsal side. The axis grows out for some distance above the carpels and unites above them so that only the similarly free but mutually adherent styles project. Obviously in the apple, as also in *Stratiotes*, we have a case of a local and limited degree of fusion of the carpel sides (semi-syncarpy). The flesh of the fruit which surrounds the glumaceous and cartilaginous carpels also originates from the axis (Fig. 93 VI, VII).

1.6.14   The pseudo-monomerous ovary

Amongst the reduction phenomena which can occur in the region of the gynoecium we have already met with the decrease of the number of carpels in an apocarpous gynoecium to one, and the reduction of the ovules. The latter can lead to the formation of only one fertile ovule, whether this be in the individual carpels of an apocarpous gynoecium or in a syncarpous ovary. It is also possible for the development of carpels in syncarpous gynoecia to be so retarded that only one remains complete, so that an apparently monomerous gynoecium is created. It is not uncommon for such **pseudo-monomerous** gynoecia to be described as though they were genuinely monomerous, because although the more or less rudimentary carpels can take part in the structure of the ovary to varying degrees, this can often only be detected by means of ontogenetic and comparative research. Some of the numerous examples investigated by Eckardt (1937a) may make this clear.

In the order Urticales the Urticaceae used to be thought of as the only family where the flower is provided with just a single carpel, differing from the other families which have two (Eichler 1878, Warming and Möbius 1911 and later also Lawrence 1962, Porter 1967).

The gynoecium of the Urticaceae does, however, occasionally provide some external indication of the presence of a second carpel: *Girardinia zeylanica*, which normally has an ovary with a simple style (Fig. 94 VI), sometimes shows two (or rarely three) style branches. A comparative investigation of the gynoecia in the order Urticales, including ontogenetic studies, yields the conclusion that the Urticaceae represent the final member of a series, in which the second carpel is progressively reduced. But even within the Urticaceae the second carpel can be traced. In the Ulmaceae it is already apparent that only one of the two carpels is fertile, and that a single ovule develops on the cross zone. Here already the second carpel is relatively small. This is shown by the ovary of *Ulmus glabra* illustrated in Fig. 94 I, and to an even greater extent by the ovary of *Celtis australis* in Fig. 94 II. The vascular bundles are illustrated for both these formations, and for the structures of the various ovaries which continue the sequence, since these relate very well to one another and can therefore be included in the comparison. Using this procedure Eckardt has at the same time given a good example of a critical evaluation of data concerning the anatomy of the vascular bundles. However, we wish to pass over Eckardt's thorough and painstaking discussion of these results, and satisfy ourselves with the conclusion that the reduced carpel in *Celtis* is only provided with a dorsal median, so that it is further reduced in its vascular system than is *Ulmus*; it also takes but a minor part in the formation of the syncarpous portion of the ovary. The ontogeny of the gynoecium is substantially the same in both species: the ovule is initiated very early and is placed almost at the base (Fig. 94 X); "it reaches the top of the ovary locule as a result of growth which mainly takes place in the zone beneath the ovule. The youngest stages of *Celtis* greatly resemble those of the young gynoecia of the Urticaceae" (Fig. 94 XI).

In *Dorstenia psilurus* (Fig. 94 III), a member of the Moraceae, the reduction of the diminished carpel is continued still further. Its dorsal median no longer enters the ovary as an independent bundle, but is united for some of its length with the ventral median of the fertile carpel, which serves the ovule. The ovary of *Morus* shows a similar structure, particularly that of *Morus alba*, the White Mulberry. However, the two style branches in *Dorstenia psilurus* are united to a considerable extent. In the ovary of *Ficus carica*, the fig tree, the second style branch can be reduced completely, as well as the dorsal median (IV, V), which otherwise is only weakly developed. Thus the ovary is already pseudo–monomerous here; corresponding to the number of style branches either one or two carpels are described for *Ficus carica*. It is only a short step from this to the pseudo–monomerous gynoecium of the Urticaceae, which would be hard to detect there, if that family were only considered in isolation. The series of forms, for which Eckardt provides many further examples, and a comparison of the ontogeny of their ovaries shows, however, that a second carpel is involved in the structure of the gynoecium in the Urticaceae. This is not, however, any longer recognisable externally (for this see the sequence of forms in Fig. 94 VII to IX).

Eckardt also recognised the ovary of the Thymelaeaceae-Thymelaeoideae as pseudo–monomerous, which was formerly (and is still to some extent today) described as genuinely monomerous. Externally the ovary certainly gives the impression, as in the case of *Daphne mezereum* (Fig. 94 XII), that it is formed from only a single carpel. It can be proved, however, that a second carpel is involved in the formation of the ovary, which can in fact only be discovered in the part which is cross-hatched in diagram XIII. Eckardt was also able to demonstrate the presence of pseudomonomery in Globulariaceae, Phrymaceae, Sparganiaceae, and a series from the Araceae and in many cases in individual genera from other families. It turns out that there are genuinely monomerous ovaries amongst the Lauraceae (Endress 1972a, b) and the Berberidaceae. According to Eckardt, *Bouchea* in the Verbenaceae forms a carpophore from the sterile carpel together with parts of the placenta of the fertile carpel, and in *Broussonetia* it forms a false style which is 1cm long and densely set with

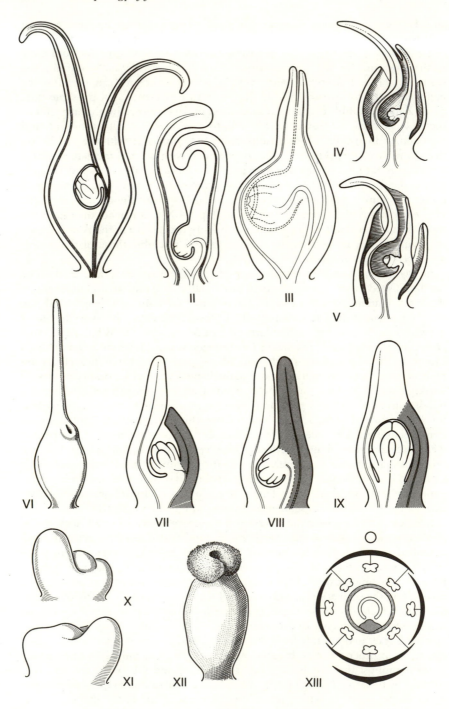

stigmatic papillae. Cases of pseudomonomery were also found by Junell (1938) in the genera *Rochelia* and *Harpagonella* in the Boraginaceae, in which the posterior carpel atrophies. According to Weber (1971), the gynoecium of the genus *Hemiboea* in the Gesneriaceae can "almost be described as pseudo-monomerous".

1.6.15   Polymerous gynoecia

The reduction in the number of carpels in apocarpous gynoecia and pseudomonomery in syncarpous gynoecia contrasts with the increase of the number of carpels in the formation of polymerous whorls of carpels, which can occur singly or in multiples. Apocarpous polymerous gynoecia are particularly often found in the Rosaceae-Rosoideae. According to the investigations of Kania (1973) their origin is connected with a more or less pronounced swelling at the base of the receptacle, which leads to the formation of a "carpel bearer". As a rule, some 10, 15, 20 or more carpel primordia arise on this base at the same time and at the same level. "Then follow further rings of primordia above these, with progressively fewer numbers of members for reasons of space. Lower down on the 'carpel bearer' these usually alternate with one another", whilst "further up, no alternation can be detected". According to Kania, the spiral insertion of the carpels which we know from *Magnolia* or in many members of the Ranunculaceae with compound gynoecia is not observed here.

In the tribe Malveae of the Malvaceae a very large number of carpels is often found in a circular arrangement. In the Malopeae (*Malope, Kitaibelia, Palaua*) the number of carpels which are initiated in this whorl may increase to as much as 45. This large number of carpels is "disposed of" in the confined space of the hemispherical receptacle in such a way that the sequence of carpels stands out above the sepals in radial loops, so that in the end 10 rows are formed, in which the carpels are placed one above the other in transversal position (Fig. 95 I, II). According to the old interpretations (Duchartre, Payer 1857) this multiplicity of carpels arises from 5 primary primordia, which differentiate into individual carpels. Sattler (1973) also reports that in *Malva neglecta*, the 15 carpels originate from 5 primary swellings, which likewise form 3 secondary primordia each. The accompanying photograph may however be interpreted in other ways. On this point, the principal reference is the research of van Heel (1978) on the gynoecium of the Malvaceae–Ureneae. The gynoecia of this tribe in fact possess twice as many style branches as carpels. The reason for this is that the inner of the original two 5-merous rings of carpels which are initiated one after the other is restricted in its development and remains sterile, whilst the style sections reach maturity. In the photographs given by Sattler the sequence of the two rings of carpels in the young stages of development can produce an image similar to that provided for *Malva neglecta*.

Fig. 94 Pseudomonomery. I–VI Median sections through the gynoecium of *Ulmus glabra* (I), *Celtis australis* (II), *Dorstenia psilurus* (III) and *Ficus carica* (two younger stages of development in IV, V). VI *Girardinia zeylanica*, normal ovary in a young stage. VII–IX Derivation of the *Urtica* gynoecium (IX) from *Celtis* (VII) and *Laportea, Parietaria* (VIII) by means of schematic longitudinal sections. X, XI Initialisation of the gynoecium of *Girardinia zeylanica* (X) and *Celtis tournefortii* (XI). XII, XIII Ovary (XII) and floral diagram (XIII) of *Daphne laureola* and *D. mezereum*, respectively. All after Eckardt, XIII altered.

In this context we want to mention three further cases which are already known in which a syncarpous ovary is formed from two successive whorls of carpels. These are: the well-known navel orange (*Citrus sinensis*), in which a second, smaller and often incompletely-developed ring of carpels follows the normal one, and embedded in the apex of the ripe orange like a secondary fruit of the size of a cherry or a plum; secondly the genus *Siphonodon* in the Celastraceae; and lastly, the pomegranate, *Punica granatum* (Punicaceae). In the latter, the completely inferior ovary* (Fig. 19 V) consists of an outer 6-merous ring of carpels and an inner 3-merous ring, of which, however, the outer is placed higher in the ovary than the inner and furthermore its placentas are borne on the outer wall, whilst in the inner ring a normal axile placentation type prevails. The reason for the upward displacement of the outer ring of carpels above the inner and for the peculiar placentation features was realised by Payer (1857). The unusual arrangement originates, according to Eichler (1878), "from the same processes as we have already encountered in *Mesembryanthemum*, namely from predominating longitudinal growth which takes place in the periphery of the ovary, and which as it were inverts the ovary, so that what used to be below and inside is shifted to be above and outside." Sometimes yet a third, innermost ring of carpels is added to this, "by which the second ring is also pushed upward and with its placentae is" more or less "twisted into a peripheral position".

1.6.16   Style and stigma

We must now concern ourselves with the distal end of the individual carpel or of the syncarpous ovary, which remains sterile and serves for the reception of the pollen grains and for the conduction of the growing pollen tubes, namely the stigma and the style. It is worth remarking at this point that the usual terminology for these two organ sections is in no way standard.

We define the **styles** by reference to Baum (1948e) as "the sterile ends of the carpels which are more or less clearly demarcated from the fertile section of the carpel and more or less free or mutually fused". At their tips they bear a **stigma** which is adapted for the reception of pollen brought by wind or by animals.

For more precise characterisation we distinguish the style formed from only one carpel (pseudostyle according to Hanf 1935) as **monocarpellary** (monomerous), from the style formed from several carpels which is termed **polycarpellary** (di-, tri- or polymerous; the "stylar column" according to Juhnke and Winkler 1938).

In syncarpous gynoecia also, it may happen that the fusion of the carpels only reaches over a certain distance, such as perhaps only over the fertile section, and so the sterile carpel tips can remain completely free. That is to say, just as many monomerous styles may appear as there are carpels taking part in the structure of the ovary. When there is complete congenital or postgenital fusion of the carpels, then a polycarpellary style of completely uniform structure is formed. On the other hand, the sterile ends of the carpels can also be fused in their lower part into a pleiomerous style, whilst their tips remain free; these are termed **style branches** (**stylodia**, Troll, Hanf), or in so far as only the stigmatic region

* (semi-inferior in *P. protopunica*).

is involved, **stigma branches**. We may view the most complete structure then as a uniform style with a capitate stigma, such as is found in many families of the Myrtales. In the sequence of development from apocarpy to syncarpy there are therefore several stages to follow: "fusion of the carpels, then of the styles, and finally of the stigmas" (Wettstein 1935).

In the polycarpellary styles the margins of the individual carpels are not generally closed. The styles are therefore "paracarpous", but they can also be "syncarpous", so that the individual locules are also continued into the stylar canal (*Veltheimia, Geranium*).

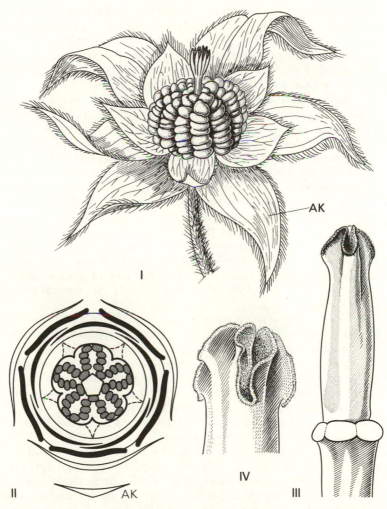

Fig. 95 I *Kitaibelia vitifolia*, flower. II *Malope trifida*, floral diagram, AK epicalyx, III *Tulipa* sp., gynoecium, IV distal section of the same with the stigma, further enlarged. I after Gürke, II after Eichler.

According to Juhnke and Winkler (1938, p. 308), the most primitive form of the stigma may be that which "in as far as it produces no ovules, the complete margin of the carpel is transformed into a stigma". This is the case for what is believed to be a primitive carpel in *Drimys winteri* or *Degeneria vitiensis* (Fig. 78 I, VI, VII): these possess neither a style nor a stigma in the usual sense. Instead the function of the stigma is taken over by the so-called "stigmatic crests", i.e. the densely papillose marginal zones of the carpel. To that extent these carpels agree with the concepts of Hallier (1912) on the origin of the stigma (see also Schaeppi 1975).

Consequently, therefore, we can refer to the simplest form of the stigma in both the monomerous styles of the carpels of an apocarpous gynoecium as well as the style branches of a syncarpous ovary as the **follicular stigma**. We want to consider first the individual carpel of an apocarpous gynoecium, namely that of *Trollius europaeus* (Fig. 96 I, II). The margins of the carpel in the fertile section are here closed ventrally and fused together postgenitally, and in what here is only presumed to be the zone of the style separate laterally from one another to form the stigma. They still remain folded here nevertheless. Their marginal zones are densely papillose. The result is therefore two stigmatic lines, which are divided in the centre by a "ventral groove", but which are united in a curve at the apex. As a rule the margins in the stigmatic zone are folded back further and expose the upper surface of the carpel to the open. The regions next to the margin can be included in the differentiation of the stigma, so that broad stigmatic lines run along both sides of the ventral groove, as is the case in *Butomus umbellatus* (Fig. 96 III, IIIa). At the same time a shortening of the stigmatic surfaces can begin from below. The result of this are the types of stigma seen in *Helleborus* and *Fragaria* (V, VI); the monocarpellary stigma from the syncarpous gynoecium of *Buxus* (IV) is included as an intermediate form. On the other hand, it is possible for the apex of the stigma to develop more broadly and become reniform (VII), or finally the stigmatic surface can encroach over the upper surface of the style to make a stigma shaped like the cap of a toadstool (Fig. 96 VIII), as we have already encountered in the Japanese Cherry (Fig. 75 VI).

Equivalent stigma forms are not only found on the style branches of syncarpous ovaries (Fig. 96 XI), but they also occur in styles which are fused as far as the stigmatic zone. The sutures left by the fusion of the stigmas can disappear to a great extent or disappear totally, while the ventral grooves remain. The result of this is a more or less uniform stigma, where the number of carpels involved can be recognised by the lobed outline of its general shape and by the number of channel-like depressions on the apex (Fig. 96 IX).

It is not uncommon for the stigma lobes to be more or less deeply emarginate at the tip. We have already come across this in *Butomus* (Fig. 78 XIV), where the papillose stigmatic tissue even extends beyond the style tip and descends the under surface of the style. This also applies correspondingly to the style branches of *Philadelphus pubescens* (Fig. 96 XI), or to the stigma lobes of the Tulip (Fig. 95 IV), which are inserted directly on top of the ovary on account of the complete suppression of the style (III). This can be generalised by the observation that the formation of stigmatic papillae is not confined to the upper surface of the carpel.

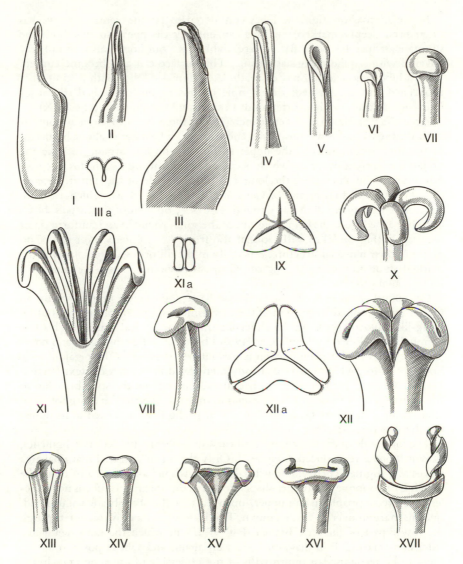

Fig. 96 Types of stigma. I, II *Trollius europaeus*, I follicle seen from the side, II upper part and follicular stigma seen from in front; III *Butomus umbellatus*, follicle in side view, IIIa schematic section through the stigma zone; IV–VII stigmas of *Buxus balearica* (IV). *Helleborus niger* (V), *Fragaria vesca* (VI), *Prunus spinosa* (VII) and *P. avium* (VIII). IX–XVII End of style with stigma of *Eucharis grandiflora* (IX), *Epilobium dodonaei* (X), *Philadelphus pubescens* (XI, XIa schematic cross-section through the stigma), *Lilium regale* (XII, XIIa cross-section through the stigma), *Bergenia cordifolia*, from in front (XIII) and from behind (XIV), *Begonia longipila*, similarly (XV, XVI) and *Begonia hybrida* (XVII). III original, VIII after Troll, the rest after Juhnke and Winkler.

In the stigmas or stigmatic lobes considered so far the ventral groove has often been deeply marked, but this can entirely disappear so that both the stigmatic stripes lie on a flat surface, which is commonly also completely covered with papillose stigmatic tissue. This surface can also become curved. This is how lingulate stigmas or spathulate stigmatic lobes with a somewhat broadened end are formed; an example of this is the four-lobed stigma of *Epilobium dodonaei*, which corresponds to the number of carpels (Fig. 96 X).

According to Baum and Leinfellner (1953), these types of stigmas contrast with **unifacial** (monomerous) styles and stigmas, which come about by means of the total suppression of the upper surface of the leaf organ, just like the unifacial excurrent processes which occur on other leaf organs. In this category Baum and Leinfellner place the long and filiform style of the Thrift, *Armeria* (Fig. 98 IX), and the stigmas of *Coriaria japonica* (Fig. 98 V), which are both covered with papillae all the way round. The development of the carpels shows that they arise "from the beginning out of the solid, rounded and uniform tip of the carpel" (Fig. 98 VI to VIII), so that also from the ontogeny there can be no doubt of their unifacial structure. Especially in the filiform styles of *Armeria*, the marked elongation of these solid carpel tips is obviously caused by additional longitudinal growth.

On the other hand, the broadening and flattening of style branches is by no means always caused by growth of the carpel margins, but is brought about by wing-like outgrowths from the underside of the carpel 'leaf' surface. This was definitely established by Leinfellner (1952c) by means of ontogenetic and comparative studies. This phenomenon is particularly striking in the petaloid style branches of *Iris*, which we have already mentioned in another context (section 1.4.4.3). Goebel (1933, p. 1922 ff.) had already compared the style branches of *Iris germanica* with those of *Crocus sativus* in order to explain the morphological relationship, but it was Troll (1957, p. 139 ff.) who first succeeded in recognising the structure clearly.

The long style of the Crocus flower bears three deeply divided style branches, of which one is reproduced in Fig. 97 I. Only the much dissected marginal area serves as a stigma. The style branch taken as a whole takes the form of a tube which is open above and funnel-shaped towards the apex. This tube is nevertheless only recognisable in the upper, open part of the style branch illustrated, since its margins overlap further down. This is shown by a cross-section through the lower part of the style branch (Fig. 97 II), in which the same wing-like extensions marked F in diagram I are seen again, and whose position allows them to be recognised as outgrowths of the underside of the style branch, i.e. from the lower surface of the carpel.

Let us now compare this with a style branch of *Iris*, whose upper and lower surfaces are shown in Fig. 97 III and IV, respectively! On the underside the stigma is recognised as a flap of tissue running right across the organ from below the end lobes. As already mentioned in section 1.4.4.3, this bears the tissue which is receptive to pollen grains on the side which faces the two end lobes. The longitudinal stripe which projects somewhat from the upper side turns out from a cross-section (V) to be a dorsal groove, whose margins overlap in this case too. A comparison of this cross-section with the cross-section through the style branch of *Crocus* indicates that here too the broadening is caused by wing-like outgrowths of the underside of the style, which was confirmed by studying the ontogeny (Leinfellner 1952c). The morphological relationship between the style branches of *Crocus* and *Iris* is illustrated by the diagrams in Fig. 97 VI to VIII. A simplified version of the structure of the style branch in *Crocus* is reproduced in diagram VI. Here the wings only reach about as far as the middle of the style branch. If the growth were restricted in the median zone of the style branch, then an organ of the form reproduced in diagram VII would result, where the apex lies at the position marked with an x and the marginal parts project a long way forwards as free lobes. For the form of the style branches in *Iris*, it is now necessary to postulate a greater degree of widening of the wings with the inclusion of the two end lobes, in which the median marginal zone is extended to the right and the left at the base of the wings to form a broad flap – the stigma.

The salverform and two-lobed stigmas of *Campsis* (*Tecoma*) *radicans* (Fig. 98 I to IV) and *Incarvillea* (Bignoniaceae) also do not arise from enhanced growth of the carpel margins, but according to Leinfellner, result from the development of wing-like structures from the underside of the carpel at a very early stage. The same is true for *Martynia proboscoidea* (Martyniaceae) and for a series of species in the Scrophulariaceae, such as the Musk (*Mimulus*), whose stigmas close up when touched on the inner surface, as also happens in *Incarvillea*, *Catalpa* and *Torenia*.

The origin of the umbrella-shaped style of *Sarracenia* (Fig. 100 I) can perhaps be explained in a fashion corresponding to that for the style branches of *Iris*. Hanf has classified this style as a type of "pollen collecting equipment". Since the flowers hang downwards, the pollen grains which fall out of the anthers of the same flower are in fact collected on the inner surface of the "umbrella" and prevented from falling out. The style is 5-lobed, with deep, rounded sinuses between the lobes, but the receptive stigmatic surfaces are only found on the upper side of small points which are inserted under the tips of the lobes, i.e. are placed on the dorsal side of the style. The position of these stigmatic points and the fact that the tips of the broadened and flattened tips of the style are emarginate suggest the interpretation that the lateral expansion of the fused style branches may come about in a similar fashion to that described above for *Iris* etc., although in a manner much modified by congenital processes. Unfortunately the only indication given by the ontogenetic researches of Hanf is that "a groove is still present at the beginning", in the middle of each stylodium during development, "which corresponds to the median of the carpel".

The emarginate condition of the tip of the stigma which we have already met in *Butomus* (Fig. 96 III), can be continued to the extent that the tips become regularly bifid by the ventral groove, as can be seen by a cross-section through the end of the stigma branch in *Philadelphus pubescens* (Fig. 96 XIa). The two halves of the stigma can become entirely separated in this way, or on the other hand they can be fused with the neighbouring half of the stigma from the next carpel, as can be seen in the 3-lobed stigma of *Eucharis grandiflora* in the general view (Fig. 96 XII) and in the cross-section (XIIa). Stigma lobes arise in this way which do not consist of the pair of lobes which belong to the same carpel, but which overlap the boundary (commissure) between two neighbouring carpels. These are therefore called **commissural stigmas**, in contrast to the **carinal stigmas** which are placed in the "normal" position, above the locules. The formation of commissural stigmas is particularly characteristic for certain families, e.g. the Papaveraceae (Fig. 99 VIII), and for the Cruciferae. In the Cruciferae all the intermediate forms between carinal and commissural stigmas can be found.

One example of a carinal stigma in the Cruciferae is that of *Hesperis* (Fig. 99 I, II). Here the stigmas of both carpels lie in the "normal" position above the locules; both are divided by a ventral groove, which separates the two swollen and papillose stigma margins from one another. In *Matthiola incana* (Fig. 99 III,

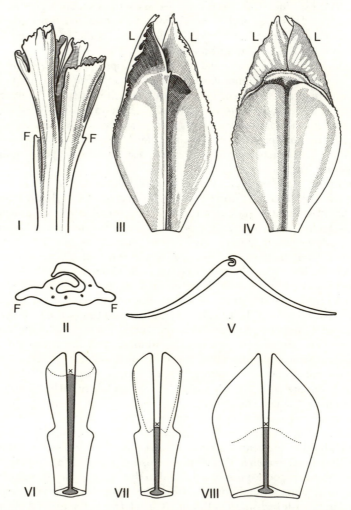

Fig. 97 Broadened surfaces of the style branches in the Iridaceae. I, II *Crocus albiflorus*, style branch, general view from above (I) and cross-section (II). III-V *Iris germanica*, upper (III) and lower (IV) surface of style branch, and cross-section (V). VI-VIII Schematic diagrams for the derivation of the form of the style branches in *Iris*, as seen from above. F Wing-like margins, L terminal lobe. All after Troll.

IV) the separation of the two carpels in the stigmatic zone can no longer be detected as such, although the loops of papillose stigmatic tissue still lie above the locules. In *Cheiranthus cheiri* (Fig. 99 V, VI) it is otherwise; here the ventral grooves are widened towards the base, so that the swollen margins of the stigmas diverge. At the same time the area of the commissural transition between the stigmas is widened out in a curve downwards. This dilatation is a great deal more obvious than the original stigma lobes: it represents the

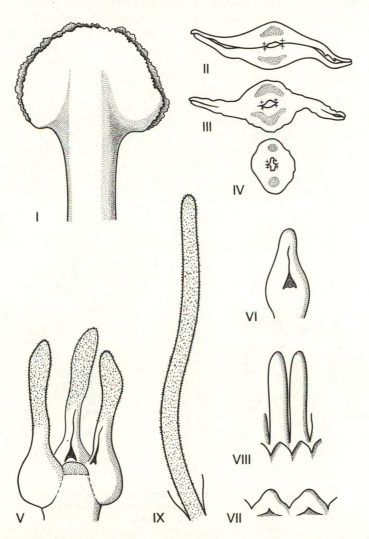

Fig. 98 I-IV *Campsis radicans* (*Tecoma radicans*), stigma (I) and a sequence of cross-sections through the stigma lobes (II, III) and the upper part of the style (IV); the + signs mark the position of the morphological carpel margins. V, VI *Coriaria japonica*, V young gynoecium, VI individual carpel at an even younger stage. VII-IX *Armeria juniperifolia*. Stages from the development of the style (VII, VIII), and mature style (IX). After Leinfellner and Baum-Leinfellner.

commissural stigmas. These are even more obviously apparent in other members of the Cruciferae. *Erucastrum gallicum* will serve as another example (Fig. 99 VII). Commissural stigmas like those in the Cruciferae are also found in the bicarpellate Papaveraceae.

On the other hand, there are some other forms of stigma whose origin lies in localised but very pronounced lateral expansion of the tip of the style. If we suppose that the limbs of the reniform stigma of *Prunus spinosa*, as illustrated in Fig. 96 VII, are pushed apart by a distal widening of the ventral groove, we obtain the form of the stigma of *Bergenia cordifolia* (Fig. 96 XIII, XIV). If we then assume more pronounced lateral growth, then this can be linked to the form of the stigma in *Begonia longipila* (Fig. 96 XV, XVI). The stigma type of *Begonia hybrida* (XVII) and other *Begonia* species is derived from this by elongation and spiral coiling of the free tips of the stigmas. There is also the possibility that the stigmatic surface may increase in area by a greater or lesser degree of branching, which often occurs by repeated bifurcation of the style branches. According to Hanf (1935), "those stylodia which resemble stag's horns are brought about by a high degree of lateral basipetal branching" (Fig. 100 III, IV). It is obvious that a high degree of ramification or pinnate branching of the style or the stigmas or the feathery type of structure caused by numerous long, simple or branched papillae, as in grasses, are principally to be understood in the context of wind pollination.

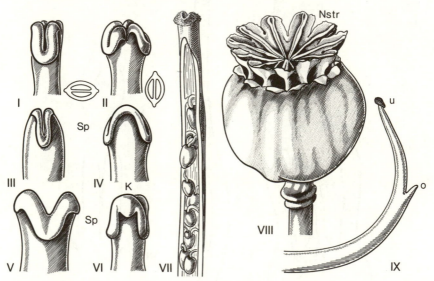

Fig. 99 I–VI, VII Derivation of the commissural stigma from the carinal stigma. I–VI Tip of the gynoecium of *Hesperis matronalis* in side view (I) and from in front (II), the cross-sections show the orientation of the carpels, III, IV do., *Matthiola incana*, V, VI do., *Cheiranthus cheiri*, VII young fruit of *Erucastrum gallicum* cut open, showing the false septum. VIII *Papaver rhoeas*, ripe fruit (poricidal capsule), the opening of the pores results from distal sections of the walls of the individual carpels between the placentas and below the deeply emarginate stigmas being bent outwards, Nstr stigmatic ray consisting of two halves of stigmas from neighbouring carpels. IX *Phlomis fruticosa*, unequal development of the upper (o) and lower (u) style branches. I–VI After Juhnke and Winkler, VII, VIII original, IX after Goebel.

As was stressed by Goebel (1933, p. 1931), the zygomorphic form of flowers is often also extended to the structure of the style and stigma. In the simplest case it is just a matter of differences of size between style or stigma branches. In *Salvia heerii* the upward-directed branch of the style is weaker and sterile, and in addition the two branches are coloured differently: the upper is white and the lower red. This difference is more marked in *Phlomis fruticosa* (Fig. 99 IX). Further examples are provided by the Acanthaceae, and also by the Polygalaceae, Proteaceae, Violaceae, Goodeniaceae and the Marantaceae, in which very complex forms of stigma often occur, as for example the "hollow stigmas" in *Viola* species (see Lange 1913) or the so-called "pollen cup" in the genus *Scaevola* in the Goodeniaceae, which we cannot attempt to describe here (but see further Goebel 1933, p. 1928 to 1939; also Fig. 100 II).

As a result of its surface structure, and often by means of sticky secretions, the stigma is specialised not only to catch pollen grains, but also retain them. Heslop-Harrison and Shivanna (1977) give a review of the surface structure of secreting and non-secreting stigmas. Above all the stigma and style have to facilitate the germination of the pollen grains and the growth of the pollen tube

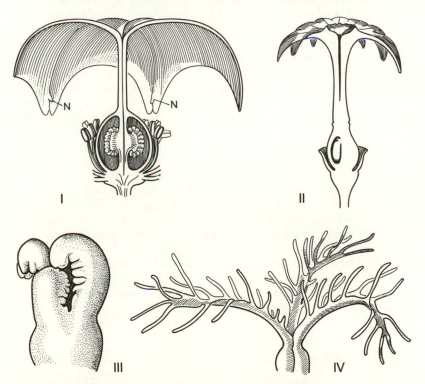

Fig. 100 I *Sarracenia purpurea*, longitudinal section through the gynoecium, which is formed of five carpels with an expanded and umbrella-like style, with the stigmas N placed below the tips. II *Hura crepitans*, female flower, the stigma is found at the tip in the centre of a broad and umbrella-shaped style (after Engler). III, IV *Acalypha obovata*, young (III) and older (IV) gynoecium with "stag's horn" style branches (with basipetal differentiation sequence), after Hanf.

from the stigma to the ovules. The means of nutrition of the pollen tube by the stigma is often provided by the fact that the secretion from the protoplasm-rich epidermal cells is usually rich in sugar. The stigma may also therefore be regarded as a nectary. Specialised tissues are also formed in the style in order to facilitate the advancement of the pollen tube and can supply it with nourishment (see Sassen 1973). This tissue which connects between the stigma and the inside of the ovary is known as **conducting tissue**.

There are two types of structure of conducting tissue to be distinguished. In so far as a stylar canal is present, which is lined with glandular tissue, this can be papillose and many-layered by periclinal divisions. This **ectotrophic conducting tissue** can cover the entire wall of the stylar canal, or be confined to one or more longitudinal bands (as in many of the Caryophyllaceae, see section 1.6.6). The placentas are also covered with this tissue, which in many plants forms a protuberance close to the micropyle, known as the **obturator**. The cuticle of the glandular epidermal cells is often lifted off before pollination, as the wall layer beneath it swells up in a slimy mass (pectin-based mucilage?), which often fills the entire stylar canal. In such cases the pollen tube can penetrate the tissue which lines the stylar canal and push its way between the cells. Usually it grows on the surface of the conducting tissue.

The majority of the Angiosperms do not possess any stylar canal, but have a solid style, which is thought to be a derived state (Coulter and Chamberlain 1912). In this case the centre of the style is filled with an **endotrophic conducting tissue** of elongated cells, whose walls swell up and whose protoplasm is strongly coloured by cytoplasmic pigments. In polymerous styles several such strands of cells usually occur. The continuity of this tissue is partially loosened, either before the intrusion of the pollen tube or during its progress. Whether the pollen tube is guided to the ovules chemotactically or by the structure and direction of the conducting tissue is still being debated. In many syncarpous gynoecia, pores, slits or canal-like openings are formed in the septa which can assist the pollen tubes to find their way to different parts of the ovary, independently of where they germinate (this system of connection of possible pathways for the pollen tubes is called the **compitum**; Carr and Carr 1961, p. 253).

With (gametophytic) self-incompatibility (see section 3.3.1), the growth of the pollen tube can be inhibited by secretions of the conducting tissue (Heslop-Harrison and Shivanna 1977, and see also the literature cited there). Information about the fine structure of the stigmatic and conducting tissues is given by various recent investigations (e.g. J. and Y. Heslop-Harrison 1975, Rosen and Thomas 1970, Dashek, Thomas and Rosen 1971, Johnson, Wilcoxson and Frosheiser 1975, Bell and Hicks 1976).

1.7   Nectaries

1.7.1   Definition and histological structure

Glandular tissues which secrete fluid containing sugars, i.e. nectar, occur in a great variety of different positions within the flower. Such glands are termed **floral nectaries** in order to distinguish them from the extrafloral nectaries which occur on organs outside the flower, such as on the stem, the petioles, the stipules, leaf blade margins, and so on.

Not all floral nectaries have a function connected with pollination biology, however. The nectaries which occur on the outside of the corolla of many members of the Bignoniaceae, or on the outside of the perigon in a number of *Iris* species are examples of this (Vogel 1977). On the other hand, for example, the glands on the cyathia of *Euphorbia* species (which are interpreted as interfoliar stipules) play an important part in the attraction of pollinators. These are of course extrafloral, but using the term coined by Delpino, can be termed "nuptial" nectaries, meaning that they "contribute to the marriage of the plants". On the other hand, the nectaries on the outside of the perianth of *Iris* or in the Bignoniaceae are not of this kind: they are certainly floral nectaries, but are **extranuptial**.

On the inner side of the perianth tube of *Iris* there are also nuptial nectaries in close association with the extranuptial nectaries, as for example in the red-flowered and ornithophilous *Iris fulva*, and these also secrete sucrose, in contrast to the extranuptial nectaries (Vogel 1967). There are also examples of change from nuptial to extranuptial function of nectaries during the process from anthesis to ripening of the fruit (see Sperlich 1939, van der Pijl 1955, Vogel 1977).

The nectaries of the jug-shaped (ascidiate) bracts in the inflorescences of the Marcgraviaceae (Fig. 175 I to II) are extrafloral, but nevertheless nuptial. In the genus *Harpagophytum* in the Pedaliaceae certain of the floral primordia do not develop into normal flowers, but become enlarged whilst keeping their embryonic form, in such a way that it is only the nectar-producing disk which becomes functional (Fig. 175 IV, see Ihlenfeldt and Hartmann 1970).

By means of histological characters, Vogel (1977) distinguishes three kinds of nectary, **mesophyllary, epithelial** and **trichomatic**. In mesophyllary nectaries the mesophyll (or axial tissue at deeper levels) is glandular in form, and the secretion of nectar takes place through the so-called **nectar slits**, i.e. stomata which have more or less lost the ability to close (Zandonella 1967). In epithelial nectaries the glandular tissue is formed from the palisade-like, protoplasm-rich epidermal cells which have large nuclei (Fig. 102 IV), and often further layers of subepidermal cells differentiate into glands and take part as well. The nectaries which occur on the upper side of the sepals in many species of the Malvales, or even on the epicalyx lobes as in *Gossypium*, will serve as examples of **trichomatic nectaries**. The section from the secretion tissue of *Abutilon striatum* which is reproduced in Fig. 102 III illustrates the generally observed fact that whilst the nectaries themselves do not have any vascular bundles, there are nevertheless many fine ramifications of the vascular system which lead up to them. Agthe (1951) describes this for *Abutilon striatum*: "The lower part of the calyx is provided with well developed vascular bundles rich in phloem. These pass along far beneath the secretion tissue and finally end in the calyx teeth. Along the whole distance which these bundles run beneath the secretion tissue, there are very numerous other bundles which branch off in the direction of the glandular tissue. These show no xylem elements except in the first section just after the branching, and come to an end finally as strands of phloem parenchyma about two or three layers below the secreting hairs" (see also Frei 1955). Further examples of trichomatic nectaries are provided by the Caprifoliaceae

(*Diervilla, Linnaea, Lonicera*), Adoxaceae and the Valerianaceae (*Valeriana, Fedia, Nardostachys* etc.; see Weberling 1977b, Wagenitz and Laing 1984).

Apart from the above types of nectary, **hydathodes** which are modified into nectaries also occur, and indeed, as reported by Frey-Wyssling and Agthe (1950) there are all kinds of intermediates (*Fritillaria, Ranunculus*, Agthe 1951), and Knoll (1921, p. 15) stresses that nectaries "only represent a special case of the general type of hydathode".

It has been established for nectaries in general that the sugar concentration in the secretion decreases as the proportion of xylem in the conducting path to the secreting organs increases (Frey-Wyssling and Agthe 1950, Agthe 1951, Frei 1955). The sugar content (saccharose, maltose, raffinose, fructose, glucose, arabinose, see Gottsberger, Schrauwen and Linskens 1973) of the nectar varies between 8 and 76% (*Origanum vulgare*), so its concentration is very variable. Nectar can also contain proteins and amino-acids (H. and I. Baker 1973), and other organic substances.

The secretion of nectar usually takes place through the cuticularised outer wall of the glandular epidermal or hair cells, and indeed through very fine secretory canals; in other cases the cuticle becomes lacerated and detached (Frey-Wyssling 1935).

The calyx glands of the Malpighiaceae, which have already been discussed in section 1.4.2.6 (Fig. 28 IV to VI), are modified in the new-world species into oil-secreting glands, the so-called **elaiophores**, and the secretion (a fatty oil) is gathered by certain bees for their brood (Vogel 1974). Vogel was able to establish close anatomical relationships between the oil-producing hairs of *Cypella* as well as *Diascia* and *Calceolaria* and the trichomatic nectaries of other members of the Iridaceae or Scrophulariaceae. A mutual transition between both forms of hairs may therefore be assumed. In Yellow Loosetrife (*Lysimachia vulgaris*), Creeping-Jenny (*L. nummularia*) and other species of the subgenus *Eulysimachia* as well as the subgenus *Seleucia* the glandular hairs on the filament tube also produce oil (Vogel 1976).

In this context, "feeding hairs" and "feeding tissue" ought to be mentioned. In the former case we are concerned with densely crowded, single-celled tubular or clavate hairs, which are rich in fat and protein (as on the labellum of *Maxillaria* species), or also with sugar-rich hairs with a single row of cells (e.g. *Sterculia javanica*, Cammerloher 1931), which are presumably "grazed upon" by insects visiting the flowers. The same applies to the "feeding tissue": "fringed scales in transverse position", "rich in protein, fats and sugar", which occurs on the labellum of *Vanilla* species (see Kugler 1970), or for example the multicellular feeding bodies, rich in protein and fats, which occur on the tips of stamens and staminodes (Fig. 103 V, VI, Daumann 1930b). Proof that cushions of hairs or other structures like this do really serve regularly for the feeding of pollinators is, however, not yet available. In addition there has to be clarification as to whether at least some of the "feeding hairs" may be examples of the elaiophores referred to above (see furthermore the work of Porsch 1905, 1906).

1.7.2　Receptacular nectaries

We have already come across floral nectaries on various floral organs in a great variety of forms. In particular we found them developed as outgrowths of the

receptacle in the form of what is termed the **disk**, as disk- or bowl-shaped structures surrounding the base of the gynoecium (Fig. 11 V), as filiform, capitate, scale-like or palmately divided glands (Fig. 11 VII), or – as for example in the Rutaceae (Fig. 11 II, III) – as a more or less inflated glandular cushion. When the ovary is inferior, "stylar cushions" of this kind can surround just the base of the style, as is the case with the "stylopodium" of many of the Umbelliferae (Figs 37 IV, V, 92 I). As the detailed investigations of Magin (1979, personal communication) have shown, the location and form of these nectaries can vary a great deal, and so – to cite two extreme cases – the nectariferous tissue can occur on the dorsal side of the style branches, and can also be shifted to the inner side of the sepal bases. In the Onagraceae, in which the receptacle is often extended above the inferior ovary in the form of a cup-shaped or long-tubular hypanthium, the inner wall of this is thickened at the base to form a ring-shaped gland (Fig. 19 III, VI). Nectaries in the form of a disk-like ring below the ovary also occur in families such as the Theaceae, Ericaceae, Polemoniaceae and the Labiatae; in the latter we saw, however, that the nectary in *Ajuga reptans* appears as a swollen appendage on the adaxial side (Fig. 86 I, II). Unilaterally developed, glandular receptacular structures, which fit into the overall dorsiventral symmetry of a flower, are by no means uncommon. Examples of this are found in the Resedaceae, and especially in the Capparidaceae. In the latter we find scale-like, swollen, bowl-shaped or even long tubular nectaries developed, usually on the adaxial side (Fig. 13 I to III). Also the free axial spurs (*Tropaeolum*, Fig. 20 II, IIa), or those which are "enclosed" in the axis (*Pelargonium*, Fig. 20 III to V, *Bauhinia*, see Fahn 1952), usually contain nectaries. In the Rosaceae the glandular tissue can cover the entire inner surface of the receptacle, or also be restricted to particular areas (Fig. 101 VII to X), and in the Papilionaceae it can even enclose the base of the gynoecium like a sheath (Waddle and Lersten 1974).

The glandular tubercles and callosities which surround the bases of the filaments on the receptacle of the Cruciferae in a great variety of different arrangements (Fig. 101 I to VI) have generally thus far been interpreted as axial structures (see Arber 1931a; Norris 1941, Dvorak 1968). Similar remarks also apply to the Polygonaceae.

The disk of the Paeoniaceae, however, does not function as a nectary, contrary to what has previously been thought. Rather, the nectar which is detectable in the flowers of certain *Paeonia* species is produced from vascular bundles which originally served certain stamens, whose primordia have, however, been smothered and suppressed by the vigorous growth of tissue from the base of the disc (Hiepko 1966).

1.7.3 Phyllome nectaries

We must distinguish the **phyllome nectaries** from the **receptacular nectaries** as the second morphologically defined group of intrafloral nectaries. They can be found on all the foliar organs of the flower. We have already got to know them as sepal nectaries in the Malpighiaceae (Fig. 30 IV to VI), where they sometimes take on a rather strange form, and in the Malvales in the form of trichomatic nectaries. The sepal spurs of *Impatiens* species (Fig. 31) should also be mentioned. These bear a nectary at the base of the hollow spur which sheds its nectar on the inside. In this case therefore the spur serves both as a **nectary** and as a **nectar container** at the same time, in contrast to the spurred sepal of the Larkspur (*Delphinium*), which serves only as a nectar container for the nectar secreted by the two nectar leaves which project far inside it (see also *Aconitum*, Fig. 47 II). In a number of other cases, such as in the saccate petals or spurs of the

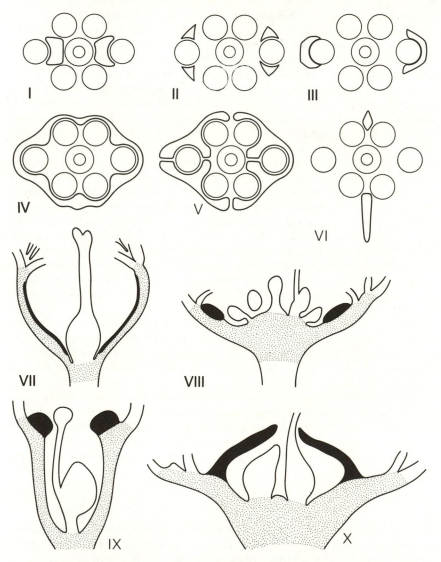

Fig. 101 I–VI Schematic diagrams for the arrangement of the nectaries in various species of the Cruciferae (from Goebel). VII–X Axial sections of the flowers of *Prunus padus* (VII), *Potentilla sterilis* (VIII), *Alchemilla vulgaris* (IX) and *Rhodotypos scandens* (*R. kerrioides*) (X), glandular tissue shown in black. After Schaeppi.

Bleeding Heart (*Dicentra*, Fig.5 XII, XVI, 38 I, III), in Fumitory (*Fumaria*, Fig. 103 XIII), in *Corydalis* (Fig. 38 IV to VI) and in other members of the Papaveraceae-Fumarioideae, we have to distinguish between the saccate floral organ or spur which contains the nectar, and the nectary which projects into it and which is derived from the axis.

In the corolla we find nectaries especially in the zygomorphic sympetalous flowers of the Caprifoliaceae, Valerianaceae, many of the Verbenaceae, Lentibulariaceae and some Scrophulariaceae (*Calceolaria*) in the form of trichomatic nectaries. These are also usually positioned in median saccate lobes or spurs of the corolla. In Snapdragon (*Antirrhinum*, Fig. 43 I to III) and in Toadflax (*Linaria*, Fig. 43 V, VI), once again the ventral saccate corolla lobe or spur only serves as a container for the nectar secreted by a receptacular nectary.

Frasera carolinensis in the Gentianaceae provides an example of free nectaries which are located on the surface of a corolla lobe. The long corolla lobes are only united for a short distance at their bases and on the midvein of each, and at about one third of the way up, is placed an oval flask-shaped nectary about 3–4 mm in diameter with a fringed margin (Fig. 103 VII, VIII). According to McCoy (1940), the glandular tissue arises from the subepidermal cell layer of the upper leaf surface. Repeated periclinal cell divisions lead to 7 or 8 layers of cells which form the base of the nectary. The walls of the "flask" arise from further periclinal divisions at the periphery, and divisions in various directions on the margins of these lead to the differentiation of the fimbriate appendages which conceal the mouth of the nectary. Later on these become sclerified and hard. According to Davies (1952) the secreting tissue is confined to the epidermal and subepidermal cell layers of the base of the cup and the thickened part of the wall.

Nectar glands are found in a variety of shapes (discoid, capitate, filiform, urceolate) on the upper and lower surfaces of the **tepals** of *Nepenthes* (Daumann 1930a). Moreover, we have already discussed the nectaries of tepals in the context of the morphology of the perigon, particularly the **nectar leaves** (section 1.4.3.10) and the way in which **staminodes** often take over the function of nectaries (section 1.5.12). We have also already mentioned the formation of nectar glands which occur, usually in pairs, at the bases of the filaments in the Caryophyllaceae (section 1.5.11). **Paired staminal glands** of this type also occur regularly in the Lauraceae in certain rings of stamens or staminodes. Fig. 103 I shows a fertile stamen of *Cinnamomum camphora*, and diagrams II and III a staminode and a fertile stamen, respectively, of *Laurus nobilis*; in II the pair of nectar glands have completed their development, whereas the anthers remain undeveloped. In the staminodes of *Cinnamomum camphora* (IV) this reduction has gone a stage further: the staminode consists almost entirely of the two nectar glands. Daumann (1930a) reports on the nectaries on the filament tube of the male flowers of *Nepenthes*.

In *Viola* there are well-developed glandular "connective appendages" of the two anterior stamens (see Leinfellner 1957a) which sometimes take the form of spurs, and which project into the spur formed from the median ventral petal (see Fig. 38 VII), in which the nectar secreted by them is collected. Here once again it is necessary to distinguish between the nectary and the nectar container. The formation of this kind of arrangement does of course require an exact mutual correspondence, or co-adaptation, in the development of the different floral organs; in the present case, the nectaries and the separately formed corolla spur. In *Pentstemon* bands of glandular hairs are found on the dorsal side of the filament bases (Hartl, personal communication).

We also encounter nectaries in the region of the gynoecium. In the simplest case, as in *Magnolia denudata* (*M. yulan*) and *M.* × *soulangiana*, nectar is secreted from all over the free upper surfaces of the carpels (Daumann 1930b). A similar situation was reported for Lilac (*Syringa vulgaris*), for Privet (*Ligustrum vulgare*) and for the Teaplant (*Lycium*) by Müller (1873), Knuth (1894) and others, and furthermore for *Tofieldia palustris* by Schniewind-Thies (1897), whereas only septal nectaries are found in *Tofieldia calyculata* (see also Sterling 1979). In the

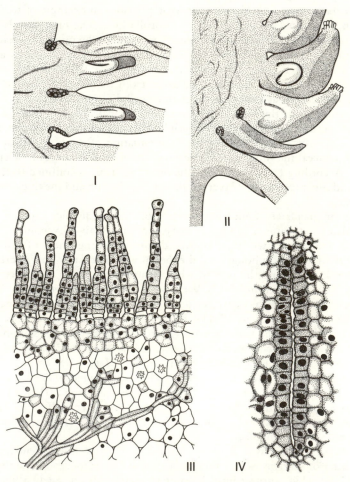

Fig. 102 I, II *Sagittaria sagittifolia*. Axial sections (segment) of the reduced gynoecium of a male flower (I) and the gynoecium of a female flower (II), showing the nectaries lying below the carpels (after Brown). III *Abutilon striatum*, trichomatic nectary on the inner side of the calyx, segment (after Agthe). IV *Hosta* species, septal nectary, cross-section (original).

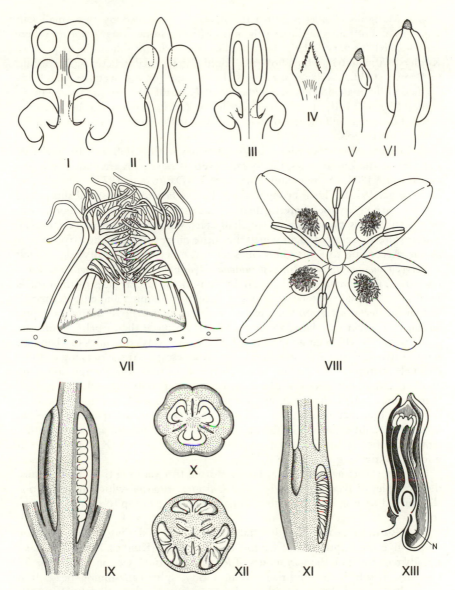

Fig. 103 I-IV Stamens of the Lauraceae with nectaries. I, IV *Cinnamomum camphora*, I stamen with two basal nectaries, IV staminode, almost completely modified by the formation of nectaries; II, III *Laurus nobilis*, staminode (II) and fertile stamen (III), after Goebel. V, VI *Calycanthus floridus*, inner staminode (V) and fertile stamen with feeding bodies (dotted) after Daumann. VII, VIII *Frasera carolinensis*, flower (VIII) and an individual nectary from the petal (after Davies). IX-XII septal nectaries, axial and cross-sections through the gynoecium of *Haworthia reinwardtii* (IX, X) and *Gladiolus* species (XI, XII), after Brown. XIII *Fumaria officinalis*, transverse longitudinal section through the flower, N nectary (after Baillon).

Marsh Marigold (*Caltha palustris*) the nectar is produced by two small hair cushions on both sides of the carpels, and in Biting Stonecrop (*Sedum acre*) there are so-called "nectary scales" placed well down on the outside of the carpels (see Kugler 1970, Fig. 63), which however could also be axial structures. According to Brown (1938), the latter may well also be true for the nectaries which are formed in the male flowers of *Sagittaria latifolia* at the bases of the rudimentary carpels (Fig. 102 I, II). These also occur in the female flowers, but between the bases of the petals and the staminodes as well as between the staminodes and the bases of the carpels.

Among the Monocotyledons, and particularly in the Liliiflorae, Bromeliales, Palmae and the Zingiberales, the so-called **septal nectaries** are widespread (see Grassmann 1884, Schniewind-Thies 1897, Daumann 1970, Silberbauer-Gottsberger 1973). This is brought about by the fact that the septal slits of the hemisyncarpous gynoecia are clothed with a glandular epithelium. For an explanation of this phenomenon we shall start with the example of *Tofieldia palustris* already referred to, in which the entire outer surface of the carpels emits nectar. According to Schiewind-Thies, in a number of Monocotyledons, such as *Polygonatum multiflorum*, *Scilla peruviana*, *Hyacinthus orientalis*, *Aloe barbadensis* (*A. vera*) *Yucca gloriosa* etc., the partial fusion of the lateral faces of the carpels results in a "division of their secretory surfaces into an inner and an outer part", so that the nectar secretion takes place "in three outer grooves along the septa, and in three inner slits, lying along the line of fusion of the carpels". From the septal nectaries which lie within the slits there are always openings or channels, which often lead upwards, so that the nectar can escape. At the opening of these channels therefore the "inner" and "outer nectaries" can merge into one another. Often, however, the glandular tissue covers only the surface inside, so that genuine septal nectaries are formed, which can furthermore become twisted or branched to a greater or lesser extent (*Hemerocallis* species, *Phormium tenax*). The above quoting of the names of the orders has already implied that septal nectaries occur not only in superior ovaries (Fig. 103 IX, X) but also in inferior ovaries (Fig. 103 XI, XII)[*, **].

Septal nectaries have not so far been reported from among the Dicotyledons; the examples of *Buxus sempervirens* and *Cneorum tricoccum* which are occasionally quoted do not, according to Daumann (1974) stand up to critical examination.

We also need to mention the "false nectaries" – "shining bodies of tissue which give the impression of having been wetted" (Kugler, 1970). We have already seen them on the staminodes of *Parnassia palustris* (section 1.5.12, Fig. 70 I to III), in which the fimbriae end in a tiny, spherical, shining head. It is generally accepted (but not without disagreement, see for example Daumann 1960), that these attract insects by simulating nectar, and that these then carry out the fertilisation (see Kugler 1970, Cammerloher 1931). "However, cleverer insects are not deceived by this, whereas stupid ones (flies and beetles), are attracted to it again and again"... (sic Knuth 1894, p. 59)!

---

[*] For the fine structure of septal nectaries and the secretion of their nectar see Benner and Schnepf (1975).
[**] The nectariferous tissue covering the basal carpel surfaces in the gaps between the free carpels of *Butomus umbellatus* could also be interpreted as a preliminary stage of septal nectaries.

2   Morphology of the inflorescence

Maximi autem momenti esse in plantarum vita distributionem florum, et optimos characteres eandem nobis offere e modo allatis exemplis videri jamquidem potest, adhuc magis tamen probabitur inflorescentarium variarum descriptione. Inflorescentiam autem eam vocamus caulium vel ramorum (axium) partem, quae nullos alios profert axes (ramos) ac florigeros. Roeper 1826.

2.1   Definition, descriptive terminology and classification of inflorescences

2.1.1   Definition and delimitation

We are in agreement with Troll (1964), who defines the inflorescence as "the shoot system which serves for the formation of flowers and which is modified accordingly". The forms of the flower-bearing branching systems and their position in the overall structure of the plant not only determine the external appearance of flowering plants to a great extent, but also at the same time provide important criteria for their relationships. The latter is of course only true if identical structural elements are compared when the floral region is considered. It must therefore be established which parts of the flowering shoot system correspond to one another within a particular plant, and between different plants, before any assertions can be made about which modifications distinguish the inflorescence of one plant from that of another, or in what respects they may correspond. The basis for such a procedure is therefore the recognition of general regularities of structure which dominate the diversity of Angiosperm inflorescences. This has become possible for the first time during recent decades by means of the fundamental principles laid down by Wilhelm Troll and his school. These enable us to understand the various forms of inflorescences within larger or smaller taxonomic groups which can be so different in their appearance as modifications of often no more than one basic plan. Quite often we can also recognise further developments of this plan in different directions, so that it is then possible to make certain inferences about the lines of evolution amongst the groups concerned. Without using a typologically based method the attention can easily be distracted during the comparison of inflorescences of any taxonomic group by flower clusters which distinguish themselves by purely external features. This might be for example by a different arrangement of leaves, a more complex branching or by the elongation of certain portions of the axes. Such flower bearing parts may then be recognised as "units" within the general architecture and then be compared to one another as "inflorescences", often of course without noteworthy success.

The distinction between the inflorescence and the "vegetative region" is very often made by means of the differences in the foliage, as for instance by Goebel

(1931, p. 1): "the floral region becomes different from the vegetative, the floral leaves are developed as bracts instead of foliage leaves, or else are completely lacking. We then use the term 'inflorescence' to distinguish the flower-bearing part of the plant from the vegetative part." On the other hand Goebel also then speaks of "flowering shoots" when the flower-bearing portion is not sharply differentiated from the vegetative zone because its leaves are all of the same type. Such a distinction is artificial however. To see this, let us examine the inflorescences of various *Veronica* species where we have: racemes, in which the floral leaves can be either bract-like (bracteose) or like normal (frondose) leaves; in addition there are species in which there is a transition from one kind to the other within the (frondose-bracteose) raceme. The distinction usually made when determining *Veronica* species: "flowers solitary in the axils or in terminal racemes", can therefore be regarded as suitable for the construction of diagnostic keys. A **typological** classification of different inflorescences is, however, not possible on the basis of differences in the formation of the leaf organs. The same is also true for the delimitation of the inflorescence and the vegetative zone: all the possible intermediates between bracteate, frondose-bracteate and frondose inflorescences can be found (Fig. 104). The so-called "flowering shoot" is therefore to be interpreted as a frondose inflorescence. This is all the more so when, as established by Troll (1950a), "the floral leaves, when they occur in the form of foliage leaves, often also exhibit characters of bracts" ("frondose bracts", Troll 1964, p. 6). It is not uncommon for the leaf organs in the inflorescence zone (e.g. as in woody plants) to occur as "mere diminutive" (frondulose) "forms of the foliage leaves".

A review of the terminology which is indispensable for the description of inflorescences and the use of these terms in the appropriate literature is given in papers by Rickett (1944, 1955).

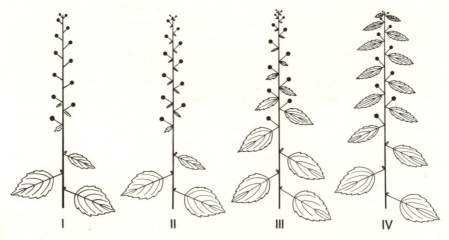

Fig. 104 Various types of subtending leaves (pherophylls) in racemose inflorescences. I, II Bracteate (I) or substantially (distally) ebracteate (II) racemes, III frondose-bracteate, IV frondose racemes. After Troll.

2.1.2 Determinate and indeterminate inflorescences, distinction according to branching types

The distinction between determinate (open) and indeterminate (closed) inflorescences, which goes back to Roeper (1826), has proved to be an important one. If the main axis of the inflorescence ends in a flower (terminal flower, Fig. 117 I, Fig. 122 II), then the inflorescence is called a **determinate** inflorescence. In this case, the growing point of the inflorescence is transformed into a floral apex when the terminal flower is initiated, and from then on only produces the organs of the terminal flower instead of further bracts and side branches. It is not unusual to find this condition expressed by having the organs of the terminal flower put in the position continuing the phyllotaxis of the preceding bracteose leaf organs (on this point see also the evidence of Wydler and the Bravais brothers which is cited by Troll 1964, p. 9). On the other hand, an **indeterminate** inflorescence is one where the growing point only produces lateral flowers or partial inflorescences*, and often the last of them never complete their development (Fig. 107 I). If all the flowers which are initiated complete their full development, then the inflorescence axis ends abruptly between the uppermost partial inflorescences (Fig. 105 I) or lateral flowers (Fig. 164 V). In this case the uppermost lateral flower, or the uppermost partial inflorescence can become erect and align itself with the general direction of the main axis, so that the inflorescence axis appears to be continued, and the presence of a terminal flower is simulated. However, the lateral origin of this flower or partial inflorescence is often still recognisable by the presence of the appropriate subtending leaf (sometimes also by the occurrence of the prophylls of the lateral axis), and often also by means of the visible remains of the rudimentary main axis which often appears as a small projection (Fig. 105 II, III). Troll (1964, p. 25) distinguishes such **subterminal** flowers, whose lateral origin is still clearly visible, or when at any rate this can be established by studying the ontogeny, from **pseudoterminal** flowers, where "the ontogeny yields no evidence" for their lateral origin, "although comparative studies give reasons why they cannot be awarded the status of a terminal flower". The presence or absence of a terminal flower is frequently characteristic even for the larger families, and – as we shall see – of great importance for the typology of the inflorescence.

The type of branching of an inflorescence is used to a considerable extent for its characterisation and classification. **Simple** and **compound** inflorescences can be distinguished, which is to say that we will describe inflorescences as **simple** when "the branching does not extend beyond the first order" (Troll). Thus inflorescences can be distinguished according to their mode of branching as follows:

I   simple inflorescences: raceme, spike, spadix, umbel, capitulum;
II compound inflorescences: double raceme, double spike, double umbel, panicle, thyrse.

---

* Here therefore only the axes of second and higher orders terminate with the formation of a flower. A distinction used to be made between single and multiple axis systems, depending on the number of axes which are formed before axes terminate with flowers (Saint-Hilaire 1840, Braun 1842, Wydler 1844).

We shall now examine these modes of branching systems more closely, using the illustrations in Fig. 106 and Fig. 109.

2.1.3   Simple inflorescences

Characteristic properties of the **raceme** (or botrys, Fig. 106 I) are that the internodes of its main axis are clearly developed and that all flowers have stalks.

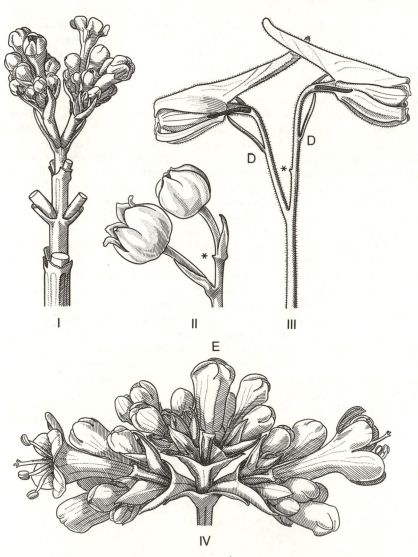

Fig. 105 I, IV *Valeriana officinalis*, tip of the inflorescence without (I) and with (IV) abnormal terminal flower E, this flower has two short spurs, instead of one as normally. II *Convallaria majalis*, tip of the inflorescence with vestigial growing point at *, III *Delphinium elatum*, similarly, the subtending leaves (D) of the upper lateral flowers are shifted upwards (recaulescent) on the pedicels. I, IV, original, II, III after Troll.

It does not terminate with a flower, so its apex remains indeterminate. The position of leaves on the rachis and consequently the arrangement of the flower-bearing lateral branches can be either opposite or alternate, as in all the other branching forms discussed below. The order in which the flowers open usually corresponds to the order in which the flowers are developed, thus starting with the lowermost flower and proceeding upwards. The pedicels are often provided with one or two prophylls, whose topological arrangement we have already discussed in section 1.2.1. It is not unusual for a progressive reduction of the bracts to be observed, which in any case are often inconspicuous. Such a reduction can be found for instance with the flowers which are inserted higher up on the axis, but also in the basal flowers. The same remarks also apply to spikes and umbels.

The **spike** (Fig. 106 II) is also without a terminal flower. It is distinguished from the raceme by the sessile flowers placed in the axils of the bracts. It does nevertheless happen from time to time that the lower flowers of a spiked inflorescence have more or less distinctly developed pedicels, as in the orchid *Spiranthes* (Irmisch 1853). There are therefore intermediates between the raceme and the spike.

The **umbel** (or sciadium) is distinguished from the raceme by the compression of the rachis (Fig. 106 IV), which is compensated here by the fact that the pedicels of the flowers which now radiate from one central point are greatly elongated. The subtending bracts of the flowers in the umbel are also found crowded together at this point. Their number is indeed often less than the number of flowers. For these reasons it is not always possible to establish without further investigations whether the umbel is indeterminate, like the raceme, or whether a terminal flower is present. In the latter case we are not concerned with a "compressed raceme", but with a compressed botryoid (see section 2.1.4) – perhaps even with a pseudosciadioid, a "contracted cymoid", as in the Jagged Chickweed (*Holosteum umbellatum*). The relationship between the form of the raceme and the umbel appear clearly in the inflorescences of the

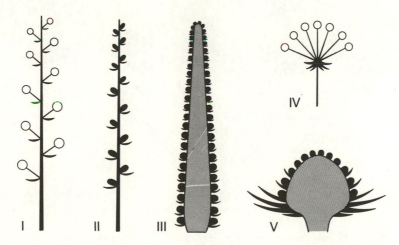

Fig. 106 Branching patterns in simple inflorescences. I Raceme, II spike, III spadix, IV umbel, V capitulum.

genus *Iberis* (Candytuft) of the Cruciferae. In this instance the internodes of the racemose inflorescence remain very short until after anthesis of the flowers, whilst from early on the pedicels are so much elongated that at the moment when the flowers open they are elevated to the same level as the not yet opened flower buds inserted further up the axis, or even somewhat exceed the apex of the raceme (Fig. 108 I). The display effect of what has become a rather umbel-like inflorescence is increased by the fact that the two petals of each flower which point downwards or outwards from the inflorescence become larger and, especially, longer, than the two which are directed towards the centre of the inflorescence (Fig. 108 II). However, the elongation of the internodes of the rachis between the individual flowers does not begin until the flowers wither, so that later on the inflorescence takes on the form of a raceme (Fig. 108 III). If the elongation of the axis does not take place, then the umbellate form of the inflorescence remains until fruiting time, as is the case in *Iberis umbellata* (Fig. 107 II).

In the spike-like **spadix** (Fig. 106 III) the axis of the inflorescence is much thickened, and in the **capitulum** (or cephalium, Fig. 106 V) it is shortened as well, and may be broadly conical or even a flattened disc. In many cases, as for example in the Compositae or the Dipsacaceae, the base of the capitulum is surrounded by an involucre of numerous sterile leaves (phyllaries) arranged in a rosette, which are not to be confused with the subtending bracts of the flowers within the head. A capitulum which is provided with an involucre in this way used to be known as a **calathidium**, in the Compositae at least. Inflorescences can of course also assume the form of a head simply because of congestion of the flowers caused by the inflorescence axis failing to elongate, without the axis being thickened as well. The capitate form of the racemose inflorescence, for

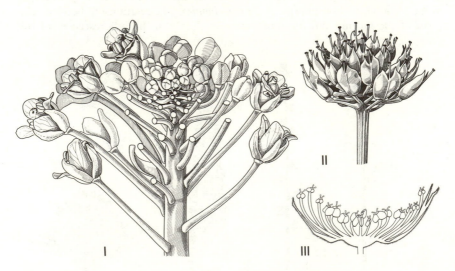

Fig. 107 I *Capsella bursa-pastoris*, apex of inflorescence. II *Iberis umbellata*, inflorescence after flowering. III *Didiscus caeruleus*, umbel in lateral section. II, III after Troll.

example, of the clovers (*Trifolium*) or of many species of *Acacia* is due to compression of this kind. The flowers can also be arranged in heads in compound inflorescences which, however, often only lasts until the beginning of anthesis, but often persists until fruiting time as well. On the other hand, a thickening of the inflorescence axis in the region where the flowers are initiated can even take place in umbellate inflorescences, as for example in *Didiscus caeruleus*, whose umbels bear a certain resemblance to the capitula of the Compositae (Fig. 107 III).

2.1.4   Compound inflorescences

If the individual flowers of the simple inflorescences (raceme, spike, umbel and capitulum) are replaced by a complete inflorescence of the same branching pattern, then certain forms of the compound inflorescences, the **double raceme** (diplobotryum, Fig. 109 I, IV), **double spike, double umbel** (Fig. 109 II) and the **double capitulum** are obtained. This type of branching can be repeated several times, so that third–order inflorescences (e.g. triplobotrya) and, finally, inflorescences of multiple order (e.g. pleiobotrya) can result. In each case we are concerned with compound inflorescences with multi-axial branching systems. The individual racemose, spicate, umbellate or capitate elements of the

Fig. 108 I-III *Iberis sempervirens*. I and II young inflorescences seen from the side (II) and from above (II). III a further stage in the development of the inflorescence; after Troll.

compound inflorescences are termed **partial inflorescences**, as are also the cymose branches of a thyrse (see below).

We encounter double racemes after the style of Fig. 109 I in many species of the Papilionaceae. The Melilot (*Melilotus officinalis*) and the Hop Trefoil (*Trifolium campestre*) will serve as examples of this. In *Trifolium campestre* (Fig. 110 II) the lateral position of the long-stalked, capitate, compound racemes in the axils

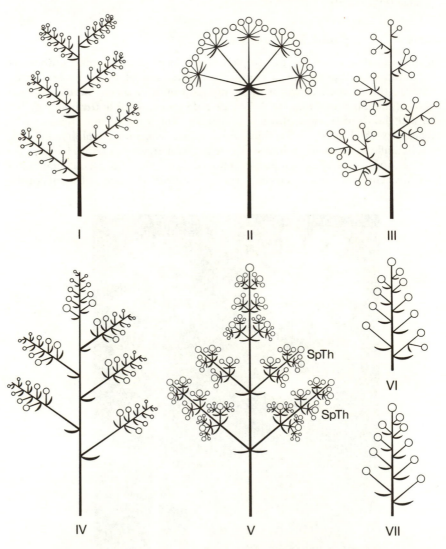

Fig. 109 Types of branching in compound inflorescences. I Homothetic double raceme, II double umbel with 5 partial umbels, III panicle, IV heterothetic double raceme, V determinate (double) thyrse or dithyrsoid, SpTh partial thyrsoids, VI, VII highly depauperate forms of panicle, VII botryoid. By reference to Troll.

of the upper leaves is readily recognised. Evidently the growing point of this double raceme exhausts itself in the differentiation of lateral racemes and their bracts, without itself terminating in a racemose inflorescence.

There are, however, many clover species which give the impression that the inflorescence axis terminates with a capitate, condensed raceme, as for example in the Crimson Clover (*Trifolium incarnatum*, Fig. 110 III). This comes about because the completion of a double raceme is limited to the formation of a single raceme of lateral origin, whose hypopodium becomes erect in such a way that it appears to continue the main axis of the flowering shoot, whilst the growing point is displaced laterally and becomes rudimentary (IV). Sometimes also such apparently terminal heads consist of two lateral racemose partial inflorescences, which have been brought together by compression of their hypopodia and of the internode of the main axis between their branching points. This regularly occurs, for example, in the inflorescences of Red Clover (*T. pratense*, V); for further details see the commentary by Troll (1957, p. 295ff., 1964, p. 43ff.).

The forms of the second- or multiple-order racemes illustrated here in the diagram in Fig. 109 I and by the inflorescence of *Trifolium campestre* (Fig. 110 II) are characterised by the fact that they are composed only of lateral racemes (or double racemes), without the main axis of the inflorescence itself ending in a raceme. Forms of inflorescences such as these are termed **homothetic** by Troll, in contrast to those which have terminal racemes as well as lateral ones, which are called **heterothetic compound racemes** (Fig. 109 IV). Examples of this are found in the genus *Hebe* (Scrophulariaceae, e.g. *H.hulkeana*), or in several members of the Cruciferae, such as *Isatis tinctoria* and species of *Crambe*. The differences mentioned here between homothetic and heterothetic branching systems in compound racemes are again found in corresponding modifications of inflorescences constructed on the models of the double spike, double umbel and double capitulum.

A more detailed discussion is necessary for the panicle and the thyrse. The **panicle** (Fig. 109 III) is characterised by the fact that the main axis of the inflorescence is terminated by a flower, and similarly also for all the lateral axes. The degree of branching increases more or less regularly downwards from the uppermost lateral single flower below the terminal flower, so that the complete inflorescence has a conical outline, or at least primarily so. The inflorescence of *Polemonium caeruleum* (Fig. 134 I) may be used as an example of this.

In a panicle, or in inflorescences derived from a panicle, as well as in a determinate thyrse (see below), the terminal flower assumes a dominating position. One way in which this can be expressed is when this is the only flower which comes to maturity, in cases where the inflorescence is impoverished or reduced, which may be characteristic of the species, or be determined by modifications, such as those caused by starvation (section 2.2.7), or by inadequate stimulus for flowering (section 2.2.7). Further evidence of this fact is that the terminal flower is often the first to open, or at least that it opens before its neighbouring lateral flowers. This "precursive" opening of the terminal flower is at least partially explained by the fact that its development precedes that of the nearby laterals. As we have already established in section 2.1.2, the growing point of the inflorescence undergoes a conversion when the terminal flower is initiated as a floral apex, and now directly produces primordia of floral organs, whereas previously the organs of the lateral flowers had to be developed from growing points which originated in the axils of the bracts differentiated from the same apex as the organs of the terminal flower. Ontogenetically therefore the terminal flower holds the lead over the upper lateral flowers.

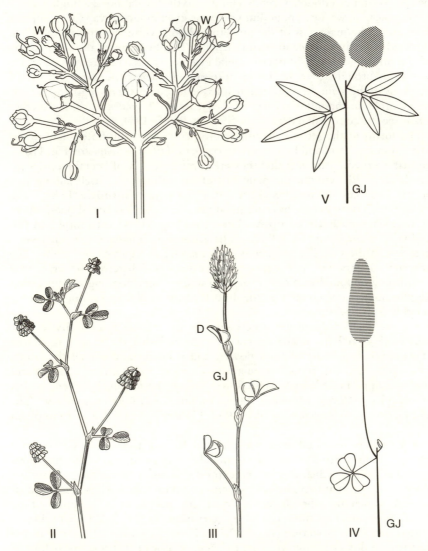

Fig. 110 I *Scrophularia umbrosa*, dichasial partial inflorescence, the branches partly tending towards a scorpioid cyme. II-V inflorescences of *Trifolium*, IV and V schematic. II *Trifolium campestre*, double raceme, III *T.incarnatum*, flowering shoot with racemose partial inflorescence, D subtending bract, GJ basal internode, IV corresponding branching scheme, V *T. pratense*, branching scheme of a double-racemose inflorescence with two racemose partial inflorescences approaching one another. After Troll, V modified.

The terminal flower can also differ in form from the lateral ones, particularly in respect of the number of its parts. We may recall the well known example of *Ruta graveolens* with its 5-merous terminal flower and 4-merous lateral flowers (Fig. 11 II, III). The Bogbean, *Menyanthes trifoliata*, possesses a 6-merous terminal flower and 5-merous laterals.

According to whether the arrangement of the pherophylls and their corresponding panicle branches is decussate or alternate, decussate and alternate types of panicle can be distinguished. Those panicle branches just below the terminal flower which are limited to a single terminal flower are often called **monads**, and the 2- and the 3-flowered branches **dyads** and **triads**, respectively. The branches which include numerous flowers, and which are themselves further paniculate-branched, can also be referred to as **sub-panicles**.

The conical outline of the panicle can be modified in such a way that all the flowers are elevated to lie at or approaching the same level as the terminal flower by corresponding elongations of the lateral axes of different orders, so that taken collectively they form an umbellate flat or curved surface (**corymb**, Fig. 111 I). As examples of this we have the inflorescences of the Rowan (*Sorbus aucuparia*) and of many species of *Viburnum*. If there is a preferential development, especially of the lowest and outermost inflorescence branches, so that all their terminal flowers "overtop" their parent axes, the inflorescence can even become completely "inverted" (Fig. 111 II), as for example in the Meadowsweet (*Filipendula*). Such an inflorescence is termed an **anthela**.*

An impoverished panicle which still possesses its terminal flower can resemble a "determinate raceme" (Fig. 109 VII), but should better be termed a **botryoid** (the equivalent for a spike is called a **stachyoid**). Its paniculate nature can make itself apparent by the fact that its lowest branches sometimes bear two or three flowers instead of one (as for example in Bogbean, *Menyanthes trifoliata*, or as in *Amelanchier alnifolia*, see Fig. 109 VI).

The **thyrse**, by contrast to the panicle, is defined as an inflorescence "with cymose partial inflorescences". By "**cymose branching**" is meant a branching exclusively from the axils of the prophylls, which are developed as the only leaf organs preceding the individual flowers. They usually, as in the dicotyledonous plants (and in some monocots), occur in pairs and inserted in a more or less transverse fashion (opposite or alternate).

The branching type of the thyrse may occur either in a determinate form, where the inflorescence is provided with a terminal flower (**determinate thyrse, thyrsoid**, Fig. 109 V, 112 I), or in an indeterminate form (**indeterminate thyrse**, thyrse s. str., Fig. 112 II).

The terminal flower of a cymosely branched partial inflorescence may also be referred to as the **primary flower** (P in Fig. 113 I), and those ending the branches which arise from the axils of its prophylls are the **secondary flowers** (S).

In more detail, it is necessary to distinguish a number of very characteristic forms from among those with cymose branching. If the branching takes place regularly out of the axils of both prophylls, then the result is a **dichasium** (Fig. 113 I, II), where the branches originating in the axils of the prophylls continue the same manner of branching and mostly exceed the parent axis.

---

* The German words for this are rather apt, viz. "Trichterrispe" = funnel-panicle (Radlkofer 1891 p. 180), and "Spirre", which is an anagram of "Rispe" = panicle (F. K. Mertens in Meyer 1819).

Instead of this dichasial ramification, the branching at higher orders is often **monochasial**. This means that the ramification continues out of only one of the axils of corresponding prophylls, as can be seen to some extent in the partial inflorescence of *Scrophularia umbrosa* illustrated in Fig. 110 I. When the branches which grow out of one another develop alternately from the left and the right hand prophyll axils, then a **cincinnus** (or scorpioid cyme, Fig. 113 III) is formed. If the branching always proceeds from the left or right hand axil only (in each case related to the median which passes through the subtending leaf and the (respective) original axis), then the result is a **bostryx** (or helicoid cyme, Fig. 113 IV). When the corresponding arrangement occurs on both branches of a partial inflorescence which has dichasial branching to begin with, then this is referred to as a **double cincinnus** (or double scorpioid cyme, as in many Labiatae or the Boraginaceae) or a **double bostryx**, or double helicoid cyme (*Hypericum perforatum*).

In many monocotyledonous plants, and in several dicots as well, only a single so-called **"addorsed" prophyll** occurs, which is positioned on the adaxial side of the axillary shoot, i.e. between the parent axis and the lateral axis (see section 1.2.1). In this case the branching mode gives the form of a cymose partial inflorescence known as a **rhipidium** (Fig. 113 V, Fig. 114 I), as occurs for example in *Iris* species. The axes of this are all oriented in the same vertical plane, which is the median plane corresponding to the relevant subtending bract and the axis of the inflorescence. The latter is also true for the **drepanium** (Fig. 113 VI, Fig. 114 II) as produced by the rushes (*Juncus*), in which, however,

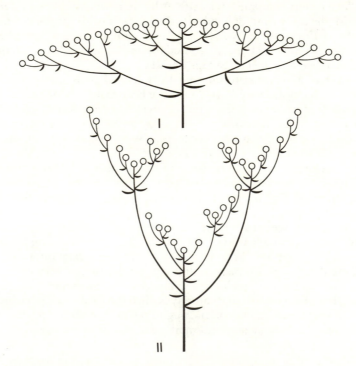

Fig. 111 Schematic outlines of a corymb (I) and an anthela (II) (after Troll).

the continuing shoot which ends in a flower proceeds from out of the axil of a second leaf opposite to the addorsed prophyll (not depicted!), so that all the axes are placed on the side which is turned away from the primary axis.

If it is desired to express the difference between the cincinnus and the bostryx graphically in the plane of a sheet of paper, a ground plan diagram has to be used, as is done in Fig. 113 III, IV, in order to reproduce the different spatial arrangement of the successive axes. In the usual longitudinal diagram (Fig. 114 II), however, there is no remaining difference between these two types of branching, and furthermore, without considering the number of leaves, the difference from the drepanium would not be evident either. It is possible, of course, to try to illustrate the zigzag branching of the cincinnus by introducing the ground plan into the vertical plan (Fig. 114 III), then, however, a configuration is obtained which corresponds to the side view of the rhipidium.

These attempts at illustration may suggest to us relationships between the different cymose branching types, as was already partly recognised by Buche-

Fig. 112 Diagrams of a thyrse, I determinate thyrse (thyrsoid) with decussate, II indeterminate thyrse (s. str) with alternate arrangement of the leaves. E terminal flower, vb prophylls, h hypopodium, e epipodium, P primary flower, S secondary flower. After Troll, modified.

nau (1865), who also originated the German terms for rhipidium and drepanium. Whereas Buchenau compared the rhipidium with the cincinnus and the drepanium with the bostryx, Eichler (1875, p. 39) saw the relationships between the rhipidium and the bostryx on one hand, and between the cincinnus and the drepanium on the other, as illustrated in Fig. 114 IV to VI.

According to Müller-Doblies (1977) not only the possibility of a gradual transition from the cincinnus to the drepanium can be demonstrated by a straightforward geometric elongation (as was done by Eichler, in Fig. 114 VI; Fig. 114 IXc to e), but also, on the other hand, a transition between the cincinnus and the rhipidium, by compression of the cincinnus, as in Fig. 114 IXc to a. Since it is also possible to arrive at a rhipidium via a tightly coiled bostryx from a normal bostryx (Fig. 114 VIIIc to a), the overall relationship between the branching types can be illustrated by means of the diagram in Fig. 114 VII.

According to Müller-Doblies, certain stages in the transition from the bostryx to the drepanium can be seen in the monochasial partial inflorescences of some genera of the Amaryllidaceae (progressive "opening" of the bostryx as in Fig. 114 VIIIc to d in species of *Crinum*). On the other hand, *Clivia* and *Lapiedra* (Müller-Doblies 1978) form a flat cincinnus (Fig. 114 IXb). Corresponding transitions from the rhipidium to the cincinnus can be observed amongst the Amaryllidaceae in the branching of the inflorescence of *Leucojum autumnale*, in which each flowering shoot (paracladial scape with coflorescence, see section

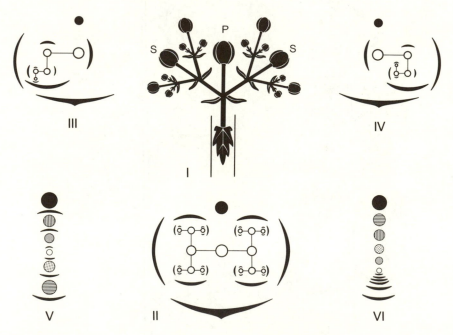

Fig. 113 Forms of cymose branching. I, II Dichasium, general view (I) and in a diagram (II), III cincinnus, IV helicoid cyme, V, VI rhipidium, and drepanium respectively; however, for the outermost branches (flowers) the bract is still shown, but without any continuation in the axil. By reference to Troll.

2.2.1) proceeds from the axil of the single prophyll of the preceding axis. Each of these prophylls is "displaced from its usual addorsed position in such a way that a flat cincinnus is created" (Müller-Doblies 1977, p. 358; see also 1972, p. 678, Fig. 9); the branching of the inflorescence of *Elegia capensis* (Restiona-ceae) is interpreted in the same way. According to Müller-Doblies, flat bos-tryces occur also in *Allium*.

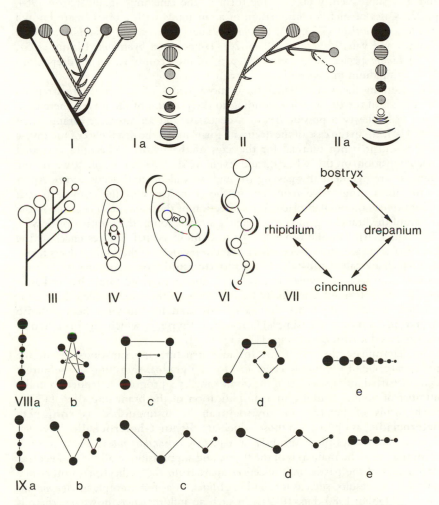

Fig. 114 Monochasial branching in the inflorescence, I, Ia diagrams of rhipidium and II, IIa drepanium, seen from the side and from above, respectively. III Lateral view of a cincinnus, in which the imposition of the zigzag branching of the ground plan makes it appear in the same form of a rhipidium in side view. IV Ground plan of the rhipidium, V helicoid cyme in transition to a rhipidium, VI cincinnus in transition to a drepanium (IV–VI after Eichler). VII Scheme showing the geometrical relations between the four types of monochasium VIIIa–e sequence from the rhipidium (a) to the drepanium (e) via the helicoid cyme (bostryx; c, d). IX Sequence from the rhipidium (a) to the drepanium (e) via the scorpioid cyme (cincinnus; b–d); VII–IX after Müller-Doblies, drawings partly altered.

Müller-Doblies found examples of the transition from the rhipidium to the bostryx before anthesis in the floriferous bulbs of many Scilloideae e.g. in the genera *Scilla, Ledebouria* and *Drimiopsis* and in the tuber of *Hypoxis* (Hypoxidaceae).

Apart from these transitions between the cymose branching types, there are also examples of so-called "mixed sympodia" (Wagner 1901), in which "an inflorescence which begins in the form of a cincinnus is modified to a bostryx, and may subsequently take on the form of the cincinnus again" (Troll 1964, p. 32). Cases of such a combination of a cincinnus with a bostryx are known in *Phlox paniculata* (Wagner), *Allium* (Weber 1929, p. 5), *Leucojum aestivum* (Luyten and van Waveren 1938, p. 26), *Hippeastrum hybridum* (Blaauw 1931, p. 80), *Linum* species (Schumann 1889, p. 573), and a combination of a drepanium with a rhipidium in *Saruma henryi* (Wagner).

Concerning the way in which the cymose partial inflorescences are arranged on the axis of the thyrse corresponding to the position of the leaves, there exist once again the two possibilities of a decussate and an alternate arrangement (Fig. 112 I, II). In the case of the decussate form and in the distal part of the thyrse or thyrsoid it is not unusual for the axes of the whorl to become separated (decomposition of the whorls), and even to show a transition towards the alternate arrangement (by passing through so-called spiral decussation). Apart from this, a progressive reduction in the complexity of the cymose partial inflorescences often takes place in the direction of the apex, i.e. a decrease in the extent of the branching and consequently a reduction in the number of flowers (Fig. 112 I, II). On the other hand the cymose partial inflorescences in the proximal zone are often preceded by branches which are themselves thyrsiform, and which are called **sub-thyrses** or **sub-thyrsoid** (corresponding to the sub-panicles, this section, above). In this way a double thyrsoid (diplothyrsoid as in Fig. 109 V) is formed, and finally by repetition of this mode a pleiothyrsoid is reached. Troll distinguishes thyrses of this kind by the term **heterocladic thyrse**, in contrast to the (simple) homocladic thyrse, in which the formation of sub-thyrses does not occur (Fig. 112, Fig. 115 I).

In determinate thyrses of heterocladic structure the sub-thyrsoids in the proximal zone of the inflorescence can form a gradual transition to the purely cymose partial inflorescences of the distal zone by a progressive reduction in the number of nodes, resulting in a simplification of the branching (Fig. 115 II). Such forms of heterocladic thyrsoid can be distinguished as **conjunct-heterocladic**, as opposed to those which are **disjunct-heterocladic**, in which this transition takes place abruptly (as e.g. in *Centranthus ruber*). In the case of compression of the main axis of the thyrse and suppression of all the consecutive lateral axes of the partial inflorescences apart from the pedicels, an umbel-like inflorescence results, such as is found in *Allium*, or as for example in *Agapanthus* (Müller-Doblies 1980, Fig. 19–21). In such an inflorescence, however, there is evidence that it is really a "compressed" thyrse (thyrsoid). This is also demonstrated by the order in which the flowers open.

2.1.5   Metatopic displacement

It is not unusual for the flowering branches to be further modified by what is called **metatopy**, i.e. displacement and fusion of organs, which can be determined to a great extent by congenital processes. The phenomena of concaulescence, recaulescence and anaphysis arise from simple changes of proportions during the axillary branching, and can be explained by reference to the diagrams in Fig. 116. If the primordium of an axillary shoot is situated somewhat more towards the main axis, it can grow upwards for a distance in common with the relevant internode of the main axis, and be moved up above the point of insertion of its pherophyll. In such cases of **concaulescence**, the lateral axis appears to arise in "extra-axillary" fashion from the main axis (Fig. 116 II). Numerous examples of this are found in the inflorescences of the Boraginaceae, Hydrophyllaceae, Solanaceae and Polemoniaceae (see Fig. 140 II, III, Fig. 157 I). In **recaulescence** the axillary bud is shifted for some distance towards the base of the subtending leaf, the insertion of which is displaced on the branch for a smaller or greater distance above its original position, after stretching of the

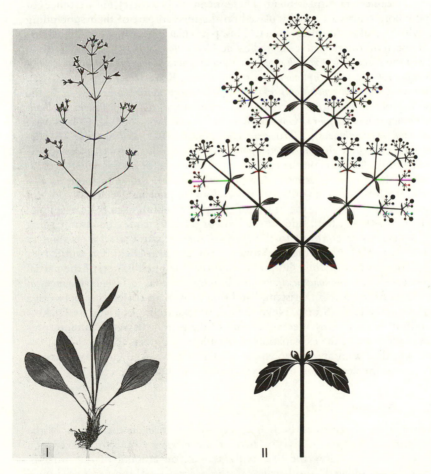

I   II

Fig. 115 I Flowering plant of *Valeriana saxatilis* as an example of a monothyrsic inflorescence without a terminal flower; II diagram of the structure of a pleiothyrsus without terminal flowers, as occurs in many of the Valerianaceae (*Valeriana officinalis*), after Weberling 1961.

common basal zone of both organs (Fig. 116 I). A well known textbook example of this is the inflorescences of species of *Thesium*, in which the degree of recaulescence (corresponding to the early initiation of the axillary buds) undergoes an increase with the distance up the axis (progressive recaulescence; regressive recaulescence occurs in *Samolus valerandi*). We speak of **anaphysis** if an axillary bud with its subtending bract "is moved up to a position above the bract which follows it genetically" (Troll 1964, p. 132, see Fig. 116 III). This is quite frequently observed in the Cruciferae in the transition region between the vegetative zone and the inflorescence, (Fig. 116 IV) and often in the partial inflorescences of the Lime tree (*Tilia*).

2.1.6   Epiphyllous inflorescences

The so-called **"epiphyllous inflorescences"** may also owe their special location to a recaulescent displacement. These intrinsically axillary inflorescences or partial inflorescences are usually placed on the median part of their subtending leaf, which is often a foliose bract. This particular phenomenon, which has already been thoroughly investigated by de Candolle (1890), can be met with in many of the Flacourtiaceae (*Phylloclinium paradoxum, Moquerysia multiflora*, Gilg 1925, Fig. 196, p. 432, *Phyllobotryum soyauxianum*, Keay 1954, Fig. 69, p. 163); Meliaceae (*Chisocheton pohlianus*, Harms 1917, where there is a review of all the cases of epiphyllous inflorescences known at that time); Dichapetalaceae (*Chailletia, Dichapetalum, Tapura, Gonypetalum*, see Stork 1956, Fig. 1); Celastraceae (*Polycardia phyllanthoides*); Cornaceae (*Helwingia*); the Saxifragaceae (*Phyllonoma*), and elsewhere. In *Chisocheton pohlianus* epiphyllous inflorescences appear as "small tufts of flowers on the upper side of the rachis of the pinnate leaves..." "between the opposite leaflets". On the other hand, the inflorescences also occur in the form of "lax axillary narrow elongate panicles" (Harms). Since in *Helwingia* and *Phyllonoma* the bracts of the partial inflorescences bear stipules which are inserted at both sides of the base of the leaf, it may be assumed that the recaulescence is confined to the median zone of the bract (Weberling in Troll 1961, p. 116). Nevertheless, Stork (1956), and also Dickinson and Sattler (1974), argue as de Candolle did previously, that the initialisation of the partial inflorescence in *Phyllonoma ruscifolia* and *P. integerrima* (as in *Helwingia japonica*) is produced from the leaf meristem, "and that there is no coalescence between a floral axis and a leaf" (Stork). Nevertheless, the photographs given by Dickinson and Sattler illustrating median sections of the primordia permit an interpretation in the sense of the explanation of recaulescence given above.

We will deal with an additional form of metatopy, known as **syndesmy**, later on after the description of the typology of the inflorescences.

2.1.7   Accessory buds

The formation of accessory buds is especially common in the region of the inflorescence. As is well known in dicots these accessory buds are placed in a sequence above or below the first axillary buds, on account of the order in which they are initiated. In other words, the order is serial, upwards or downwards. In monocots it is always lateral (collateral) beside the first bud (for further details see the thorough treatment by Sandt 1925). These accessory buds may also develop into flowers, or remain purely vegetative. In as far as they come to flower during the process of flowering they may alter the overall appearance of the inflorescence to a great extent (see Fig. 140, 141).

2.1.8 "Racemose" and "cymose" inflorescences

Since Hofmeister (1868) it has been customary to classify inflorescences as **racemose** or as **cymose**. Those inflorescences which are characterised by a continuous main axis are termed **racemose**. The cymose inflorescences on the other hand lack "a *homogeneous* main axis which dominates the lateral branches" (Fitting in Strasburger 1942) since this completes its development with a terminal flower, whilst the lateral axes continue the branching system. Its branching system is therefore constructed with a sequence of axes which each terminates with a flower ("centrifugal inflorescence"). If it continues with more than two distal lateral branches, each inserted below the terminal flower, then this is called a pleiochasium (*Sedum* species, Fig. 117 III; *Sempervivum*), and if there are

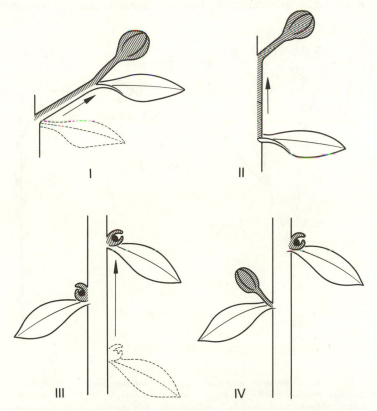

Fig. 116 Diagrams for the explanation of recaulescence (I), concaulescence (II) and anaphysis (III); anaphysis makes it possible for vegetative buds to still appear in the floral region (IV). After Troll.

only two branches (usually opposite one another) then the branching system is called a dichasium (Fig. 117 I) and, if there is just one, then it is a monochasium (Fig. 117 II). In the latter case the consecutive branches are often aligned more or less in the direction of the main axis, so that their respective terminal flowers are pushed to the side, hence creating the impression that these are solitary axillary flowers.

The straightening of such a sympodium, which leads to the simulation of a homogeneous main axis, begins to take place secondarily after the differentiation of the individual members of the cincinnus, each of which consists of one flower with its prophylls. In this case therefore it takes place postgenitally. Straightening can also, however, take place much earlier, namely at the same time as the initialisation of the individual flowers, or even almost preceding this, so that the impression of a monopodial structure is already given in the early stages. In this **"homaxone"** or **"congenital cincinnus formation"** (Troll) there occurs an abbreviated process of development, in which the flowers still of course appear in the arrangement typical of the cincinnus. Since this type of cincinnus formation is very characteristic of the inflorescences of the Boraginaceae, the term **boragoid** has been coined for it.

A distinction as made above between racemose and cymose inflorescences

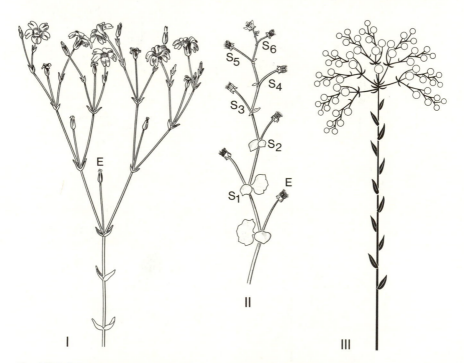

Fig. 117 I *Cerastium arvense* as an example of a cymoid of the dichasial type; II cincinnal cymoid of *Saxifraga cymbalaria* (monochasium), E its terminal flower, S^1, S^2, S^3, etc., are the "sympodial members" of the cymosely branched partial inflorescence growing out of each other, and each provided with a pair of prophylls and a terminal flower (after Troll, I altered). III Branching pattern of an inflorescence of *Sedum glaucum* as an example of a pleiochasium (original).

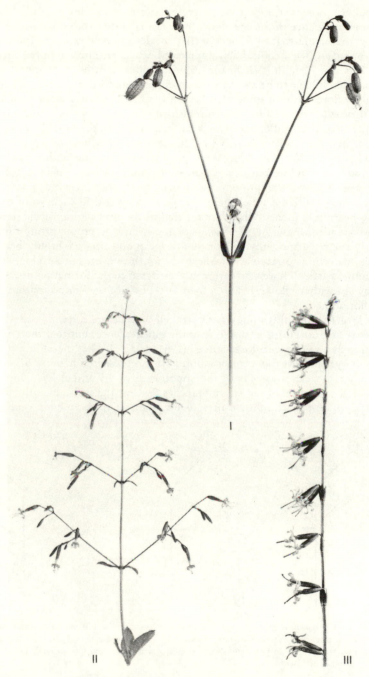

Fig. 118 Inflorescences of *Silene* species, I *S.vulgaris*, II *S.nutans*, III *S.tatarica*. After Troll.

does not, however, reflect the natural relations. According to this classification the inflorescence of *Silene vulgaris* (Fig. 118 I), a dichasium, would be classified as cymose, and that of *S. tatarica* (III), a "determinate raceme" (botryoid) and of *S. nutans* (II), a conical diplothyrsoid, both as racemose inflorescences.

It is bad enough, that in this case the inflorescences from members of the same genus have to be associated with different inflorescence types! The "dichasial" inflorescence of some members of the Caryophyllaceae or as in the genus *Valerianella* (Fig. 119 II), is even transformed into a thyrsoid (or diplothyrsoid) – i.e. a racemose inflorescence – when in addition to the distal pair of flowering branches, the axillary branches of the next lowest whorl of leaves manage to flower (Fig. 119 I, Fig. 120). What we are really dealing with in the difference between the inflorescences of *Silene nutans* and *Silene vulgaris* is a difference of degree of promotion in the development of the ramification of a thyrsoid: in *Silene nutans* there is a basi-mesotonic promotion (the lowermost of the flowering-bearing branches, which are situated in the mid region of the stem are the most vigorous) and in *Silene vulgaris* an extreme acrotonic promotion, so that only the uppermost thyrse branches develop and become highly branched. In the "determinate raceme" of *Silene tatarica*, however, the development of the cymose partial inflorescences remains restricted to the formation of the terminal flower [primary flower]. This is therefore a case of the "impoverishment" of an inflorescence.

Troll has therefore quite rightly stressed that: "there certainly exists a **cymose form of branching** in the zone of the inflorescence, but that there is **no such thing as a cymose inflorescence**" (1964, p. 33). He has therefore proposed the term **cymoid** for the form of inflorescence represented here by *Silene vulgaris*.

Cymoids can also achieve the appearance of an umbel by compression of the hypopodia in all successive branches (except the pedicels of the flowers) of the cymose partial inflorescences, as for example in the pseudosciadioid of the

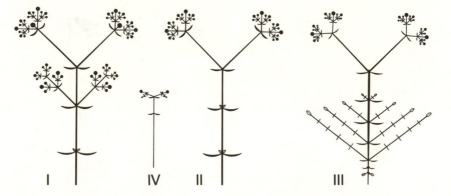

Fig. 119 Valerianella. I Inflorescence which has developed a whorl of thyrsoid partial inflorescences below the uppermost favoured whorl of cymose partial inflorescences, II with formation of the distal whorl only, III development of branches in a richly ramified specimen, IV inflorescence of a starved form of *V.locusta.*

Fig. 120 I *Valerianella discoidea,* a much-branched flowering plant. II-IV *Valerianella tuberculata,* fruiting plants with different degrees of branching, IV depauperate form.

Jagged Chickweed (*Holosteum umbellatum*, see Troll 1957, p. 392 ff.) already mentioned in section 2.1.3, or also in African Hemp (*Sparmannia africana*, see Troll 1957, p. 395 ff., 1964, p. 123; and also Fig. 172 III).

Undoubtedly the branching pattern of the flowering systems provides valuable characters, which are important to consider for the descriptions and recognition of individual taxa. There is no way therefore that we can abandon the distinctions and terminology explained above, which have turned out to be useful and sufficient for purposes of descriptive recognition. The determination of the different forms of branching cannot, however, serve as the sole basis for the elucidation of the fundamental types and the recognition of homologous and non-homologous structural elements.

2.2   Typology of inflorescences

If we were only to consider the branching patterns, then the "long-stalked flower pairs" which occur in the axils of leafy pherophylls in Fly Honeysuckle (*Lonicera xylosteum*) or in *Lonicera tatarica* (Fig. 121 II), and the similarly twin-flowered inflorescences of *Linnaea borealis* (Fig. 121 III) would all be interpreted as "two-flowered inflorescences". Each, however, achieves a completely different position in the overall structure of the plant. On the other hand, it is scarcely possible to recognise a structural relation between the externally very different forms of the inflorescence in the Caprifoliaceae by means of the forms of branching alone – we only need to compare *Sambucus nigra* (Fig. 161 I), *Symphoricarpos rivularis* (Fig. 121 IV), *Lonicera periclymenum* (Fig. 121 I) and *Linnaea borealis*. This is only possible by a comprehensive comparison including the consideration of the position of the flower-bearing branching systems within the architecture of the whole plant. There have been several noteworthy initial attempts to advance to such a comparative-morphological interpretation and thus to go beyond a purely descriptive standpoint (see Čelakovský 1893, Parkin 1914, Pilger 1921, 1922, 1933, etc.), but nevertheless their empirical basis remained too narrow. Troll was the first to succeed in recognising a general typology of inflorescences which satisfies the above requirements. In this it turns out that it is necessary to distinguish according to the fundamental plan between inflorescences of the **monotelic** and **polytelic** types.

2.2.1   Structure of polytelic synflorescences

We wish first of all to explain the structure of **polytelic inflorescences**, and for this we shall use the diagram in Fig. 122 I. In this complex inflorescence the cluster of flowers which terminates the main axis assumes a special position. It will therefore be called by a term of its own, a **florescence**. The fact must be stressed in this case that the main axis does not terminate with a flower. Hence the apex of the florescence remains indeterminate, and it forms exclusively lateral flowers.

Similarly, the lateral branches which proceed from the main axis below this florescence terminate in florescences having the same structure and likewise not being terminated by a flower. We distinguish these from the **main florescence** which terminates the main axis by the term **coflorescence**. The lateral branches which end in coflorescences repeat the structure of the main axis to some extent, and are therefore termed **paraclades**. They also repeat the architecture of the main axis in that they can produce further paraclades of second order (Pc') below the coflorescence themselves; the branching can also continue still further.

The complete flowering region is therefore represented here by a system of florescences: a **synflorescence**, in which the main florescence assumes a dominating role. The flower forming function of the latter is enhanced to a varying degree by the paraclades. For this reason the paraclades (or "repetition shoots") had already been termed **enrichment branches** by Alexander Braun (1851, p. 41). Troll has also adopted this term, but with the qualification that it is only to be used for paraclades which have "developed fully and reached flowering" (1964, p. 231), and not for paraclades which have remained completely (ananthic paraclades) or partially sterile. Troll calls the complete zone of the main shoot in which the enrichment branches are fully developed the **enrichment**

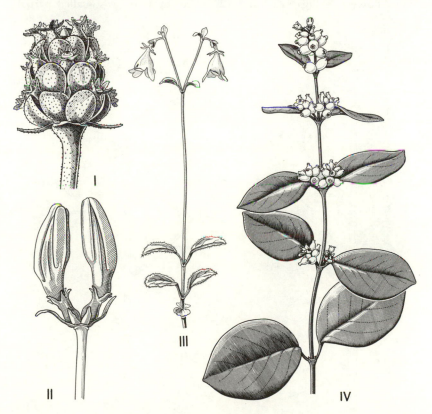

Fig. 121 I *Lonicera periclymenum*, head-like contracted florescence, II *L.tatarica*, two-flowered partial florescence, III *Linnaea borealis*, flowering shoot, IV *Symphoricarpos rivularis*, first-year shoot with terminal and axillary (indeterminate) spikes (heterothetic double raceme). After Troll and Weberling.

zone. The synflorescence is therefore divided into the main florescence and the enrichment zone. The border between the two is formed by the **basal inter-node**, which is the internode preceding the main florescence. This is often considerably elongated, and in some cases is even developed like a scape.

Simple examples of synflorescences of this kind are found in Charlock (*Sinapis*) or in Shepherd's-purse (*Capsella bursa-pastoris*); a flowering plant of the latter is illustrated in Fig. 127 I. This exhibits an extended racemose main florescence whose basal internode is not especially conspicuous. In the enrichment zone below the main florescence five paraclades have developed, each ending in a racemose coflorescence. These, like the main florescence, are clearly demarcated from the preceding leafy zone, since they are ebracteate, as is the case in most of the Cruciferae.

Since the pedicels of the Cruciferae never possess prophylls, we only find racemose florescences in this family. When, however, the flowers are provided with prophylls, as in the diagram of Fig. 123 I, then further flowers or sympo-dially branched flowering systems can appear out of their axils. Hence the individual flowers in Figs 122 I and 123 I are replaced by so-called partial

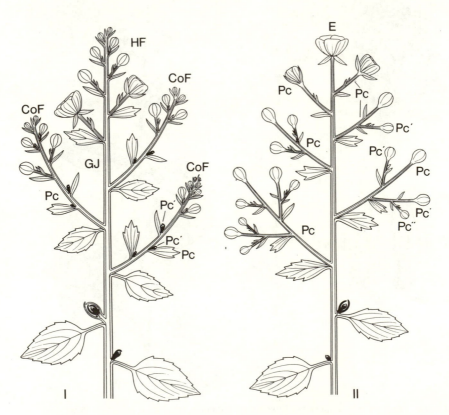

Fig. 122 Schematic representation of a polytelic (I) and a monotelic (II) synflorescence. HF main florescence, CoF coflorescence, Pc paraclades, Pc', Pc'' paraclades of the second and third orders, E terminal flower, GJ basal internode. After Troll.

florescences* (Fig. 123 II). A good example of this is provided by the florescences of the Labiatae, which usually occur as indeterminate thyrses. However,
just within this family and even in the different species of the same genus are
found not only thyrses with much-branched cymes but also impoverished and
therefore racemose florescences whose (originally cymose) partial florescences
remain limited to no more than the primary flower. On this point, the Whorled
Clary, *Salvia verticillata* (Fig. 124 I) should be compared with *Salvia patens* (Fig.
124 II). Corresponding examples of thyrsic florescences with alternate leaves are
provided by Common Figwort, *Scrophularia nodosa* and by many other members of the Scrophulariaceae.

2.2.2   Structure of monotelic inflorescences

In the second and no less widespread type of angiosperm inflorescence, the main
and lateral axes terminate with flowers. The panicle in Fig. 122 II may be used
as an example of this. We are therefore dealing with a **determinate** inflorescence. Hence in this case the main axis terminates with a single flower, instead
of a many-flowered ("polytelic") florescence, and this is why inflorescences of
this **monotelic** type are distinguished from polytelic inflorescences. In such a
system the flower-bearing lateral axes which are produced from the main axis
below the terminal flower are all to be interpreted as homologous elements
which are equivalent to one another, whether they are branched or not. They
are accordingly all termed **paraclades**, since each in a certain sense repeats the
architecture of the main shoot, and similarly for their branches from the first to

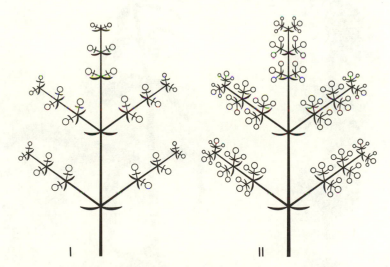

Fig. 123 Typological correspondence between racemose (heterothetic dibotric, I) and
indeterminate thyrsic (diplo-thyrsic, II) synflorescences. After Troll, redrawn.

---

* In contrast to the term **partial inflorescence**, which denotes neither the type of branching nor its
  morphological position in terms of defined inflorescence types, the term **partial florescence** is
  morphologically defined and refers to a cymosely branched element of a florescence.

the nth order (Pc', Pc'', Pc''' etc.). It should be stressed that with the term paraclade (i.e. repetition shoot!) only the respective elements of one or the other inflorescence type which are equivalent to one another may be included. The paraclades of the monotelic type can also be referred to as enrichment branches, if they succeed in flowering (see section 2.2.3), and correspondingly the collective paraclade-bearing part of the inflorescence may be called an enrichment zone. The internode which precedes the terminal flower in the monotelic type of inflorescence is termed the **final internode**, by analogy (!) with the "basal internode" of the polytelic inflorescence. Therefore in this instance the complete flowering region consists of the terminal flower of the main axis, the terminal internode which precedes it and the **enrichment zone** comprising the monotelic paraclades.

2.2.3   General zonation in the architecture of flowering plants

In the inflorescences of both the monotelic and polytelic types there is usually a zone which precedes the **enrichment zone** in which the paraclades are re-

Fig. 124 Inflorescences of *Salvia verticillata* (I) and *Salvia patens* (II). After Troll.

tarded (Fig. 122 I) or do not manage to develop at all (Fig. 125). Troll (1964, p. 231) ascribes this to the "effect of an inhibiting factor, whose influence increases in proportion to the distance from the floral region and which finally increases to such a degree as to inhibit floral development altogether", and therefore designates this as the **inhibition zone** (HZ). The inhibition of the development of paraclades in a basipetal direction can appear in a progressive fashion, so that the transition from the enrichment zone to the inhibition zone is gradual, or it can also take place suddenly, so that a sharp boundary between the two zones is created. A gradual inhibition is sometimes expressed by paraclades being transformed into thorns, as is the case in the transition zone of *Hydrolea spinosa* (Hydrophyllaceae) given by Troll (1954, in 1950–75, p. 232).

If the inhibiting influence is removed by cutting off the main axis below the enrichment zone at an early stage, or is (spontaneously) relaxed at fruiting time, then some or all of the buds which are located in the inhibition zone may then develop and reach maturity ("second flowering").

It is necessary to point out that by no means all the paraclades which are

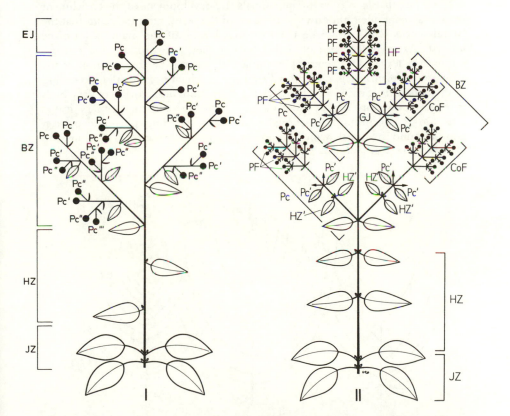

Fig. 125 Zonation of a flowering shoot in the structure of a monotelic (I) and polytelic (II, thyrse) synflorescence. T Terminal flower, EJ terminal internode, Pc, Pc', Pc", etc., paraclades of the first and successive orders, BZ enrichment zone, HZ inhibition zone, JZ innovation zone, HF main florescence, CoF coflorescence, PF partial florescence, GJ basal internode.

thrown out necessarily come to flower. In fact there are quite a number of examples, especially in monotelic inflorescences, of facultative or obligatory sterility of paraclades ("ananthic" paraclades). These do not therefore represent enrichment shoots, but remain purely vegetative (at least in the case of obligatory "ananthy") and in the end serve only for the purpose of assimilation. Their occurrence is allegedly confined to herbaceous perennials. Troll (1964, p. 203ff.) cites species of *Adonis* (*A. vernalis!*), *Ranunculus*, *Caltha*, *Drosera*, *Euphorbia* (*E. cyparissias!*), *Ruta* and *Linum* (*L. perenne!*) as examples.

In perennial plants the inhibition zone is preceded by a section at the base of the main shoot whose axillary buds function as **innovation buds**. These remain dormant until the following vegetative period and then develop innovation shoots, which in herbaceous perennials usually provide the renewal of the complete branching system above ground. The enrichment zone, the inhibition zone and, in perennial plants, the **innovation zone** (Fig. 125 JZ), together form what is called the **hypotagma** ("Unterbau"), which therefore comprises the whole of the more or less characteristically vegetative portion of the main shoot.

It is a fact that in very luxuriant annuals the development of the enrichment shoots can continue right down to the base of the plant, with the result that the lateral shoots which originate from the basal zone of the main axis are the longest. As a consequence of their position they also produce the greatest number of leaves. Characteristic examples of this can be found amongst the semi–rosette plants such as Shepherd's-purse, in which occasionally, according to the data of R. Schroeder (thesis, Mainz 1950, unpublished), all axillary buds manage to develop, except those of the cotyledons and the primary leaves (Fig. 126 I). According to Schroeder, the corresponding phenomenon can also be found in annuals which do not have any basal rosette, as in Common Hemp–nettle (*Galeopsis tetrahit*, Fig. 126 II).

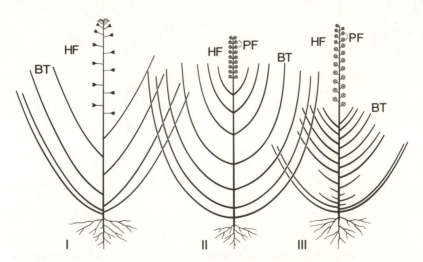

Fig. 126 Branching schemes of I *Capsella bursa-pastoris*, II *Galeopsis tetrahit* and III *Verbascum pulverulentum*. HF Main florescence, PF partial florescences, BT enrichment branch. After R. Schroeder (unpubl.).

In many hapaxanthic* semi-rosette plants the paraclades which arise from the axils of the rosette leaves at the base of the flowering shoot may even be the ones whose development is preferred. Since these **rosette shoots** are produced from the zone in which the primary growth of the primary shoot is at its maximum, it is not surprising that these lateral shoots are not only the longest, but also the strongest. When completely developed they may fall only a little short of the main shoot in respect of flower formation (Fig. 126 I). Since below the main florescence or the terminal flower of the main shoot there are usually also enrichment shoots which develop in basipetal order, a considerable gap in the branching of the main stem is produced between the zone of the **stem shoots** and the basal rosette shoots due to the intercalation of the inhibition zone (Fig. 126 III).

A similar situation can also occur in subrosette-forming hapaxanthic plants which do not possess a definite rosette, but in which the lower leaves are crowded together in a rosette-like manner, as shown in Fig. 142. Troll (1950, 1964, p. 251) denotes the enrichment shoots which occur here as **basal shoots**, in contrast to the rosette shoots of the semi-rosette plants.

Among the basal shoots, the cotyledonary shoots which arise from the axils of the cotyledons assume a special significance in this respect, since it is not uncommon for these to be the only ones to develop and to form strong shoots (see Fig. 155 I). The formation of vigorous cotyledonary shoots can be found in a very pronounced fashion among the annual species of Spurge, such as in Sun Spurge (*Euphorbia helioscopia*, Fig. 179 I), where they terminate in a pleio-chasial inflorescence (monotelic pleiothyrsoid) as does the main stem. This remains the case even though the main shoot did not undergo any primary growth in the zone of their insertion. The strong development of the cotyledonary shoots can therefore not be explained by the general rule which applies to the rosette shoots, that the formation of the lateral organs depends to a great degree on the vigorous growth of the main shoot. The same is true for the often considerable vigour of the basal shoots in subrosette-forming plants, which often does not correspond to the degree of primary growth of the main stem in the zone of their insertion. Troll (1964, p. 255) assumed that there might be a "modifying influence of a precocious and preferential secondary thickening at the base of the main shoot".

2.2.4 Woody plants

Woody plants are distinguished from herbaceous perennials by the extensive and persistent vegetative basal framework, which is raised above the ground to a greater or lesser degree. This is termed the **phanerocorm**, in contrast to the **cryptocorm** of a herbaceous perennial, which spreads out in the ground (Troll 1964, p. 282).

The seasonal shoots of a tree[†] correspond to the innovation shoots of a herbaceous perennial. Whilst the innovation shoots of the latter normally also

---

* **hapaxanthic** plant – an annual or perennial plant which only flowers once, in contrast to pollacanthic plants, which flower repeatedly.
† The difficulties which can arise in recognising the seasonal shoots in many tropical woody plants have been especially remarked upon by Pilger (1922).

come into flower, those of woody plants remain partly vegetative. The flowering shoots of the herbaceous perennials are therefore equivalent to only the fertile innovation shoots of the woody plant. The position of these flower-bearing shoots in the overall branching system is therefore of particular interest. According to Troll (1960, in 1950–75, p. 87) there are mainly two possibilities to consider here: "either the current main shoot comes into flower, or the formation of flowers occurs on lateral short shoots which arise from long shoots, usually of the previous year".

Although the structure of the inflorescences themselves ought to follow a pattern of monotely or polytely according to the taxonomic group to which the plant belongs, and in so far as no substantial modifications take place (apart from the proliferation which is frequent in woody plants, see section 2.2.9) the inflorescence shoots as such often prove to be highly derived. It is well-known of course that even in many herbaceous perennials we find pronounced differences between the flowering and the vegetative shoots, as for example in the purely bracteose foliation and the precocious development of the flowering shoot in the Colt's-foot (*Tussilago farfara*) or in species of Butterbur (*Petasites*). Similar phenomena can be observed in numerous woody plants. According to Troll (1958, in 1950–75, p. 134) a simple example is provided by the Bog Bilberry (*Vaccinium uliginosum*) by the differentiation of its shoots into sterile leafy shoots and bracteose flowering shoots which are reduced to a few-flowered raceme. This phenomenon may be classified in the following way: in the innovation shoots of herbaceous perennials the foliation usually starts with scale leaves, which are replaced further upwards by foliage leaves and in the flowering region, by bracts. Fundamentally the same thing applies to the seasonal shoots of woody plants. In the flowering short shoots however the region of foliage leaves is often lacking, so that in the leaf sequence, the basal scale leaves pass over directly into the bracts, with the result that the foliage of these shoots is purely bracteose. In extreme cases even the prophylls can be fertile (Troll 1960, p. 87, 1965, in 1950–75, p. 116/117), as for example in the Ash (*Fraxinus excelsior*). According to Troll, the correctness of this interpretation is confirmed by cases "in which the short shoots are abnormally elongated, so that foliage leaves are the first to be produced above the scale leaves, before the onset of flowering", as can happen in Lilac (*Syringa vulgaris*).

2.2.5   Cauliflory

In **cauliflory** ("flowering on the stem"), which is the "effusion of the flowers from the older parts of the trunk", the flowering shoots also have bracteose leaves throughout, which further increases the peculiar impression made by this phenomenon. Cauliflory is best known in the cocoa tree (*Theobroma cacao*), but is also typical of numerous other tropical woody plants, as in members of the Bignoniaceae (*Crescentia*, Calabash Tree, *Parmentiera*, Candle Tree), Lecythidaceae (*Couroupita*, Cannon-ball Tree), Moraceae (*Artocarpus* species, *Ficus* species), Meliaceae (*Dysoxylum*), Papilionaceae (*Mucuna*), many members of the Menispermaceae, etc. It depends on the delayed development of the inflorescences, which consequently break out "through the old wood". There are, however, forms which show a transition to this highly derived phenomenon:

"in some plants the inflorescences partly occur in the axils of existing leaves, and partly from those of fallen leaves; in extreme cases" ... "a number of years can pass between the shedding of the subtending leaf and the development of the inflorescence, during which the secondary thickening takes effect. Thus it is that the flowers are found breaking out of a thick trunk" (Wagner in Linsbauer 1917). On the subject of the derivation of cauliflory Troll (1967, p. 93) comments on the property of *Cercis siliquastrum* (Judas Tree), in which the development of the constantly bracteose flowering shoots (which resemble short shoots) takes place over a longer period of time, so that these often come to be positioned on older branches, and finally partly even on the trunk. On these there can even occur "branching for which the development of innovation buds of the shoot may be responsible". In other cauliflorous plants such as *Couroupita* or *Theobroma* the relation between the formation of inflorescences and the original arrangement of the leaves on the axes which bear them is also often still clearly recognisable.

In typical cases of cauliflory, however, there is characteristically a great distance, often amounting to as much as 20 metres, between the flowering shoots and the vegetative shoots. Furthermore, the often very close relationship between the formation of flowers and the phases of leafing which applies to woody plants of the temperate zones does not necessarily occur here. This may be the reason why some of the cauliflorous species flower throughout the year. On the other hand, the periodic occurrence of a sudden onset of flowering may have its origin in a delay of development, such that there is a buildup in the number of flower buds which are all at the same stage. These may possibly even be held at rest below a thin layer of bark, until they all blossom at once in response to some minor stimulus, such as a sudden temperature change of 2 to 3 °C (Bünning 1953).

The spatial separation between the leaf- and flower-bearing shoots can be continued even further, as in *Ficus ribes*. According to the observations of Koorders (1902), the shoots which bear inflorescences ("receptacula") "are almost 2.5 metres long where they arise from the basis of the stem and are partly appressed to the ground, and partly buried beneath the soil level, and in the latter case are partially provided with short, normal roots. These 'fruiting' shoots which radiate out from the base of the stem exceed both in length and number those receptacula-bearing branches which are inserted above the base of the stem and on the branches". He made similar observations on *Ficus geocarpa* (fruits "partly buried on cable-like, horizontal leafless shoots") as well as in *Cyrtandra geocarpa* and *C. hypogaea* (Gesneriaceae), *Saurauia callithrix* (Saurauiaceae), *Sageraea cauliflora* (Annonaceae) and *Diospyros cauliflora* (Ebenaceae). This should also be compared with the data given by Bureau (1888). The species of the genus *Paraphydanthe* (Flacourtiaceae), a small rain forest tree of West Africa, also flower from underground inflorescences of several metres in length, where it is only the flowers which are raised above ground level (see Letouzey 1970).

On the other hand, there are definite relationships between cauliflory and some forms of **flagelliflory**, in which the flowers or inflorescences hang down from the crown of the tree on long rope-like branches, and so are brought into an exposed position which allows the flowers to be visited by bats. Besides the well-known example of *Kigelia africana*, Bignoniaceae, the "sausage tree", there exist numerous other examples from different families, such as in the Annona-ceae (*Polyalthia*), Chrysobalanaceae (*Couepia*), Mimosaceae (*Parkia*), Caesalpini-aceae (*Hymenaea, Eperua falcata, Elisabetha*), Papilionaceae (*Mucuna*), Sonneratia-ceae (*Sonneratia*), Sterculiaceae (*Dombeya*), Solanaceae (*Markea, Trianaea*); in addition see van der Pijl 1930, 1949, and Vogel 1958, 1968/69.

2.2.6   Forms of polytelic synflorescences

According to the experience gained so far, which is based on extensive material covering members of nearly all Angiosperm families, the inflorescences of all flowering plants can be assigned to one or other of the two fundamental types described, the monotelic or the polytelic. The astonishing variety of forms of inflorescences which confronts us is determined by nothing more than the fact that the formation of the individual structural elements can undergo a great many variations of a quantitative kind according to the principle of variable proportions. To begin with we want to acquaint ourselves with some of these possibilities for variation in the structure of polytelic inflorescences by means of a few species of the Cruciferae which have heterothetic di-botric to pleiobotric

Fig. 127 I *Capsella bursa-pastoris* and II *Isatis tinctoria*, flowering plants. After Troll.

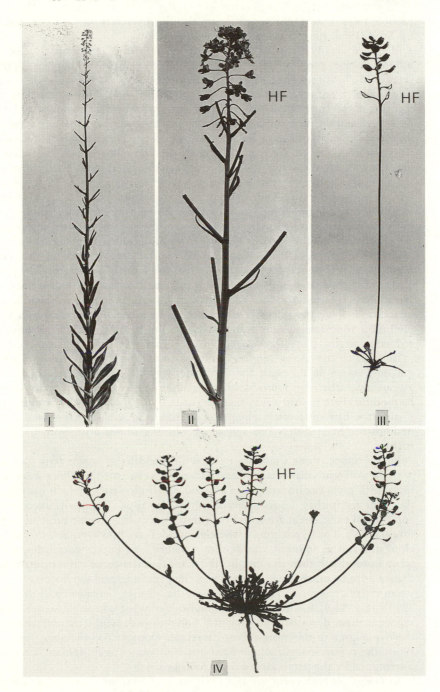

Fig. 128 I, II *Isatis tinctoria*, inflorescence after the excision of all paraclades, with the terminal section further enlarged in II, HF main florescence. III, IV *Teesdalia nudicaulis*, flowering plant confined to its main florescence (III) and greatly branched (IV). After Troll.

(double racemic to multi-racemic) ramification by referring to investigations by Troll (1960, 1964; personal communication).

We have already discussed the most important features of the structure of the inflorescence of the Cruciferae by means of the example of Shepherd's-purse (Fig. 127 I). In comparison with this the inflorescence of *Isatis tinctoria* (Woad) is much more extensively and richly branched (Fig. 127 II). This is caused by the fact that in the enrichment zone there are numerous paraclades developed. On the other hand the main florescence which terminates the principal axis is far less important in terms of its size and number of flowers. This is made obvious if all the paraclades are removed, as is the case with the inflorescence illustrated in Fig. 128 I, II. From this it is also evident that in this instance the basal internode remains very short. Whereas the lower paraclades are greatly branched, particularly so in their distal regions, and here bear paraclades from the second to the fourth order, a higher proportion of those inserted in the distal zone of the inflorescence are confined to their florescence only, and also lack the leaf organs which precede the florescence. Thus the formation of the hypotagma in these paraclades is suppressed. The same is true for the distal regions of the more highly branched paraclades, where moreover the subtending bracts of the paraclades of second and higher order remain small, or in any event may be **frondulose**. Only in the lowermost paraclades can some larger foliage leaves be found. In *Isatis* the extensive enrichment zone is usually preceded by a long inhibition zone (and an innovation zone at the base), in which an obvious reduction in the size of the leaves in the distal direction can be observed.

In order to consider a number of other modifications of the basal structure we once again begin with the properties of Shepherd's-purse, for which a branching scheme is reproduced in Fig. 129 I. The ramification can in fact be modified here in such a way that the entire hypotagma remains compressed, so that the first moderately elongated internode to appear is the basal internode of the main florescence and all enrichment shoots arise from the basal rosette.

In this respect some similarity exists with *Teesdalia nudicaulis* (Fig. 129 III), with the difference that here the basal internode is elongated to a long scape. It is this which gives the impression of a naked, i.e. leafless stem, which has earned the plant its epithet. Usually several paraclades appear from the rosette-like compressed enrichment zone, and their florescences are likewise on long scapes (Fig. 128 IV). The hypotagma of these paraclades can, however, be more or less elongated and in addition, can develop and bring to flower paraclades of the second order. It is then not rare for the main florescence which terminates the main axis to be much less important than the paraclades and the majority of the stronger coflorescences. This is even more the case in *Coronopus didymus* (Fig. 129 IV, Fig. 130), in which the basal internode of what is in any case not a very vigorous main florescence remains still undeveloped, whilst the internodes in the hypotagma of the more robust paraclades elongate considerably. Furthermore, the coflorescences, whose basal internodes once again remain short, are overtopped by the paraclades of the second order.

In *Lemphoria procumbens* the axis of the main florescence has been shortened greatly, corresponding to the diagram in Fig. 129 V. This is, however, compensated by a corresponding increase in length of the pedicels, so that the completely compressed main shoot of the plant terminates directly above ground level in an umbellate florescence (Fig. 131). This is much exceeded by the elongated, leafy and much-branched paraclades, whose florescences are all of racemose form.

Some members of the Cruciferae are distinguished by the fact that the zones of their florescences are also provided with foliage leaves. *Buchingera axillaris* belongs to this group. Its branching diagram in Fig. 129 II corresponds in other respects to that of Shepherd's-purse. If it is assumed that, similarly to *Lemphoria procumbens*, the complete main axis with the inclusion of the zone of the main florescence remains compressed, while the main florescence bears foliage leaves like the hypotagma, then the branching scheme in Fig. 129 VI is obtained,

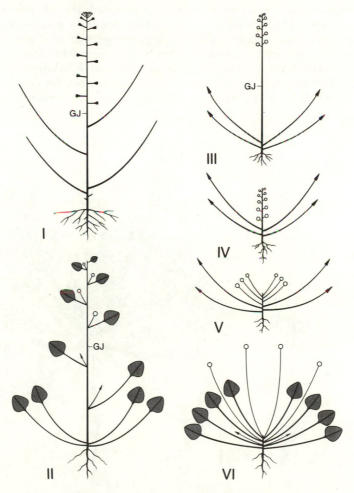

Fig. 129 Diagrams of the growth forms of I *Capsella bursa-pastoris*, II *Buchingera axillaris*, III *Teesdalia nudicaulis*, IV *Coronopus didymus*, V *Lemphoria procumbens*, VI *Ionopsidium acaule*. GJ, principal axis. After Troll.

which corresponds to the properties of *Ionopsidium acaule* (Fig. 132 I). The compression of the inflorescence axis is once again compensated for by a considerable elongation of the pedicels, which exceed the long-petioled leaves when they are fully developed. The plant illustrated in Fig. 132 II, where the paraclades have been cut away for clarity, shows that in *Ionopsidium acaule* an enrichment zone is formed as well.

2.2.7   Impoverished forms

Extremes of variation in the inflorescence structure occur in the so-called 'starved' forms. These occur primarily in the hapaxanthic plants, and especially in annuals, which are particularly sensitive to conditions of undernourishment. By comparison with perennial plants of course, annuals are only provided with scanty reserves, which are only sufficient to take care of the juvenile stages. This explains the tendency, under unfavourable growing conditions, to bring development to a conclusion by producing ripe fruit within a short time. This is also the explanation for the German name of the Common Whitlow-grass (*Erophila verna*), which they call "spring hunger-flower". This is another case of a rosette plant in which the only part of the primary axis which elongates is the basal

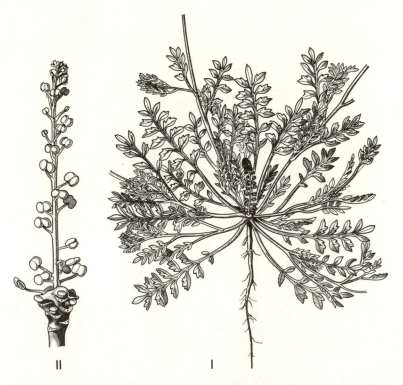

II   I

Fig. 130 Coronopus didymus, I flowering plant, paraclades partially removed, II primary axis (after abscission of all rosette leaves) with main florescence, more greatly magnified. After Troll.

internode, forming a scape similar to that in *Teesdalia*. Besides the main florescence there are also, in favourable conditions, paraclades which come into flower. These arise from the enrichment zone, which is compressed into a rosette (Fig. 133 I). The hypotagma of the paraclades, in which only two leafy prophylls or one or two more foliage leaves manage to develop, remains foreshortened, but under favourable conditions it can even produce further paraclades of the second order. In this way numerous racemose florescences can arise from the basal rosette. The plant often occurs as a so-called spring ephemeral in places with especially shallow soil, as for example on walls or on the tops of rocks. In such localities it is of course necessary for the plant to complete its development quickly before the thin soil dries out, which is only possible by reducing the extent of its overall growth. The development of the rosette hypotagma is therefore limited to a few foliage leaves, paraclades do not occur, and only a few flowers in the main florescence reach complete maturity. Not uncommonly there may be only one flower, namely the lowest, which is the first to open in the acropetal sequence of development. This often assumes an erect habit and then simulates a terminal flower, while the vestiges of the apex of the florescence are pushed to one side (Fig. 133 II). However, by closer inspection (Fig. 133 III), and by ontogenetic means, the lateral origin of this

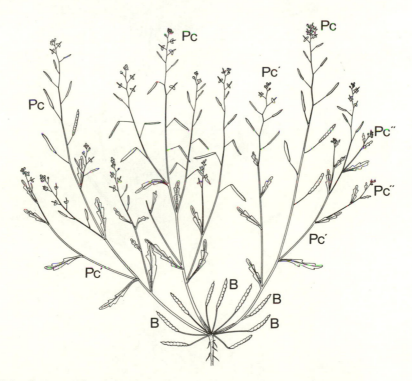

Fig. 131 Lemphoria procumbens, habit, rosette leaves already withered, B flowers of the main florescence, already ripened into fruit. Pc, Pc′, Pc″ Paraclades of different orders. From Schulz, simplified by Troll.

flower is readily established. The starved form of *Raphanus raphanistrum* (Fig. 133 IV) is somewhat less reduced, since it nevertheless succeeds in producing three flowers, or even manages to ripen their fruit; it does, however, turn out to be very much reduced in comparison with the normal form of the plant. However, even under extreme conditions of development the fundamental plan of the structure of a polytelic synflorescence is still preserved throughout. The same remarks also apply to starvation forms of plants with monotelic

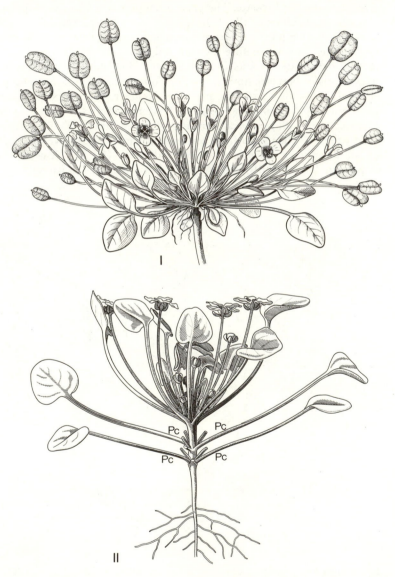

Fig. 132 Ionopsidium acaule, habit of a flowering plant (after O. E. Schulz), II flowering plant after excision of the paraclades Pc (after Troll).

synflorescences. As an example of this the plants of *Microcala filiformis* (Gentian-aceae) reproduced in Fig. 133 V, VI will be sufficient for the time being. The characteristic structure of the inflorescence of the Valerianaceae, recognised by the consistent lack of the terminal flower (which will be discussed in more detail later on), is also still present even in extremely reduced forms of *Valerianella* (Fig. 119 IV, 120 IV).

The characteristic differences between monotelic and polytelic inflorescences are also apparent in the way they are affected by photoperiodic influences (see further the remarks by Troll, 1964).

2.2.8   Forms of monotelic synflorescences

We also encounter a great variety of different modifications of the same basic plan often even within one and the same family with monotelic synflorescences. A series of examples, which have almost all been selected from the *Polemonia-ceae*, may make this clear.

We have already referred to the inflorescence of *Polemonium caeruleum* (Fig. 134 I) as a classic example of a panicle where the branching increases in a regular fashion towards the base and whose outline is at least primarily conical (see section 2.1.4). The lowest panicle branches show not only the most com-

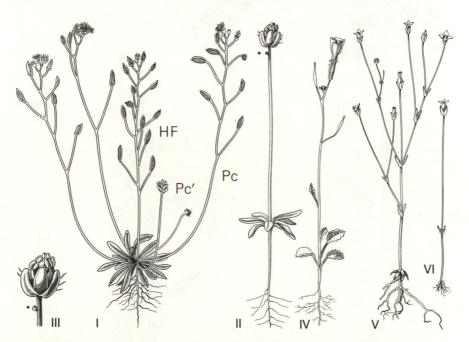

Fig. 133 I-III *Erophila verna*, I well-grown plant with main florescence (HF) and the (co)florescences of paraclades of first (Pc) and second (Pc′) order. II Impoverished form, III with its distal portion further enlarged, ∗ marks the laterally displaced rudiments of the florescence apex. IV *Raphanus raphanistrum*, starved form. V, VI *Microcala filiformis*, a normally developed plant (V) and a starved form (VI). All due to Troll.

plex branching, but are also the most robust. Hence all in all, the formation of the branching system is characterised by a basal-mesotonic direction of development. The unfolding of the paraclades proceeds in a basipetal direction as far as the inhibition zone. The flowering sequence begins here with the terminal flower which "closes" the main axis and continues downward with the neighbouring lateral flowers. The same pattern repeats in the more highly branched paraclades in a corresponding fashion. As the development of further paraclades continues the number of flowers in such a panicle can grow to many times the number in the inflorescence illustrated here. On the other hand, the panicles of other *Polemonium* species are characteristically few-flowered. This is shown in the example of the inflorescence of *Polemonium pauciflorum* illustrated in Figs 135 II and 155 II. In *Polemonium viscosum* (Fig. 135 I) the panicle is reduced to a botryoid, whose flowers are compressed into what is almost an umbel, which is raised by a few only slightly elongated internodes of the main shoot.

The inflorescence of *Polemonium carneum* presents the form of a broadly conical panicle from the beginning until anthesis. Very soon, however, the

Fig. 134 I *Polemonium caeruleum*, II *P. carneum*, synflorescences. Due to Troll and Weberling.

terminal flower is overtopped by the middle and upper paraclades (Fig. 134 II), and the panicle is inverted into an anthela by intensified longitudinal growth of the paraclades. This growth is the most marked in the lowest paraclades. A distinctly acrotonous form of development is found in *Polemonium micranthum* (*Polemoniella micrantha*), as illustrated in Fig. 136. This is observed repeatedly within different genera of the family Polemoniaceae, as for example in *Linanthus bolanderi* (Fig. 137 I) with its decussate leaves. In *Linanthus liniflorus* the vigorous development of the favoured distal lateral branches can make such demands on the meristem of the inflorescence that the terminal flower of the main axis is no longer properly formed (Fig. 137 II).

The form of the inflorescence of *Bonplandia geminiflora* (and also that of *B.linearis*) which appears to diverge from that of all other members of the Polemoniaceae (Fig. 138), is also due to marked **acrotony**. It looks as though it were an indeterminate inflorescence, in that paired flowers are produced from the axils of leafy bracts. In reality this is a case of a monotelic **thyrsoidal** synflorescence, whose structure may be explained by the diagram in Fig. 139. It must be realised that each of the consecutive lateral axes of the much-branched cymose distal paraclades is aligned in the direction of the main axis and so its respective terminal flower is pushed to one side, together with that which arises from the axil of the β-prophyll, so that both simulate axillary flowers (see Weberling 1957).

In the inflorescence of *Bonplandia geminiflora* we also observe the occurrence of serial accessory buds, which are constantly found in association with the more highly branched paraclades (and which can make the survey of the branching pattern more difficult). Serial accessory buds of this kind often occur in the inflorescences of the Polemoniaceae and when they develop, they can

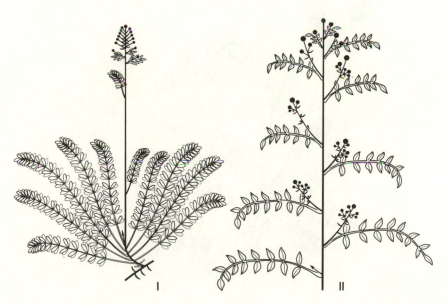

Fig. 135 I *Polemonium viscosum*, II *P. pauciflorum*, branching of the inflorescences; true to scale.

radically alter the appearance of an inflorescence. The inflorescence of *Ipomopsis rubra* is a case in point. At the beginning of anthesis this presents the form illustrated in Fig. 140 I, whose structure is explained by the branching diagram in Fig. 141 I. What is noticeable about this is the relatively long botryoid terminal section of the panicle as well as the recaulescent displacement of the bracts in the distal lateral flowers. Already at this stage a (phylloscopic) serial accessory bud can be observed in the axils of all subtending bracts of the paraclades of the first order, inserted between the paraclade and the subtending bract. The accessory buds all begin to develop and flower in the distal region during the further stages of anthesis ("accessory paraclades"), at which point it can be seen that they bear many more leaf organs than the primary paraclades (Fig. 140 II, III). At least the terminal flowers develop from among the numerous flower buds initiated on these accessory shoots in this region. If the state of development of the individual accessory shoots is examined in the subsequent stage of anthesis, then at first in the distal paraclades of the panicle and from there downwards an increase in the development and branching (Bs!) is observed, and then a continually increasing inhibition of development beginning above the dyadic paraclades, until finally the accessory buds persist in the bud stage. Instead of these, in the dyadic and triadic paraclades those accessory buds manage to develop, which are associated with the paraclades of second order

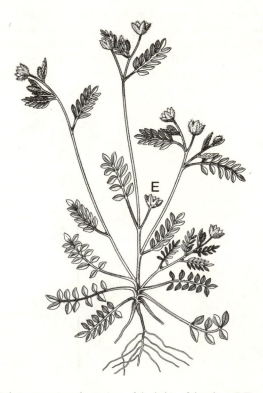

Fig. 136 Polemonium micranthum, view of the habit of the plant. E Terminal flower.

(Fig. 141 II). – Furthermore, it is possible for the accessory buds in the inflorescence region to develop in a totally vegetative form (e.g. *Caryopteris incana* = *C. tangutica*, Verbenaceae, see Troll 1972, p. 108).

Very often a capitate condensed form of part or all of the paraclades occurs in the inflorescences of the Polemoniaceae, so that the flowering shoot, and also the ends of the stronger lateral shoots, bear a head-like flower cluster (cephaloid, Troll 1964, p. 53). This is the case, among others, in many species of the genus *Collomia*. Amongst these in particular, the most highly branched flower heads occur in *Collomia grandiflora* (Fig. 142), and information about the branching properties of these is given in the outline diagram in Fig. 157 III. What is noticeable here is the extended botryoid terminal section and the accessory buds which occur in serial descending order in association with the stronger paraclades, and whose unfolding here also establishes various maximum points along the inflorescence axis. In spite of the large number of flowers in the botryoid terminal section the flowering sequence of the panicle begins with the

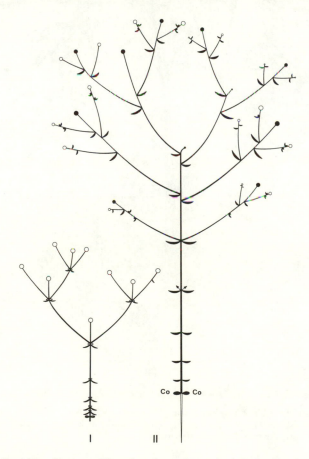

Fig. 137 Branching diagrams of flowering plants of *Linanthus bolanderi* (I) and *L. liniflorus* (II). Co, cotyledon.

Fig. 138 Drawing of the habit of *Bonplandia geminiflora*. The uppermost and normally one-flowered enrichment shoot (★) below the terminal flower E of the main shoot (marked with 1 in Fig. 139) is here abnormally two-flowered. After Brand.

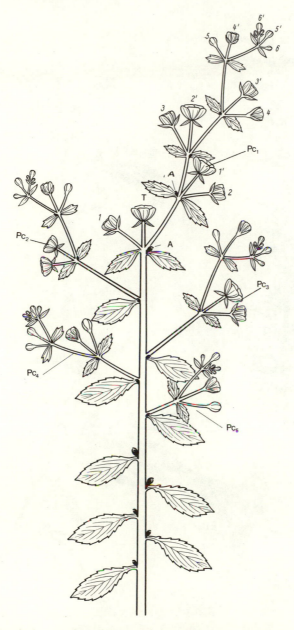

Fig. 139 Bonplandia geminiflora, diagram of the structure of the synflorescence. PC$_1$-PC$_5$ enrichment branches in descending order; that of the lateral flower 1 (= single-flowered Pc) beside the terminal flower is not counted. 1′ terminal (primary) flower of the enrichment branch (paraclade) PC$_1$, 2′ the terminal flower of the consecutive branch which proceeds from the axil of one of the prophylls of 1′, opposite flower 2, etc. PC$_1$, PC$_2$ etc., accessory buds.

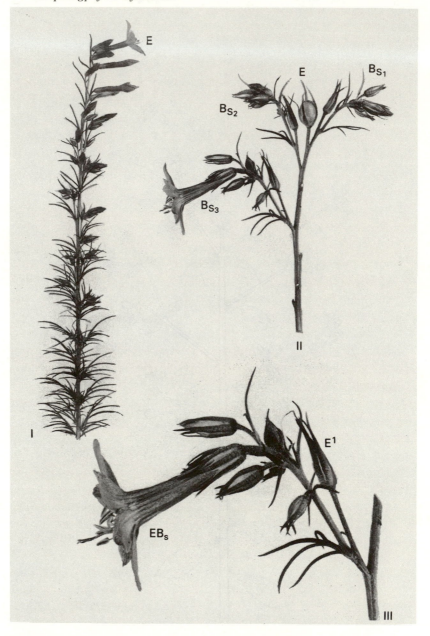

Fig. 140 Ipomopsis rubra, I terminal section of the synflorescence with terminal flower E, II terminal flower E with three postfloral monadic paraclades at the time of anthesis of the accessory buds (Bs₁-Bs₃), III postfloral monadic paraclade (terminal flower labelled with E') with flowering accessory shoot, EBₛ its terminal flower. Due to Troll and Weberling.

terminal flower of the complete inflorescence (Fig. 142 I). The plant reproduced in Fig. 142 I, II will be able to bring many of the flower heads terminating the lateral branches to maturity later on, after the large terminal head. Those which are already visible below the terminal head develop to small heads only, whilst the branches which are inserted further down the main axis can terminate in extensive flower clusters corresponding to the basal-mesotonic sequence of development of longer and more robust lateral axes.

In other species of *Collomia* which form flower heads, as for example *Collomia linearis*, the number of flowers in the head is smaller (Fig. 143 I, II). Various highly reduced impoverished forms, which can be met with in *Collomia linearis*,

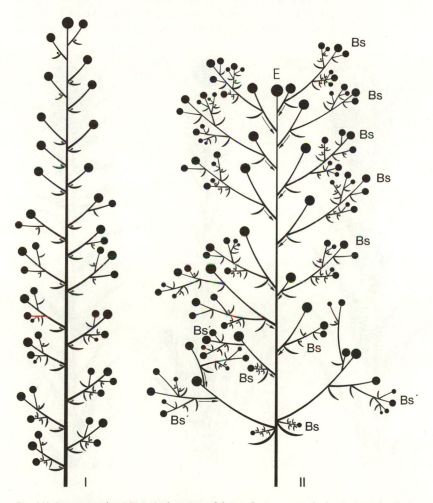

Fig. 141 Ipomopsis rubra. I Terminal section of the synflorescence at the beginning of anthesis, accessory buds still undeveloped, II after development of the accessory buds, Bs (Bs′ accessory buds of paraclades of the second order) E terminal flower. The arrows indicate the normal position of the displaced recaulescent subtending bracts. Original.

C. biflora, C. grandiflora and others species which normally form heads, lead finally to forms in which only the terminal flower is developed (Fig. 143 III to VI). However, in cases where the starved forms do produce several flowers, these still occur in head-like clusters.

If it is possible to speak of the reduced forms of *Collomia*, as well as the starved forms illustrated in Fig. 133 as "adaptive dwarfism", which is ascribed to unfavourable growing conditions alone, then in the case of the genus *Gymnosteris* the dwarfism is clearly determined intrinsically. In fact both the two species of the genus, *Gymnosteris nudicaulis* and *G. parvula*, apparently always adhere to the growth form shown in Fig. 144 II (which is for *G. nudicaulis*). This is characterised by the fact that only two internodes develop and elongate, namely the hypocotyl and the scape-like prolonged epicotyl. Apart from the rather small cotyledons, which persist in these species, as they do in the stunted

Fig. 142 Collomia grandiflora. I, III Upper part of a flowering plant with the capitate and compressed terminal section of the synflorescence, I at the beginning of anthesis, the terminal flower E shortly before opening, III capitate terminal part in full flower, the lateral heads further down already well developed, II complete plant with basal shoots GT, which already show the beginnings of the flower heads at their distal ends. Due to Troll and Weberling.

forms of *Collomia* already described, the only leaves borne by the main axis are several rather larger bracts in the region of the terminal flower head. The assimilation function of these bracts must doubtless compensate for the lack of foliage leaves. Thus the life cycle is extremely foreshortened, in such a way that

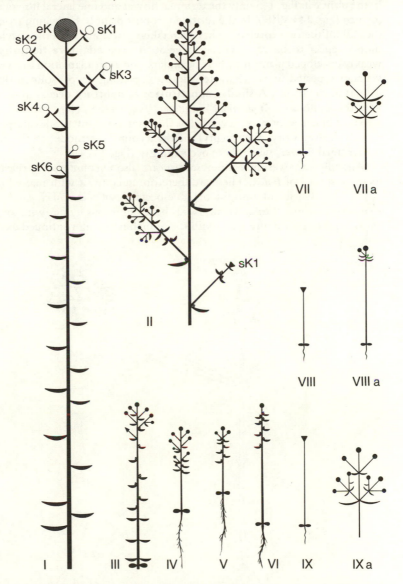

Fig. 143 I-IV *Collomia linearis*, I schematic outline of a flowering plant with terminal (eK) and lateral (sK1–sK6) flower heads, II branching diagram of the terminal head in I. III, IV Stunted forms. V, VI stunted forms of *Collomia grandiflora*. VII-IX *Gymnosteris nudicaulis*, vertical diagrams of flowering plants (all to the same scale) and VIIa-IXa branching diagrams of their terminal heads.

the plant already reaches the flowering stage in its first juvenile phase, which must here surely be ascribed to endogenous factors. The plants reach only to a height of a few centimetres, and the flower heads bear only a few flowers, in the arrangement as illustrated in the branching diagrams reproduced in Fig. 143 VIIa to IXa. In the plant c in Fig. 145 only the terminal flower and one lateral flower are fully formed (Fig. 143 VIIIa). In the relatively robust plant b, branching from out of the axils of the two cotyledons has even taken place, but both the cotyledonary shoots appear to be very weakly developed. The **adaptive neotony** of the weakly developed plants may here be understood as an adaptation to a very short vegetative period in the characteristically very long dry seasons in the native habitat of the plants. A similar **constitutional neotony** occurs in connection with the uniflory of *Argostemma concinnum* (Rubiaceae). Here also only a single whorl of bracts is formed, nevertheless this time three internodes are produced, namely the hypocotyl, the epicotyl and the terminal internode, i.e. the pedicel of the terminal flower, which here occurs singly (Fig. 144 I).

Markedly impoverished inflorescences are also encountered in the cushion-forming species of *Phlox*. These are therefore in contrast with many "herbaceous" Phlox species of upright habit, especially such as *Phlox paniculata* and *P. maculata*, whose floriferous thyrsoids bear numerous cymes with an equivalently repeating structure – i.e. with the branching being continued exclusively

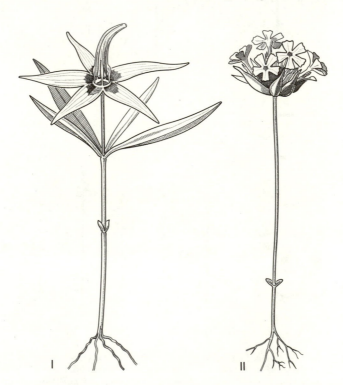

Fig. 144 I *Argostemma concinnum*, II *Gymnosteris nudicaulis*. I Due to Hooker (Icon. Pl.), II Due to Brand (Englers Pflanzenreich).

from the axil of the β-prophyll. On the other hand the relatively richly-flowered inflorescences of *Phlox amoena* (Fig. 146 I) consist only of a few whorls of one-to three-flowered paraclades, which are all raised above the vegetative zone with its characteristically compressed internodes by two scape-like elongated "segregation internodes" of the inflorescence axis. The number of flowers can be reduced even further; in *Phlox amoena* almost to a botryoid (Fig. 146 II), and in *Phlox subulata* (Fig. 146 IV, V) to only two or three flowers or even to the terminal flower alone. The latter is normally the case in *Phlox bryoides* (Fig. 146 III), but here the only remaining terminal flower is not preceded by any elongated internode.

The innovation, or perhaps in this case better to say, the continuation of the vegetative branching system, takes place in all these species from the axils of the foliage leaves which directly precede the inflorescence zone, or in the case of

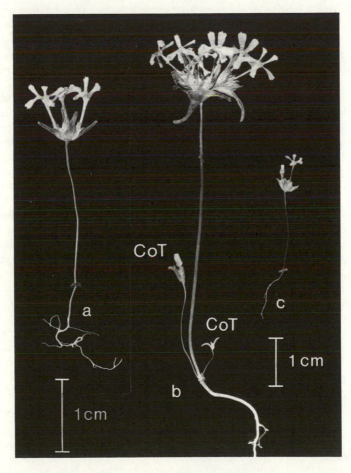

Fig. 145. *Gymnosteris nudicaulis* (W. M. C. Cusick 2366, Z), plant b with two cotyledonary shoots CoT.

P. bryoides, the terminal flower. In this process an acrotonous preference of development is usually observed. The initiation of these continuation shoots begins in the same year, often before the terminal inflorescence has finished flowering, which Troll (1964, p. 304) calls **prolepsis***. In *Phlox subulata* the distal continuation shoots detailed in the branching diagram in Fig. 146 IV often themselves succeed in flowering only a few months later; they therefore exhibit the phenomenon of **syllepsis*** (Troll 1964, p. 306).

The stoloniferous continuation shoots to which *Phlox stolonifera* (Fig. 146 VI, 147) owes its name have a proleptic* character. The inflorescence of this plant looks like an umbel because of an impoverishment of the paraclades and compression of the internodes in the floral region whereas the preceding internodes show a pronounced elongation. The ramification of the lowest "branch of the umbel" proves nevertheless that the inflorescence is an impoverished and compressed panicle (sciadioid).

A further variation of the monotelic synflorescence type consists of the preferential, or even exclusive elongation of the terminal internode of the inflorescence axis and the paraclades. The extent to which this influences the habit of flowering plants may be seen in many species of Poppy, as for example the Common Poppy (*Papaver rhoeas*, Fig. 148), in whose description it is stated: "flowers solitary, terminal" (Hegi 1919, IV/I, p. 29).

Underlying this, however, is the branching mode of a panicle, which of course is modified not only by the considerable elongation of the terminal internodes, but also by the specific manner in which the paraclades of second and higher order develop. The unfolding of the latter is often confined to the prolongation of the distal internode and is much delayed, so that the foliage leaves in the hypotagma of the paraclades come to be more in evidence. The paniculate branching mode is more clearly recognisable in *P.caucasicum*, in which the terminal internodes are less highly elongated and the development of the paraclades takes place more uniformly (Fig. 148 II). On the other hand in the alpine poppy species, such as *P.rhaeticum*, the complete hypotagma of the primary shoot remains compressed as a rosette; the same is true to a great degree for the hypotagma of the paraclades. Only the terminal internodes of the main axis and the paraclades are elongated like scapes, and raise the flowers far above the foreshortened vegetative zone – a character which finds its analogy in the polytelic synflorescences of *Teesdalia nudicaulis*. – In the Oriental Poppy (*P. orientale*), we have by way of contrast a species which grows as a semi-rosette plant, in which the paraclades all persist undeveloped and only the terminal flower matures. What we have here is different from the starvation forms of monotelic plants, in which the property of having a single flower is adaptively determined (see section 2.2.7), since this is a case of obligatory uniflory. This is encountered within many monotelic families as a character typical of a species, or even as a generic character (e.g. *Dryas*).

In this context we still have to consider the reduction of a paniculate inflores-

---

*For a more consistent terminology the development of the continuation shoots of *Phlox bryoides* should be characterised by the term **sylleptic-vegetative**, whereas the continuation shoots of *Phlox subulata* which come to flower in the same season as their mother axis are **sylleptic-anthetic**. The stolons of *Phlox subulata* (see below) which bear bud-scales at the bases, but also develop in the same season as their mother axis without coming to flower, are **proleptic-ananthic** (Müller-Doblies and Weberling 1984).

cence to the terminal flowers of the main shoot and the paraclades. An example is the inflorescence of *Cantua buxifolia* (Polemoniaceae, Fig. 149 I), in which, apart from the flower which terminates the main axis, only the terminal flowers of the paraclades of the first order reach anthesis. In the axils of the leaf organs which precede these flowers further flower buds are indeed initialised, but these

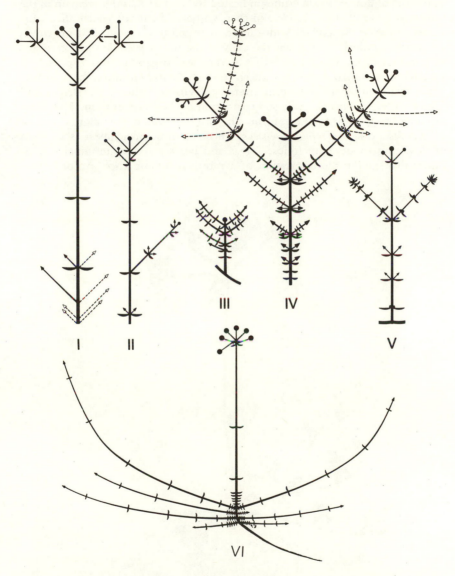

Fig. 146 Branching diagrams of flowering shoots of I, II *Phlox amoena*, III *P.bryoides*, IV *P.subulata* var.*nivalis*, next generation of shoots shown dotted, V *P.subulata*, VI *P.stolonifera*, a stolon which has formed a rosette at the distal end and which has already succeeded in flowering again and developing new stolons.

wither sooner or later (with perhaps the exception of the uppermost bud of the lowest paraclade). This phenomenon can perhaps be explained by an unusual degree of dominance of the terminal flower of the paraclades. We have already discussed (section 2.1.4) the fact that the development of the terminal flower takes the lead over that of the lateral flowers. "The delayed initiation of the upper lateral flowers is sometimes indicated by the fact that they remain in the bud stage" (Troll, 1964, p. 14) and then wither. The delay which affects the development of the axillary buds can be increased to the point that growth is suppressed, with the result that the axils of the bracts preceding the terminal flower remain empty (Fig. 149 II). 'Sterile' leaf organs of this type which precede the terminal flower have been termed **Zwischenblätter** by Nordhagen (1937, p. 12), and the term **metaxyphylls** introduced by Briggs and Johnson (1979) seems to be a good translation of the German term. They may be either bracteose or foliose in form. The latter is found in the inflorescences of *Nigella damascena* or *Trollius europaeus*, whose inflorescences appear as a system of flowering shoots with foliose leaves and bearing only a terminal flower, rather than giving the impression of a ramified inflorescence. According to

Fig. 147 Phlox stolonifera, flowering shoot with stolons in the rosette region. Due to Troll and Weberling.

Holthusen (1940, p. 619), it is often not possible to detect any meristematic cell tissue in the axils of metaxyphylls.

2.2.9   Proliferation phenomena

The possibility that the inflorescence apex may be able to grow out and to return to vegetative growth, which is termed **proliferation**, is of far-reaching significance for the understanding of many inflorescences of the monotelic as well as the polytelic type. What is meant here is not so much the cases where the inflorescence grows out occasionally, as can be observed in many plants, e.g. in *Lupinus polyphyllus, Cheiranthus cheiri* (see Hegi 1919 IV/1, p. 445) or in *Matthiola incana* (Weber 1860), but the **habitual** proliferation which is typical of quite a number of plant groups. From amongst these Troll (1960: Komm. Ber. 1959, p. 116; 1964: Komm. Ber. 1963, p. 123) has distinguished between **early proliferation** (premature proliferation) and **late proliferation**.

Late proliferation is confined to polytelic synflorescences, "in which it consists of the return of the inflorescence apex to a vegetative condition"

Fig. 148 I *Papaver rhoeas*, II *Papaver caucasicum*, flowering plants. Due to Troll.

(proliferation of the main florescence). An important prerequisite for this is that the subtending leaves of the individual flowers or of the partial florescences should be of the foliose type – a reason why a proliferation of the main florescence is rarely a typical character of the Cruciferae (as for example in the Corsican rosette plant *Morisia monantha* = *M. hypogaea*). On the other hand, numerous examples of this are found in the Onagraceae, of which only the genus *Fuchsia* may be cited here, as well as in the Primulaceae. *Lysimachia*

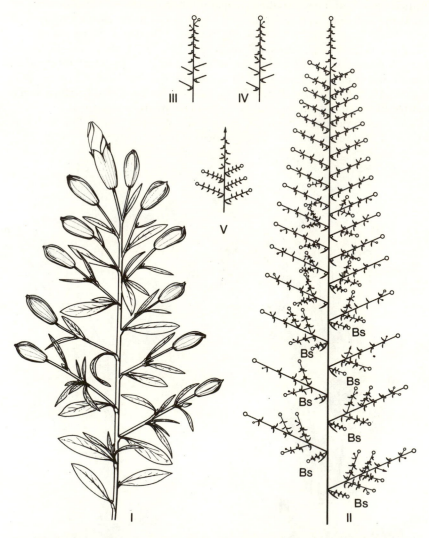

Fig. 149 Cantua buxifolia, inflorescence. II-V *Loeslia mexicana*, II branching diagram of an inflorescence, III, IV terminal portions of the more robust paraclades with terminal flowers and numerous sterile bracts, some of the distal ones, however, still bear rudimentary flowers in their axils; V diagram of the top of an inflorescence where the terminal flower has failed to form. Bs Accessory shoots.

punctata, a familiar herbaceous perennial of gardens, may be quoted from the latter family. Its many-flowered racemes are provided with leafy bracts and regularly change over to vegetative growth again at their tips after the end of flowering. Another species which regularly proliferates is Creeping-Jenny (*Lysimachia nummularia*), where the shoot as a whole, including the foliose-leaved main florescence persists, in growing with a creeping habit. In this case the proliferating shoot soon passes over to flowering again and even produces paraclades from the section below its flowering region, which can likewise

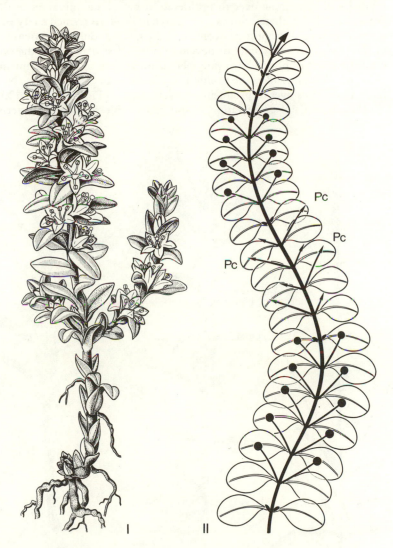

Fig. 150 I *Glaux maritima*, flowering plant (after Garcke, modified), II *Lysimachia nummularia*, shoot with repeated proliferation of the racemose main florescence, Pc marks where the paraclades will grow out.

come into flower. Hence, on the uniformly foliose-leafy creeping shoots one often finds a successive alternation of racemose flowering sections and vegetative sections, whose axillary buds can develop into paraclades (Fig. 150 II). Whilst *Lysimachia ephemerum* (Fig. 151 I) and the other species of the section Ephemerum exhibit a racemose and bracteose main florescence and similar coflorescences, exclusively lateral racemes are found in *L. thyrsiflora*, which arise from the axils of leafy bracts on the main axis, which itself remains vegetative (Fig. 151 II, III). This divergent character may be explained by the fact that the apex of the synflorescence has already gone over to proliferation before the initiation of the main florescence. This is therefore a case of the what Troll has termed **early proliferation**, whose essential property consists in the fact that only the enrichment zone of the synflorescences comes to develop – regardless of whether the synflorescence structure is monotelic or polytelic (proliferation of the enrichment zone).

In *Lysimachia thyrsiflora* the proliferating section can even branch. "Nevertheless it never goes over to producing flowers itself". (Troll 1972, p. 113.)

Obligatory (late)-proliferation is also responsible for the characteristic ap-

Fig. 151 I *Lysimachia ephemerum*, II, III *L.thyrsiflora*, inflorescences. Due to Troll.

Fig. 152 Eucomis punctata, I apex of inflorescence seen from above, II terminal section of an inflorescence which is not yet fully in flower, III inflorescence in full flower. Due to Troll.

pearance of the flowering shoots of *Glaux maritima* (Fig. 150 I). The proliferating shoot section also remains sterile in this case, and can form paraclades in its proximal section.

Late proliferation (retarded proliferation) is particularly frequently met with in the **monocotyledons**. Troll (1962) has given a detailed account of it for *Chlorophytum comosum*, in which the thyrsic florescences of the flowering shoots regularly form rosettes of leaves at their tips. Since these are drooping they come into contact with the ground, where they form roots, and in this way bring about vegetative reproduction of the plant. The luxurious inflorescence apex maintains "its size during the course of the development of the synflorescence, instead of being consumed, which is really the prerequisite for the fact that it is finally able to rather suddenly change over to vegetative growth". In the process the bracts of the florescence are modified through transitional forms to larger and larger foliage leaves, which is made possible by the primary growth of the axis which finally leads to the formation of a rhizome. Apart from this "apical proliferation" it is also possible for the partial florescences (chiefly in the zone of the main florescence) to change over to proliferating growth. This takes place so that the formation of flowers within a bostryx is prematurely interrupted, "in that a vegetative bud is created in the place of a floral primordium".

The initial phase of proliferating growth can also be regularly observed at the apex of the racemose main florescence of *Eucomis punctata* (Fig. 152), where the bracts are changed into foliose leaf organs, so that the inflorescence is crowned with a tuft of green leaves. This is also reminiscent of the familiar image of the infructescence of *Ananas comosus*, which is likewise brought about by retarded proliferation. In this case the proliferating shoots can be used as cuttings. Retarded proliferation also takes place in *Tillandsia latifolia*, in the course of which the axis of the inflorescence becomes more and more pendulous and its tip becomes closer to the substratum, so that the proliferating shoot can take hold in it with its roots.

Early proliferation (premature proliferation), i.e. proliferation of the apex before the formation of the main florescence takes place, is found in the Liliaceae, as for example in the genus *Polygonatum* (especially marked in *P.verticillatum*, but also in other species). Troll (1961) has given detailed descriptions of premature proliferation in another monocotyledonous family, namely several genera of the Commelinaceae.

Whilst the majority of the Commelinaceae possess synflorescences, which terminate in a main florescence and which represent in that respect "**closed synflorescences**", in *Cochliostema odoratissimum* the shoot apex of the main axis remains vegetative, and the formation of flowers remains confined to thyrsic enrichment shoots which arise from the axils of the foliose leaves of the rosette-forming main shoot. In that respect, therefore, we are concerned with an "**open synflorescence**" (see Troll 1958, p. 133).* The same is true of *Rhoeo* and *Coleotrype*, with the difference that the main axis is not shortened. According to Troll's later results (1961, p. 116), *Campelia zanonia* is also comparable with *Cochliostema*.

---

* The concepts of "open" and "closed synflorescence" are not therefore synonymous with the terms "polytelic" (ending not with a terminal flower, but with a many-flowered open florescence) and "monotelic" (ending with a terminal flower)!

A consequence of proliferation is that a vegetative section follows after a flowering one on one and the same axis, and this alternation can repeat many times. This is particularly true of many woody plants, as for example in many members of the Myrtaceae. Especially noteworthy amongst these are *Melaleuca* and *Callistemon*, and also *Calothamnus*. The inflorescence of *Melaleuca genistifolia* reproduced in Fig. 153 I, II takes the form of a terminal spike (I) before it flowers, but the inflorescence apex reverts to vegetative growth as soon as anthesis begins (II). This alternation of flowering and vegetative regions is even more conspicuous in *Callistemon coccineus* (Fig. 154 I), where the inflorescence apex has already begun to grow out vegetatively at an earlier time. The shoot of *Callistemon rigidus* shown in Fig. 154 II is continuing to grow on again after alternating twice already between flowering and vegetative zones. There is then always the impression given as though the flower-bearing zones were being "inserted" in the system of vegetative shoots, which doubtless led Parkin (1914) to classify them as "intercalary inflorescences". In fact what appears to us at first sight to be an unusual kind of arrangement for the flowers is just the result of premature proliferation.

Fig. 153 Melaleuca genistifolia inflorescence before (I) and (II) during anthesis. Due to Troll.

2.2.10   Truncate synflorescences

(a) Monotelic truncate synflorescences

There are also in fact non-proliferating inflorescences in the Myrtaceae which
lack a terminal flower. These can still, however, be recognised as monotelic
systems, since either the loss of the terminal flower is merely facultative – as
happens for example in *Syzygium jambos* (= *Eugenia jambos, Jambosa vulgaris*),
according to Troll (1969, p. 258) – or else at least the paraclades, with their
paniculate, botryoid or stachyoid branching, produce terminal flowers. Troll
(1961, p. 116) gives the term "**monotelic truncate synflorescences**" to such
inflorescences. Further specific examples of this phenomenon are given by Troll
(1969, p. 255), partly by reference to other authors, as for instance the inflorescence of the Horse Chestnut *Aesculus hippocastanum*, and the inflorescences of
Cimicifuga spp., *Ligustrum, Schinus molle*, etc.

However, with monotelic synflorescences it can also happen that not only the
terminal flower is abortive, but also the most recently differentiated lateral
flowers, briefly: by means of a *"complete reduction of the end of the synflorescence"*

Fig. 154 Callistemon coccineus (I) and *C.rigidus* (II), flowering shoots. Due to Troll.

(Troll). *Macleaya cordata* (*Bocconia cordata*), a member of the Papaveraceae-Papaveroideae with paniculate inflorescences, will serve as a good example of this. According to the investigations of Troll (1961, p. 118) the apex of the synflorescence often remains undeveloped; "on the weaker shoots it does often however develop completely, in order to terminate with a flower in the usual manner of members of the Papaveroideae".

It is appropriate at this point to cite the species of *Loeselia*, section *Loeselia*, from the monotelic family of the Polemoniaceae. To this belongs not only *Loeselia coerulea* as reproduced in Fig. 155, but also *L. mexicana*, whose inflorescence structure is illustrated by the branching diagrams in Fig. 149 II to V. It is characteristic of all the species of this genus that numerous sterile bracts precede the terminal flowers, both on the main axis and on the paraclades. These metaxyphylls are probably a sign of the weakening of the inflorescence apex, which has previously initiated numerous lateral axes. This is therefore consistent with the fact that in very long inflorescences it sometimes does not manage to form a terminal flower (Fig. 149 V). Since the differentiation of the uppermost lateral flowers has already been inhibited beforehand, it is also possible to say that the apex of the synflorescence has become abortive. In *L. coerulea* and in *L. pumila* this seems to be the rule. In all these cases, however, the paraclades are always provided with terminal flowers, which therefore indicates the monotelic nature of these inflorescences (Weberling 1963). – Reduction of the apex of the synflorescence also occurs in the Rubiaceae, which mostly show monotelic synflorescences (Weberling 1977); *Petunga racemosa* and *Gonzalagunia spicata* may be quoted as examples of this (Fig. 156).

The inflorescences of the Valerianaceae are characterised by the constant lack of the terminal flower (Weberling 1961). Only in a single case from out of more than several thousand plants of *Valeriana officinalis* was an inflorescence found which had a terminal flower, and that was admittedly abnormal in form, i.e. with two spurs, and in that respect disymmetric (Fig. 105 IV); a similar special case was observed in *Valeriana pauciflora* (see also Cejp 1926, p. 113; concerning abnormal and multiply-spurred terminal flowers in *Valeriana* and *Centranthus*). The distal end of the inflorescences of the Valerianaceae is normally represented by a "fork", formed from the two uppermost, dichasially branched paraclades (Fig. 105 I).

The way in which this phenomenon comes about will be made clear by considering how the inflorescence develops. If the lateral differentiation of the paraclades by the apex of the inflorescence is examined, it will be observed that the initialisation of the paraclades occurs earlier and earlier in relation to their subtending bracts. They do of course become gradually weaker further up the stem, but do not apparently keep in step with the decreasing dimensions of the stem and the apex of the inflorescence, respectively. Therefore in the end the primordia of the uppermost paraclades attain such predominance, that their precocious initiation completely consumes the apical meristem for their own formation, and the latter no longer regenerates fully. This is also evident from the anatomical investigations of Philipson (1947, p. 413); "Indeed, the buds come more and more to dominate their bracts ... eventually they arise so early and involve so much of the apical meristem that no residuum of the apex remains between them to continue the main axis".

Indications that this interpretation is correct are provided by the differing ways in which the primary flowers occur within the cymose partial inflorescences in the acrotonous branching of the inflorescences of *Valerianella*, *Fedia* and several species of *Valeriana* (*V. alliarifolia*). These are often lacking on the branches of lower order, and the later they appear within the sequence of cymose branching, the more the acrotonous development of the partial inflorescences asserts itself in the vigour of the lateral axes and the extent of their branching. On the other hand, the reverse happens in the starvation forms, where the lateral axes of the first order often terminate with primary flowers (Figs 119, 120).

(b) Polytelic truncate synflorescences

The loss of the terminal flower in a monotelic truncate synflorescence corresponds to the loss of the main florescence in a polytelic truncate synflorescence.

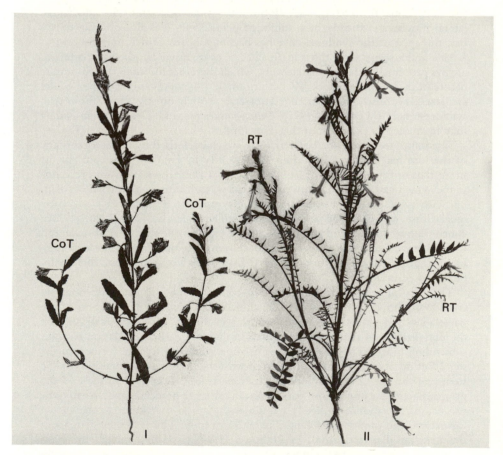

Fig. 155 Loeselia coerulea (Schaffner, Mexico; Goet), complete reduction of the apex of the inflorescence. CoT Cotyledonary shoots. II *Polemonium pauciflorum*, flowering plant, RT rosette shoot.

In both cases this is brought about by a modification of the structural plan in such a way that only the enrichment zone of the inflorescence is formed. This was termed "truncation" by Troll (1969, p. 255).

According to Troll (1963, p. 120; 1964, p. 100 ff.; 1966, p. 125/126, 130; 1967, p. 100) the families in which polytelic truncate synflorescences can be found include the Balsaminaceae, Scrophulariaceae, Papilionaceae, and the Acanthaceae (Sell 1969). This last turns out to be very informative in respect of the derivation of the truncate synflorescence from the complete synflorescence.

In the Papilionaceae the overwhelming majority of main shoots terminate in a double raceme of homothetic structure, as we have already described in the

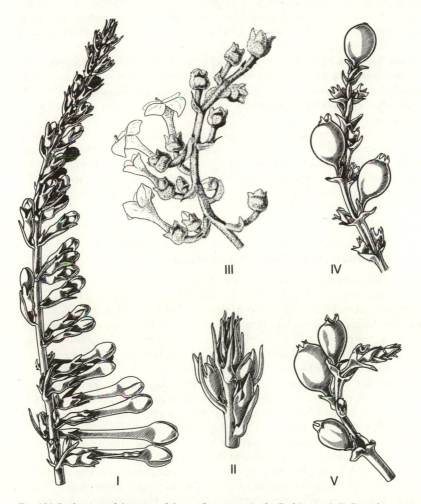

Fig. 156 Reduction of the apex of the synflorescence in the Rubiaceae. I, II *Gonzalagunia spicata*, terminal thyrse (I), and its terminal part more greatly magnified (II). III *Gonzalagunia dependens*, thyrsoid inflorescence with a terminal flower (Schimpff 660, Ecuador, M). IV, V *Petunga racemosa*, axillary botryoid inflorescence (IV), V without terminal flower.

examples of *Trifolium campestre* (Fig. 110 II) and *Melilotus officinalis* (also see section 2.1.4 and Fig. 109 I). In the Sophoreae, Podalyrieae, Genisteae and several genera of the Galegeae-Psoraleineae (e.g. *Amorpha*) on the other hand the main shoot terminates with a simple raceme, of which the Lupin may be cited as a well-known example. A general plan can be seen "if it is assumed that in forms where the flowering shoots end in a double raceme, " (homothetic branching, author's comment) "the main florescence is suppressed, which is equivalent to saying that a truncate synflorescence is present" (Troll 1963, p. 120). In fact in some of these cases, according to Troll's data, the main florescence sometimes appears in the form of a simple raceme, as for example in species of *Astragalus* and *Glycyrrhiza* (Galegeae) as well as in species of *Trifolium*, *Melilotus* and *Medicago*. This is particularly true of *Medicago lupulina*, in which this character is said to appear so often, "that in some places it is difficult to find specimens with truncate synflorescences".

In those plants which develop with polytelic truncate synflorescences the hypotagma is usually differentiated into two regions: a distal section, in which the paraclades are reduced to the coflorescence, and a proximal section which Troll calls the "**special hypotagma**, in which the paraclades are provided with a hypotagma".

The formation of truncate synflorescences can generally be considered to be a derived character. This can also be made clear on the basis of the Papilionaceae: the Sophoreae, Podalyrieae and the Genisteae are provided with main florescences and according to the interpretation of Engler, have been shown to be relatively primitive members of the family on account of the morphological characters of the flower and fruit.

2.2.11   Systematic significance of synflorescence structures, transition from monotely to polytely

It may have already become perfectly clear that the consideration of the inflorescence structure can be highly useful for the characterisation of natural families, provided that Troll's method for the typology of inflorescences is employed. The larger families also often behave uniformly by showing either the monotelic or the polytelic type of inflorescence structure. This is often also true for the specific expression of one or the other type, but also particular lines of development make themselves evident by the modification of particular elements. Since the clarification of the typological relations usually requires an analysis of the complete growth form, this results in a more precise understanding of that otherwise often only vaguely-defined character complex which determines it, commonly known as the "habit" of a plant.

The families of the Caryophyllaceae, Geraniaceae, Melastomataceae, Oleaceae, Gentianaceae, Hydrophyllaceae and Boraginaceae are amongst those which are characterised by the constancy of monotelic synflorescences. On the other hand, polytelic synflorescence structure is the rule in all members of the Cruciferae, Resedaceae, Leguminosae (exception: *Gleditsia*, *Gymnocladus*), Malvaceae, Primulaceae, Labiatae, Scrophulariaceae, Gesneriaceae, Compositae and – apart from a few exceptions, particularly in the Alismataceae, Scheuchzeriaceae, Juncaginaceae, Burmanniaceae, Restionaceae (female *Leptocarpus*), possibly some Liliaceae (*Trillium, Paris, Tulipa, Aspidistra*) and Commelinaceae, and perhaps also isolated cases in the Gramineae-Bambusoideae – in all the monocots. There are also not a few monotelic families in which there occur taxonomic groups with proliferating or truncate synflorescences as well as related groups with polytelic synflorescence structure; good examples of this are probably the Papaveraceae with the almost entirely polytelic Fumarioideae or the Campanula-ceae with the polytelic Lobelioideae.

Although most of the monotelic truncate synflorescences may be unambigu-ously classified with the monotelic synflorescence type, in spite of the failure of the terminal flower to develop, there is the obvious question as to whether or not monotelic and polytelic inflorescences are linked to one another by inter-mediates. It may even be postulated, on the basis of general systematic and phylogenetic considerations, that the two inflorescence types cannot exist in complete isolation from one another.

If we once again compare the two types on the basis of the preceding discussion by means of the diagrams in Fig. 122 I, II and Fig. 125 I, II, then we can confirm that the polytelic synflorescences are essentially different from the monotelic synflorescences, not only in the lack of the terminal flower, but also in the formation of the lateral flower-bearing branches. In contrast to the structural plan of the monotelic synflorescence, in which all the lateral axes represent elements which are completely equivalent morphologically, i.e. the paraclades, a differentiation of the lateral shoots takes place in the polytelic synflorescence. In this there occur single flowers or cymose partial florescences in the terminal section which is defined as the main florescence, and then in the region below the main florescence there are branches which terminate in a florescence (co-florescence). The latter repeat the structure of the main shoot, and are not limited by a terminal flower, and thus represent the paraclades of this system.

In order to derive this type of polytelic synflorescence, which is already more highly differentiated in its fundamental plan, from the less differentiated plan of the monotelic synflorescence, only two steps need to be postulated: 1) the **loss of the terminal flower** and 2) a **specialisation** of the lateral shoots. The latter would occur in such a way that a) the lateral shoots in the distal zone would form elements of the main florescence as individual flowers or partial flores-cences; and b) the lateral shoots of the proximal part would form the paraclades, which differ in their branching mode from the lateral axes forming the elements of the main florescence, since they now come to resemble the main shoot by ending in a florescence themselves, and being able to produce paraclades of the next highest order in their proximal zone. The loss of organs and the specialisation of what were originally equivalent structural elements are processes which we encounter again and again in the evolution of organisms. It is therefore no surprise to discover that forms which correspond to transitions between the two types of inflorescence can be identified. The transition between one type to the other has evidently taken place repeatedly and independently within different families, which makes it possible to distinguish between taxa which are primitive or more advanced in this respect from amongst members of a family or other larger or smaller related groups.

The specialisation of the flower-bearing lateral branches which is typical of polytelic synflorescences is also quite often found to be predisposed in a panicle or in a thyrsoid, in the form of a disjunct-heterocladic formation of the overall branching. We have already become acquainted with this in the thyrsoid shown in Fig. 109 V. Here it is a result of the isomorphic formation of numerous cymosely branched paraclades in the terminal section of the inflorescence and the repetition of this homocladic-thyrsoid structure in the branching of the proximal paraclades. In the panicle this process of homogenisation of the paraclades within the distal zone lies in the fact that they all remain limited to their terminal flowers, so that more or less extended botryoid terminal sections are formed, as we saw, for example, in species of *Collomia* (Fig. 143, Fig. 157 III). Nevertheless, diadic or triadic panicle branches still occur here, as a transition between the monadic paraclades of the distal section and the botryoid or paniculate proximal paraclades, so that the change of branching pattern does not occur so abruptly here as it does in the disjunct-heterocladic thyrsoid in Fig. 109 V. In the inflorescence of *Gilia capitata* (Fig. 157 I, II), whose panicle branches are all condensed into terminal or lateral heads (cephaloids), the botryoid terminal sections of the main axis and the special panicles are even more extended, and the intervening diads and triads less numerous. It is easily possible to imagine that these intervening

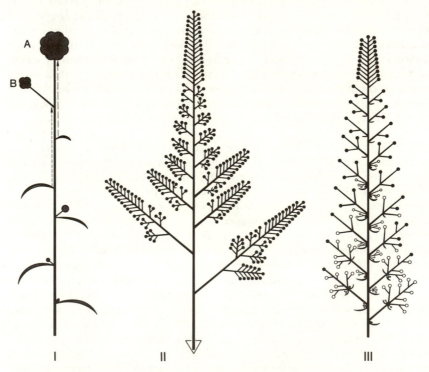

I   II   III

Fig. 157 I, II *Gilia capitata*, I Outline diagram of a flowering shoot, with the capitate flower clusters shaded; the arrows indicate the concaulescent displacement of the axillary shoots; II branching diagram of the terminal head A in I; as a result of concaulescence the lowest branch is included within the flower head. III *Collomia grandiflora*, branching diagram of a terminal flower head.

intermediate branches are also omitted, so that the disjunction between the botryoid terminal section and the botryoid proximal paraclades becomes complete. It then only needs the terminal flowers to become subject to reduction to complete the transition to a polytelic synflorescence.

The observations made by Wydler as early as 1843 (p. 183) and 1860 (p. 156) as well as those of Braun (1853, p. 52) on *Agrimonia eupatoria* (Agrimony) are noteworthy in this context. The monotelic and distinctly disjunct heterocladic inflorescence of this plant ends in a stachyoid to botryoid section several decimetres long, which can include well over 100 flowers, each provided with a pair of prophylls. This botryoid is preceded by one or several more or less well-developed branches, which according to the level of their insertion may still bear foliage leaves in their proximal part. They also terminate in a long botryoid (Fig. 158). The terminal flower of the main axis often fails to develop any more, especially in luxurious specimens, whereas it does occur in weaker specimens or on the botryoid paraclades. Inasmuch as the terminal flower does manage to develop, it is often preceded by a number of sterile bracts. In other cases the tip of the inflorescence atrophies along with the terminal flower and the primordia of those lateral flowers which are placed uppermost. – Similarly few-flowered inflorescences of *Campanula rapunculoides* also terminate with a flower, whereas in vigorous specimens the apex of the long botryoid section becomes abortive. This gives the impression that the apex of the inflorescence finally becomes exhausted by the continued differentiation of the lateral flowers (see Braun 1853; Troll 1964, p. 22). In so far as the terminal flower is completely developed, this happens at the expense of the uppermost lateral flowers, so that several metaxyphylls then precede the terminal flower. In *Campanula rapunculoides*, however, and in contrast to *Agrimonia eupatoria*, several paraclades with triads or with even richer ramification occur below the botryoidal terminal section of the inflorescence, and form a transition to the proximal paraclades, which repeat the structure of the distal section. In *Agrimonia eupatoria*, however, it needs only the loss of the terminal flowers not only from the main axis, but also from the botryoidal paraclades, to achieve the status of a polytelic multiply-racemose synflorescence.

The transition from monotely to polytely is very readily explained by reference to several members of the Papaveraceae-Fumarioideae which were investigated by Troll (1964, p. 9 ff.; 1969, p. 293 ff.), namely species of the genera *Dicentra* and *Corydalis*. Whereas the inflorescences of the Papaveroideae are monotelic in structure throughout – we have already mentioned the truncate synflorescences of *Macleaya* in section 2.2.10 – the Fumarioideae nearly all possess polytelic inflorescences with heterothetic-racemose branching, and appear to be rather more derived also on account of their disymmetric or even transversally-zygomorphic flowers (section 1.4.3.4). *Corydalis glauca* makes an exception, because its inflorescences are usually paniculately branched and have terminal flowers (Fig. 159 I). It is interesting to note that its terminal flowers show the same transverse-zygomorphic structure as the lateral flowers (Fig. 159 VI). The transition to polytely is to some extent already prepared for here in the sense that the terminal flower is often preceded by a more or less pronounced botryoid section (Fig. 159 II, III). Since the same can also occur in the more highly branched paraclades which precede this region, a conspicuous disjunction is produced in the enrichment zone. This already anticipates the differentia-tion which has taken place: in the other species of this genus and the majority of the other genera of this sub-family: on the one hand, the formation of single flowers as elements of a racemose main florescence and, on the other, the organisation of the racemose florescences of the polytelic paraclades. With the reduction of the terminal flowers this differentiation and the transition to polytely becomes complete. An example of this is shown by the inflorescence of *Corydalis lutea* in Fig. 160. Among the species of *Dicentra*, which mostly possess monotelic-thyrsoid (or di-thyrsoid) synflorescences (as for example, *D. eximia*), *Dicentra spectabilis* and *D. cucullaria* as well as *D. canadensis* are characterised by the fact that their inflorescences are reduced to racemose branch-ing, in the terminal section at least, and that the apical region becomes abortive; the transition to polytely has therefore taken place here also.

Along with the '**homogenisation**', i.e. the increasing equivalence in the formation of the lateral shoots of the florescence-like section which appear as individual flowers or cymes, and the '**truncation**' (in French, 'troncature'), i.e.

Fig. 158 Agrimonia eupatoria, inflorescence. Due to Troll.

the reduction of the terminal flower(s), another change is in progress. According to the extensive investigations of Maresquelle (1970 etc.) and Sell (1969, 1976 etc.), this is a reversal in the order of flowering so that it now proceeds towards the tip, which is usually termed "**racemisation**" (Fr. racémisation).

We have already established above that the transition from monotelic to polytelic inflorescence structure must have occurred more than once and independently within different families. Further examples of this can be found in the

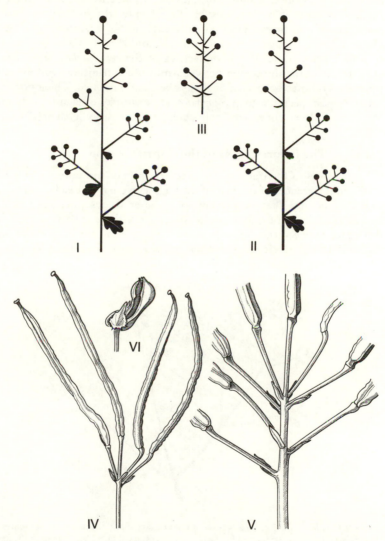

Fig. 159 Corydalis glauca, I, II vertical diagrams of synflorescences, III botryoid terminal section of another synflorescence; IV, V 4- and 8-flowered synflorescences in fruit; VI terminal flower in bud. All due to Troll.

Rubiaceae, and from among these the genus *Perama* from the tribe Spermaco-
ceae, as well as the genera *Crucianella* and *Relbunium* from the Rubieae and the
Cruciata-Valantia-Meionandra group of genera are characterised by polytelic
synflorescences (Weberling 1977a). The same is true of the Ranunculaceae, in
which in fact only the two genera *Delphinium* and *Aconitum* of the Delphinieae,
which also occupy a special position because of their highly modified zygomorphic flowers, are characterised by polytelic inflorescence structure. Furthermore, species of *Cimicifuga* and *Isopyrum* and *Thalictrum alpinum* occupy an
intermediate position with the formation of monotelic truncate synflorescences.
In the Rosaceae it is quite possible to interpret the inflorescences of several
members of the Spiraeoideae as polytelic, as for example those of *Sibiraea
altaiensis* (= *S. laevigata*), *Holodiscus discolor* and *Aruncus dioicus* (= *A. sylvestris*);
to which perhaps also the inflorescences of *Spiraea douglasii* should be added,
although here the distinction between monotelic truncate synflorescences and
polytelic synflorescences cannot always be clearly defined. The inflorescences of
other *Spiraea* species, such as *S.japonica*, are examples of facultatively monotelic
truncate synflorescences (see Troll 1969, p. 320 ff., also Schaeppi 1970).

2.2.12   The inflorescences of the Caprifoliaceae

Various stages of the transition between monotely and polytely are also found
in the inflorescences of the Caprifoliaceae, as we have already mentioned (section 2.2), because they do not appear to be at all congruent in terms of their
branching and their external appearance. Hence there appears to be a profound
difference between the highly branched corymbs of our Elder (*Sambucus nigra*,
Fig. 161 I) and species of *Viburnum* (Guelder-rose, *V.opulus*, Wayfaring-tree,

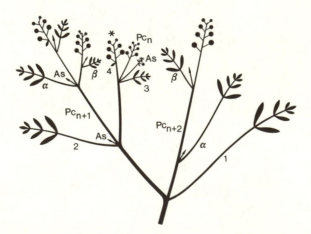

Fig. 160 Corydalis lutea. Terminal section of the primary shoot, * is the terminal inflorescence.
The paraclades Pc_{n+2}, Pc_{n+1} and Pc_n, with their accessory shoots (As), belong in the axils of the
leaf organs 1 to 3. The accessory shoot of Pc_n is confined to the florescence, as is Pc_n itself, but
only its proximal flower has been formed and the rest are rudimentary. The paraclades Pc_{n+2} and
Pc_{n+1} have pushed the main shoot aside and aligned themselves in the direction of the main axis.
After Troll.

Fig. 161 Inflorescences of *Sambucus nigra* (I) and *Lonicera periclymenum* (II). I from Troll and Weberling, II photo by Th. Arzt.

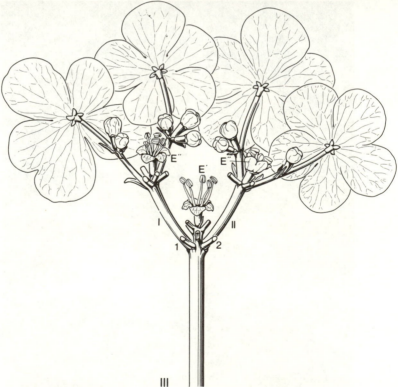

Fig. 162 I *Viburnum opulus*, II *V.lantana*, corymbose inflorescences seen from above, III *V.opulus*, top view of special panicle, the marginal flowers bent back somewhat for clarity, with all the paraclades except the two proximal ones cut away, E′ terminal flower of the special panicle, E″ of the axes labelled I, II. I, II from Troll and Weberling, II due to Troll.

V. lantana, Fig. 162), and the two–flowered inflorescences of *Linnaea borealis* (Fig. 121 III) or the "long–stalked flower pairs" of the Tartarian Honeysuckle (*Lonicera tatarica*, Fig.121 II) and by comparison with the last again the flower heads of Honeysuckle, *Lonicera periclymenum* (Fig. 121 I, Fig. 161 II), or the whorls of flowers in *Lonicera caprifolium* (Fig. 168 IV), which appear to be placed upon saucer-shaped whorls of bracts. However, the relationship between these apparently different forms of inflorescence becomes apparent as soon as the structure of the shoot as a whole is considered, without paying undue attention to what appear at first sight to be units of flower clusters, as for example in the "flower pairs" of *Lonicera xylosteum*. We do not wish to repeat all the details here, but the results of a comparison on a larger scale reveal that a paniculate inflorescence, or a thyrsoid, is the starting point for the derivation of the different forms of inflorescences.

Within the Caprifoliaceae, panicle-like, monotelic inflorescences only appear in the genera *Sambucus* and *Viburnum*. The species of the section *Botryosambucus*, to which the Red-berried Elder (*Sambucus racemosa*) belongs, are characterised by conical, decussate panicles where the degree of branching increases progressively towards the base, according to the model in Fig. 163 I. The same is true for the majority of the members of the section *Thyrsosma*, the most primitive section of the genus *Viburnum*. (Regarding their position, the inflorescences of *Sambucus racemosa* must really be regarded as derived: they terminate axillary short shoots, which arise from the previous year's [or older] long shoots, whereas the inflorescences in section *Sambucus* terminate long shoots.)

Nevertheless, the prevailing type of inflorescence in the genus *Sambucus* and in *Viburnum* as well is the corymb, which is fundamentally derived from the conical panicle in the manner already described in section 2.1.4. In detail of course the formation of the branching in the corymbs of *Sambucus nigra* and related species is extremely complex. In fact the two lowermost and the two following paraclades of the first order, respectively, are brought into proximity and form false whorls. Furthermore, the paraclades of second order in the lower of these panicle branches which arise from the axils of the prophylls are strongly developed and directed somewhat towards the underside, while the following and weaker paraclades of second order are alternately arranged (for further details see Troll and Weberling, 1966). The corymbs of *Viburnum opulus* have an essentially similar structure. In this case there are also further modifications, as in many other *Viburnum* species, namely 1) a multiplication of the paraclades of the first order by means of genuine pleiomery in the proximal (and middle) whorl branches, which continues to a multiple of six in the proximal paraclades of *Viburnum opulus*, and 2) the enlargement of the peripheral flowers of the corymb in many species, as already discussed in section 1.4.3.9. These corymbs are therefore to be regarded as highly derived, on account of their special structure.

To a greater or lesser degree, the inflorescences of the other genera of the Caprifoliaceae present modified thyroids or thyrses, respectively. The relationship with the paniculate (or thyrsoid–paniculate) structure of the inflorescence illustrated in Fig. 163 I is easily recognised when it is assumed that the branching in the paraclades of the first to the nth order is continued exclusively from the axils of the prophylls.

In the genus *Diervilla* the inflorescences are homocladic thyrsoids with rapidly increasing impoverishment of the cymose partial inflorescences towards the apex. The terminal flower can be present in a completely normal state (Fig. 164 III, IV), or can also be lacking (Fig. 164 V). What we are dealing with is therefore the facultative formation of a monotelic truncate synflorescence (Fig. 163 II).

In the closely related genus *Weigela*, which is also characterised by zygomorphic flowers and bicarpellate many-seeded capsules, the terminal flower is always absent. A flowering branch from a species such as *Weigela floribunda* (Fig. 165) may even create the impression that it is a disjunct-heterocladic thyrse of polytelic nature. However, in this case we ought not to overlook the fact that not only the terminal but also the (impoverished) lateral thyrses form the apices of short shoots, which arise from the terminal and axillary buds of a long shoot from the previous year, which can be easily seen from the large number of bud scales which occur at its base. Since we must consider each of these short shoots as an independent flowering shoot, we can only characterise its inflorescence as a monotelic truncate synflorescence with homocladic-thyrsoid branching, and cannot describe the whole branching system illustrated by Fig. 165 as a dithyrsic polytelic inflorescence.

Similarly, the inflorescences of *Symphoricarpos* represent monotelic truncate synflorescences, but differ from *Weigela* in that not only the terminal flower but the entire terminal portion of the synflorescence has become rudimentary. Since in contrast to *Weigela*, and particularly *Diervilla*, it is extremely rare that further flowers arise from the axils of the prophylls of the partial inflorescences, the result here is that spicate (to racemose) inflorescences are formed.

At first sight the same mode of branching as in *Weigela floribunda* appears to

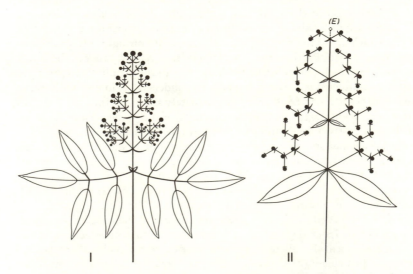

Fig. 163 I Diagram of the thyrsoid-paniculate branching of a synflorescence in the Caprifoliaceae. II *Diervilla sessilifolia*, branching diagram of the inflorescence, a terminal flower occurs occasionally (E).

predominate in the flowering shoots of many species of *Abelia*; compare it with the branching diagram of *Abelia engleriana* which is reproduced in Fig. 166 II. On closer inspection, however, it becomes evident that in the lateral axes which terminate with a flower each flower bears two further whorls of bracts besides the two prophylls, and that these are united into a ring of four like an epicalyx at the base of the flower. The same is true for the flowering shoot of *Abelia rupestris* (Fig. 167 I), where the main and lateral shoots belong to the same season. These so-called "supernumerary" bracts appear not only in most species of *Abelia*, but also in the whole tribe of the Linnaeeae, and are also characteristic of the genera *Dipelta*, *Kolkwitzia* and *Linnaea* (Fig. 121 III, Fig. 167 III). The axils of these bracts do not usually produce any further flower-bearing branches, although this can be the case occasionally. In *A. achersoniana* (Fig. 166 I) we even have a lateral flower marked B with its two prophylls which regularly appears in one of the pair of axils below the flower A, and similarly in

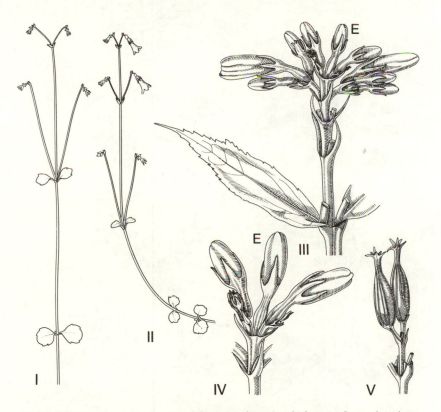

Fig. 164 I, II *Linnaea borealis*, exceptional flowering shoots in which 4 or 6 flowers have been formed (after Wittrock). III-V *Diervilla sessilifolia*, III thyrsoid inflorescence with terminal flower (E), the lower paraclades partly cut away for clarity, IV distal section of III more highly magnified, V distal portion of another inflorescence with an abortive terminal bud; in yet further instances the inflorescence axis terminates blindly between the two uppermost paraclades. From Troll and Weberling.

Kolkwitzia amabilis. These sterile whorls of bracts do not therefore in any sense represent merely accessory elements, which are unimportant for the appraisal of the branching properties. Rather they indicate that in the partial inflorescences which at first sight appear to be cymose we are not dealing with cymes, but with more or less severely reduced thyrsoids, and moreover thyrsoids which still end with terminal flowers! In this respect the branching of the two lower whorls of branches of the flowering shoot reproduced in the branching scheme of *A. aschersoniana* of Fig. 166 I is remarkable: in these partial inflorescences the terminal flower has become lost; the two proximal whorls show a thyrsic branching, and the two placed above these appear to be cymosely branched.

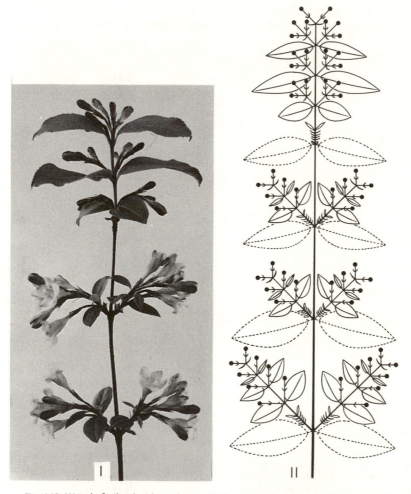

Fig. 165 *Weigela floribunda*, I long shoot with flowering short shoots, II branching diagram of the long shoot with flowering short shoots as in I; the leaves which subtended the short shoots have already fallen, and are shown with a dotted outline.

Nevertheless they still bear on their main axes one or two whorls of scale-like leaf organs which precede the branched region, as do the proximal partial inflorescences.

The tendency towards a reduction in the number of flowers which is indicated by the occurrence of whorls of sterile bracts has evidently increased in *Abelia engleriana* (Fig. 166 II), which is reckoned to be in the section *Uniflorae*, and has evidently reached the point where all partial inflorescences have only one flower. Furthermore, an impoverishment of the flowering branches can also occur, because the number of flower-bearing whorls of bracts in the lateral short shoots is reduced to a single distal whorl. This is observed within the genus *Abelia* in the subsections *Serratae* and *Biflorae*; *A. biflora*, whose branching

Fig. 166 Branching diagrams of *Abelia* inflorescences. I *Abelia aschersoniana*, II *A.engleriana*, explanations in the text. After Troll and Weberling.

diagram is represented in Fig. 167 II, belongs to the latter. Apart from the fact that no "supernumerary" whorls of bracts occur here, the branching system of this species conveniently leads us to an understanding of the shoot architecture of *Linnaea borealis*.

The branching system of *Linnaea borealis* is differentiated into a creeping long shoot which remains purely vegetative and which continues to grow during several vegetative seasons, and sterile or fertile short shoots (Fig. 167 III). Similar to *Abelia biflora* and related species, the formation of flowers is here usually restricted to two one-flowered partial inflorescences, which arise from the axils of a single distal whorl of bracts on the fertile short shoot. Here, however, the internode which precedes the whorl of bracts is elongated into a long and scapelike segregation internode (see Fig. 121 III). Each of the two nodding flowers is

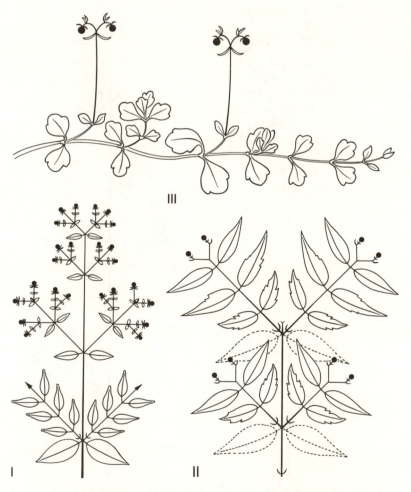

Fig. 167 Branching diagrams of flowering shoot systems of I *Abelia rupestris*, II *A. biflora* (subtending bracts displaced and recaulescent), III *Linnaea borealis*. After Troll and Weberling.

preceded by the whorl of prophylls and two more whorls of sterile bracteoles, and these form a 4-merous persistent epicalyx by fusing with the inferior ovary to a greater or lesser degree. This epicalyx is of importance for the dispersal of the fruit which becomes attached to animals by means of its sticky glandular hairs. Evidence for the correctness of our morphological interpretation is found in the fact that flower shoots can occasionally be found in which 4 or 6 flowers are superimposed in 2 or 3 stories (Fig. 164 I, II).

In the "flower pairs" of the Tartarian Honeysuckle, *Lonicera tatarica* (Fig. 121 II) and related species, we have quite a different situation from the two-flowered fertile short shoots of *Linnaea borealis*. The inflorescences of all species of *Lonicera* are polytelic synflorescences with thyrsic florescences. This is very clearly recognisable in the species of the subgenus *Caprifolium*, to which *Lonicera etrusca* also belongs (Fig. 168 I). The partial florescences here are usually 3-flowered, as in all the species of this subgenus, but they can also be 7-flowered and then are clearly recognisable as cymes. Their subtending leaves as well as their prophylls, are bracteose, and the internodes between the individual whorls of the cymes are scarcely developed, so that all the partial florescences of a florescence are condensed into a head. The capitate form of the inflorescence of *Lonicera periclymenum* (Fig. 121 I) comes about in the same way. In other species, however, the internodes of the florescence axis are clearly elongated, creating the impression of a "false spike", as in *Lonicera sempervirens*. If we then assume that the whorls of subtending leaves of the partial florescences are developed as almost orbicular and saucer-like frondose bracts, then the well-known characters of the Perfoliate Honeysuckle (*Lonicera caprifolium*, Fig. 168 IV) are the result.

The species of the subgenus *Lonicera*, to which the Fly Honeysuckle *Lonicera xylosteum* belongs, are divergent from the subgenus *Caprifolium* in respect of the formation of their inflorescences, in that the primary flowers of the partial florescences are lacking (Fig. 168 II, IV). Occasionally nevertheless plants are found in which rudimentary or even fully developed primary flowers occur in individual partial florescences (Fig. 168 III). In the majority of species the two-flowered partial florescences are placed in the axils of frondose whorls of bracts, which are frequently separated from one another by a considerable distance, so that the overall impression of a florescence becomes lost (Fig. 168 V). (Only in species such as *L. affinis*, illustrated in Fig. 168 II, is the impression of a florescence still preserved.) This is all the more true if the partial florescences appear to be long-peduncled as a result of pronounced elongation of the hypopodium, as is the case in *Lonicera xylosteum*. The "long-stalked flower pairs" of Fly Honeysuckle therefore represent partial florescences of a thyrsic polytelic inflorescence, whereas in the similarly twin-flowered inflorescences of *Linnaea borealis* we have reduced thyrses, and moreover monotelic truncate thyrsoids which terminate short shoots. Both of these, however, can be accommodated in the overall scheme for the inflorescence of the Caprifoliaceae without further ado.

The multiplicity of modifications in the inflorescences of *Lonicera* is still by no means exhausted by the formation of two-flowered partial florescences; the prophylls of both first and second orders can become fused together and modified in many ways, and the reduction of the flowers can also include one of the two secondary flowers, so that – in *L.gracilipes* – single flowered partial florescences are produced. On the other hand the final result in many species can be the congenital fusion of the ovaries of both secondary flowers (Fig. 168 VI). This is also the case to a limited extent in *Lonicera xylosteum*, and more so in *L. alpigena*, in which this synanthy, or more precisely syngyny, becomes more obvious at fruiting time by the formation of the so-called double berries. Syngyny is also found in *Kolkwitzia amabilis*, but here it is between the terminal flower of a partial florescence and the neighbouring lateral flower. It may also be mentioned that the inflorescences of *Leycesteria* and *Triosteum* are polytelic

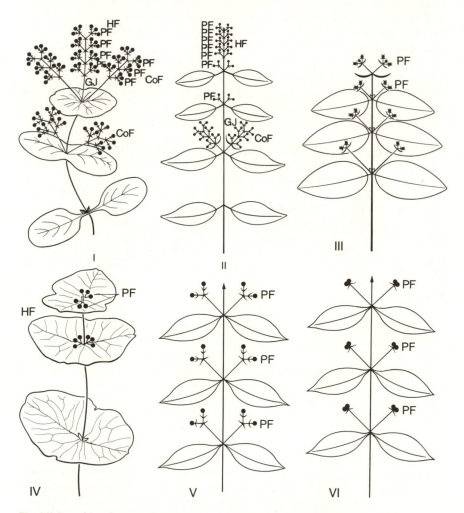

Fig. 168 Branching diagrams for the inflorescences of various species of *Lonicera*. HF Main florescence, CoF coflorescences, PF partial florescences, GJ basal internode. I *L.etrusca*, II *L.affinis* (coll. J. M. Dalziel, Thai-Yong, 1901; E); III *L.tatarica*, abnormal florescence in which the distal partial florescences occur in the axils of bracteose leaf organs and possess a more or less rudimentary primary flower, some partial florescences (PF) with acroscopic accessory buds, IV *L.caprifolium*, V *L.tatarica*, VI *L.alpigena*. From Troll and Weberling.

synflorescences, in which the florescences are represented by thyrses, or by thyrses reduced to false spikes.

2.3 Special cases of inflorescence structure

2.3.1 Anthoclades

2.3.1.1 *The concept of the anthoclade, monotelic anthoclades*

The forms of inflorescence which Goebel (1931, p. 2 ff.) has termed anthoclades have frequently attracted particular attention. According to Troll (1964, p. 157 ff.), who has defined the concept of an anthoclade rather more precisely, monotelic anthoclades are confined to thyrsoid inflorescences with acrotonic development. In the distal paraclades of such inflorescences the terminal flower is usually preceded only by the pair of prophylls (or by only one); however, the development of these paraclades – as we have already seen – can be greatly favoured, and this results in rich sympodial ramifications. If such paraclades are provided with foliose leaves and if they retain the mode of forming foliose leaf organs during the continuation of their branching, then they develop into "shoot systems which form alternately terminal flowers and foliose leaves" (Goebel). Such a shoot system is what is called an **anthoclade**. The term is not, however, confined to the distal paraclades, which branch sympodially from the beginning, but is also applied to the preceding paraclades which have a more or less extensive hypotagma, provided that these attain to the same form of branching in their apical sections. This of course is less often the case since the system as a whole develops acrotonically.

Examples of the formation of anthoclades in monotelic inflorescences are found in entirely different families, as in the Caryophyllaceae, Gentianaceae (*Exacum affine*), Hydrophyllaceae (*Nemophila*), Boraginaceae and also elsewhere – see the detailed exposition by Troll (1969). They are particularly frequent and various in the Solanaceae (*Atropa, Petunia, Hyoscyamus, Scopolia*, etc., see Danert 1958 as well as Troll 1969). Deadly Nightshade (*Atropa bella-donna*) will be chosen as a particularly appropriate textbook example, mainly because it presents an excellent exercise in the analysis of complex branching phenomena.

A vigorous flowering shoot of this herbaceous perennial is illustrated in Fig. 169. Its form is determined by the high degree of development of the distal paraclades, which often occur in fives, and follow one another closely according to the false whorl in which their subtending leaves are arranged. In this case the terminal flower of the main shoot has already fallen away. The branching of the developed paraclades is often monochasial from the very beginning, but can, however, also be dichasial at the start, or even trichasial – in the case that three leaf organs are present. The organisation of such a sympodially branched paraclade with foliose leaves, i.e. an anthoclade, is shown in Fig. 170 I.

At each individual node of the anthoclade an α-prophyll always occurs in a pair with a larger β-prophyll. It is striking that instead of being opposite one another, as would be expected in a whorl of prophylls, these two leaf organs are noticeably displaced to one side. They do not indeed belong to the same segment of the sympodium, but to two successive sections: the larger leaf is the β-prophyll of the preceding internode, and is connected to its axillary shoot in a recaulescent manner. This interpretation, which had already been proposed by Eichler (1875), may be visualised by means of the outline diagram in Fig. 170 II.

2.3.1.2   Independence of anthoclades

Anthoclades are able to a considerable extent to meet their own needs for assimilation by means of the leaves which they possess. On this basis they can even "become independent", in so far as they are equipped with the ability to produce stem-borne roots, and the hypotagma of the plant concerned remains short and insignificant. Those anthoclades which lie on the ground and then strike roots are able to separate themselves from the mother plant and grow on, continuing their creeping habit. In such cases therefore the mother plant is only significant in the juvenile stage. According to Troll (1969, p. 238 ff.), examples of this property are probably more numerous than previously thought; only *Glottiphyllum* (Aizoaceae), *Bulliardia* (Crassulaceae) and certain Solanaceae, particularly *Solanum violifolium* (which was more thoroughly investigated by Danert, 1958) were described by him in detail.

Fig. 169 *Atropa bella-donna*, flowering plant. Due to Troll.

2.3.1.3   Anthoclades in polytelic systems

According to Troll (1963, p. 123), anthoclades also occur in polytelic inflorescences, as for example in species of *Peperomia*. In particular he mentions *P. crispa*, in which the axial system "remains compressed with the exception of the basal internode of the individual members of the anthoclade, which follow each other in a spiral sequence".

2.3.2   Syndesmies

We have already mentioned the phenomenon of syndesmy during the discussion of metatopies (section 2.1.5), whilst postponing a more detailed account.

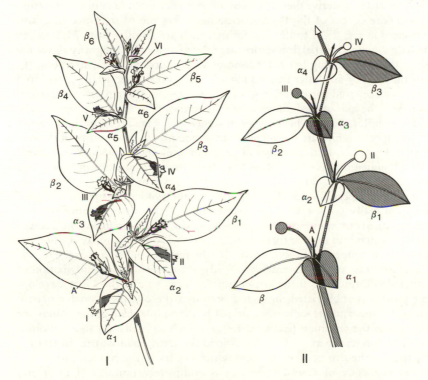

Fig. 170 *Atropa bella-donna*, I anthoclade, II diagram to explain the mode of branching and the arrangement of the leaf organs: β is the subtending leaf of the branch which terminates with the flower I and has the prophylls α_1 and β_1. The lateral branch which is emerging from the axil of α_1 and which terminates with the flower A is represented in diagram II with an arrow. It has also prepared itself for further branching from the axil of its β-prophyll. The prophyll β_1 is connected by recaulescence with its axillary shoot which terminates with the flower II; the connection is indicated by appropriate shading in the diagram. The branch which terminates with flower II bears the prophylls α_2 and β_2, which behave in the same way as α_1 and β_1. The sequential pattern of branching is emphasised by shading in the same way. The leaf organs α_2, α_3, α_4 have been moved across to the opposite side of the shoot in diagram II in order to make the branching system clearer. I is due to Troll.

Concaulescence and recaulescence are not in fact going to contribute a great deal to the explanation, because it is fundamentally a matter of the compression and fusion of an greatly branched axial system.

This phenomenon can be found in determinate, thyrsoid inflorescences, especially in cymoids. The term "syndesmy" expresses the fact that the partial inflorescences are "not freely developed, but are incorporated into the main axis, *including all their branches*, and are so to speak 'fastened' to it" (Troll 1964/69I, p. 133). Troll then distinguishes between concaulescent and recaulescent syndesmic cymoids. For the latter he quotes the inflorescences of *Elatostema* and *Procris* as examples, which we shall discuss below. As examples of concaulescent-syndesmic cymoids he describes the peculiar types of inflorescences which are encountered in a series of species from the genus *Cordia* in the Boraginaceae (Fig. 172 I, II), which had already been investigated by Baillon (1862/63) and Mez (1890, p. 569). In order to derive these forms of inflorescence from the normal structure of the inflorescences of the Boraginaceae he makes use of the three diagrams reproduced in Fig. 171 I to III. The forms which are familiar from *Myosotis*, in which the cincinnal partial florescences are moved upwards a long way above the axils of their subtending bracts by concaulescence, provide a starting point (I). With the help of an "intermediate scheme" (II) Troll now attempts to "fold down" the boragoid partial inflorescences (arrows) and illustrate their congenital fusion with the axis of the inflorescence – a "retrocaulescence", so to speak. The result is an inflorescence which occurs in *Cordia brevispicata* (III). In the majority of *Cordia* species which are distinguished by syndesmy, as for example in the two shown in Fig. 172 I, II, the number of partial inflorescences involved is of course increased, and moreover they often have the form of a double cincinnus – or better: a double boragoid – and the subtending leaves are lacking.

The above description of the derivation of the syndesmic inflorescences of *Cordia* can really only serve as an introduction to the understanding of these modified forms of inflorescence. It cannot provide any account of the growth processes which lead to their formation, since – as Troll had already realised – it is a matter of congenital fusion, in which "the features which are characteristic for the later state present themselves already at the moment of initialisation of the partial inflorescences at the growing point". In this case the entire development process is abbreviated, in such a way that the complete axial system is initialised at one and the same time, similar to the simpler case of the homaxone formation of the cincinnus (section 2.1.8). For such an "only apparently homogeneous inflorescence axis" Troll has coined the term **coenosome**. In fact the early stage of the mantle-like meristem which occurs on the convex and conical inflorescence apex of *Cordia verbenacea* is completely uniform (Uhlarz and Weberling 1977). This meristematic mantle undergoes its first division during the initialisation of the earliest floral primordia of the appropriate individual partial inflorescences, i.e. into simple or double boragoids. During this process the initialisation of the terminal flower of the complete thyrsoid inflorescence clearly precedes that of the other flowers. The further differentiation of flowers in both the proximal and distal boragoids which are initiated simultaneously takes place basipetally in a cincinnal arrangement, corresponding in this respect to the derivation according to the diagrams I to III of Fig. 171 and this means in contrast to the original polarity of the shoot apex. Floral primordia of different ages are therefore often found in close proximity in the different regions of the young coenosome. Furthermore, the vascular bundles which serve the flowers are initiated and differentiated individually in each separate cincinnus in a basipetal manner. Hence the only way in which the complex structure of the coenosome is still recognisable is by the arrangement of the flowers and in the accompanying anatomical structures. In this respect the young inflorescence of the *Cordia* species resembles the cymoid but umbellately condensed inflorescences of the African Hemp (*Sparmannia*), apart from the elongation of the pedicels which is lacking in the former case. This similarity is more obvious from the capitately expanded apex of the inflorescence axis after the flowers have fallen (Fig. 172 III). This has already been pointed out by Troll (1964, p. 14) and also later by Hagemann (1975).

Therefore, as was stressed at the beginning, the inflorescences of *Cordia* are primarily a case of compression of the complete axial system of the inflorescence. This should really give rise to a capitate form of the inflorescence. In fact there is a complete series of *Cordia* species with capitula of this sort, as for example *C. globosa* and *C. calocephala*, and of course Troll did assign something of a special position to these. In the case of the elongate coenosomes, at least in *Cordia verbenacea*, which was investigated in respect of its organogenesis and histogenesis, an early elongation of the inflorescence primordium takes place, immediately after the processes described above. During this process the young medullary cells also become capable of dividing again. Their ability to divide then ceases in basipetal order, so that the intercalary meristem at the base of the inflorescence remains active the longest and is finally used for the elongation of the "peduncle" of the inflorescence.

In the case of the thyrsic inflorescences of many members of the Urticaceae and

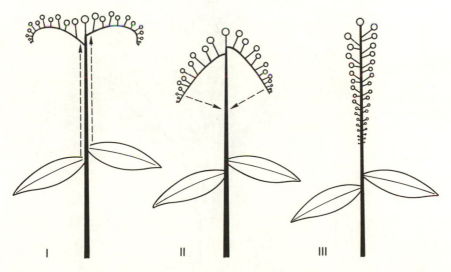

Fig. 171 Derivation of a concaulescent syndesmic cymoid according to Troll, I corresponds to the characters of a species of *Myosotis*, II intermediate situation (hypothetical), III corresponds to the characters of *Cordia brevispicata*.

the Moraceae the axial system of the dichasial and much-branched partial inflorescences is contracted into a disc-shaped structure. The fundamentally dichasial formation of such a coenosome often still becomes obvious during ontogenesis and can also often be recognised later on by means of the arrangement of the prophylls of different orders which are positioned on the margin, or near the margin on the underside (Fig. 173 I, II). Troll has classified such coenosomes as recaulescent-syndesmic cymoids on account of this prophyll position.

According to the thorough investigations of the inflorescences of the Urticaceae and the Moraceae carried out by Goebel (1931) and his student Bernbeck

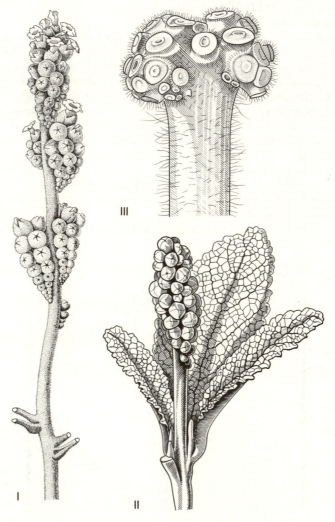

Fig. 172 Syndesmies. I *Cordia ferruginea*, II *Cordia verbenacea*, inflorescences. III *Sparmannia africana*, inflorescence after flowering, the flowers have fallen leaving the scars behind, I after Baillon, III after Troll.

(1932), what happens in these coenosomes is that there is congenital fusion of all the successive axes of the partial florescences, so that these no longer appear as such. Intermediate forms are found in the genera *Elatostema* and *Procris* in the Urticaceae, in as much as the consecutive axes in the male partial florescences are normally still freely developed, and are less contracted in the female partial florescences of several species of *Elatostema*.

In the male partial florescence of *Procris laevigata* illustrated in Fig. 173 III the consecutive branching pattern is still readily recognisable. In the female partial florescence of *Procris frutescens* (Fig. 173 I), on the other hand, all the flowers are inserted on a flat "cushion" of fleshy tissue. (These "flower cakes" emit a pleasant scent when fruiting, and the infructescences are eaten by animals which distribute the seeds.) The dichasial branching may be recognised during ontogenesis by the differentiation of the primordia, which are already broadened and cushion-like. The segments which correspond to the first order of branching are separated from one another by furrows, which are also still clearly visible in the young partial florescences of *Procris acuminata* (Fig. 173 II). In this case the primary and secondary flowers of the dichasium are not formed,

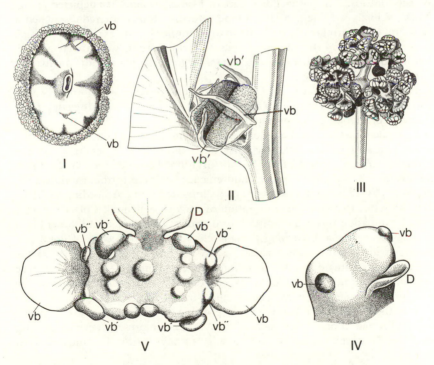

Fig. 173 I *Procris frutescens*, female partial florescence from below. II *Procris acuminata*, young female partial florescence. III *P. laevigata*, male partial florescence. IV, V *Dorstenia contrajerva*, young (IV) and older (V) primordia of a partial florescence, in IV the separation of the two complexes included in the axils of the two prophylls vb is already visible, D subtending leaf of the partial florescence, in V the primordia of primary, secondary and tertiary flowers of the cymes are visible, likewise those of their corresponding prophylls vb, vb', vb". II, III Original, the rest due to Bernbeck.

but their prophylls (vb, vb') appear well in advance of the differentiation of flowers. These are placed on the periphery of the disc-shaped partial florescence, along with the majority of the other prophylls of third and fourth order.

It is worth remarking on the stigma formation in *Procris frutescens* and *P.laevigata*, as observed by Goebel. Each individual stigma consists of a bunch of long multicellular uniseriate hairs, which are partly matted together and which serve for the collection of pollen grains. At a later stage the individual stigmas in a partial florescence are "no longer distinguishable from one another". "They merge together into a 'communal stigma' in the form of a mass of felted hairs. This receives the pollen for the complete inflorescence *collectively*, and then the pollen tubes grow further onwards in the felt and reach some female flower."

The arrangement and sequence of differentiation of the prophylls which corresponds to the dichasial branching is also often recognisable during the development of the partial florescences of *Dorstenia* species, which are similarly fused into compact bodies of tissue ("receptacles", Fig. 173 IV, V). These inflorescences, which often bear male and female flowers, show a considerable diversity of form. The partial florescence of *Dorstenia contrajerva* depicted in Fig. 173 IV, V in an early stage of development forms a four-cornered disc when it is mature. On its irregularly undulate margin numerous inwardly-curved small scales occur (Fig. 174 III) – the prophylls of various orders which were already mentioned for *Procris*. There are also several larger scales placed near the margin of the disc on the underside.

In *Dorstenia multiradiata* (Fig. 174 II), the inflorescence is completely surrounded by a border which has a fringe of longer, unequal rays projecting from it. These correspond to the prophylls of different orders.

In *D. plumeriaefolia*, on the other hand, the inflorescence takes the form of a folded disc as a result of the fact that both the second order axes and the segments corresponding to their further divisions become erect. In *D.arifolia* and several other species the partial inflorescences are even further modified: the coenosome is folded and forked into two elongated arms. According to Bernbeck this is due to a unilateral promotion of growth which takes place together with the initiation of the axes of the third order. The early establishment of a tendency to cincinnal branching is made clear by "the sequence of floral primordia at initialisation which zigzags away on both sides, starting from the primary flower situated above the inflorescence peduncle". – The same is true for the coenosome of *D. psilurus*, except that here the development of the branching is directed towards "above" and "below" (Fig. 174 I).

From these few examples it is already possible to understand that the coenosomatic formation of inflorescences in the Moraceae represents another starting point for the development of a multiplicity of new forms. The most derived

Fig. 174 I-IV Partial florescences of *Dorstenia psilurus, D. multiradiata* (II), *D. contrajerva* (III) and *D. urceolata* (IV). V *Ficus carica*, primordium of a partial florescence (later an urceolate syconium), which still has convex curvature. VI, VII *Ficus ribes*, longitudinal section through a partial florescence with (short-styled) sterile female "gall flowers" and male flowers (VI) and a female partial florescence (flowers fertile with long styles) (VII). V after Bernbeck, VI, VII after Goebel.

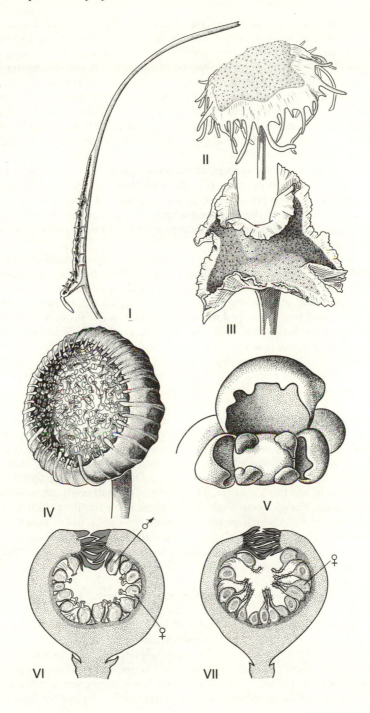

form of inflorescence in the Moraceae is undoubtedly represented by the urceo-late syconium of the figs (*Ficus*), to which the orbicular inflorescence disk of *Dorstenia urceolata* (Fig. 174 IV), with its concave upper surface and incurved margins, forms a transition. It must be presumed that the still saucer-shaped "receptacle" of *Dorstenia urceolata* (cup-shaped in *Castilla*) is further deepened by "intercalary" growth of the sides of the dish, and that at the same time the numerous scale-like prophylls which are situated on the margin in *Dorstenia urceolata* are elevated to different levels on the outer wall according to different rates of growth, with the result that they often close off the considerably narrowed orifice.

The way in which the syconia of the *Ficus* species develop is in fact closely linked with the mode of growth in the *Dorstenia* species, as is shown in the drawing of the young primordium of a syconium of *Ficus carica* in Fig. 174 V, in which the surface of the inflorescence apex is still convex. According to the evidence given by Bernbeck, the floral primordia do not appear in its interior until considerably later on.

It is of course well-known that the *Ficus* species produce different kinds of flowers in their inflorescences. In the edible fig *Ficus carica*, known in cultivation since ancient times, these are: 1) male flowers, 2) sterile female "gall flowers" with a short style, without stigmatic papillae, and 3) fertile female flowers, with a long style and well developed stigmatic papillae. These flowers are distributed in different ways between different inflorescences, which in wild forms all occur on the same tree, but generally not so in cultivars. In the cultivars are found 1) trees with "real" figs, which mostly contain fertile female flowers (see Fig. 174 VII), and 2) male fig trees ("Caprifichi"), in which inflorescences appear succes-sively in two forms, i.e. the female figs ("Mammae"), which only contain sterile female flowers, and the "Profichi", which bear sterile female flowers in the lower two-thirds of the syconium, but which have male flowers surround-ing the orifice (see Fig. 174 VI). Pollination in *Ficus carica* takes place by means of a gall wasp, *Ceratosolen arabicus* (*Blastophaga psenes*), which lays its eggs in the ovaries, but whose ovipositor can only reach the ovaries of the short-styled sterile female flowers. Eggs are therefore only laid in the sterile female flowers of the Profichi and the Mammae. The pupae of the gall wasp can overwinter in the Mammae. The gall wasps which hatch out from the Profichi have to push past the male flowers which are placed around the orifice and so load themselves with pollen. They are then able to pollinate the "real" figs, and on account of this branches with the "Profichi" have been hung in trees with "real" figs ("caprification") ever since ancient times. Nowadays of course there are also parthogenetic cultivars.

The syconia of *Ficus* species therefore represent partial florescences, in which the individual flowers are united into an assembly of higher order, which in terms of the biology of flowering and seed distribution fills the role of an individual flower. Within certain reservations this is also true for the female inflorescence of *Procris* species, and for the inflorescences of *Dorstenia*. The latter can also show some external similarity with an individual flower, as for exam-ple in *Dorstenia multiradiata* (Fig. 174 II). Hence the term **pseudanthium**, which will be discussed in more detail below, may also be applied to these inflores-cences.

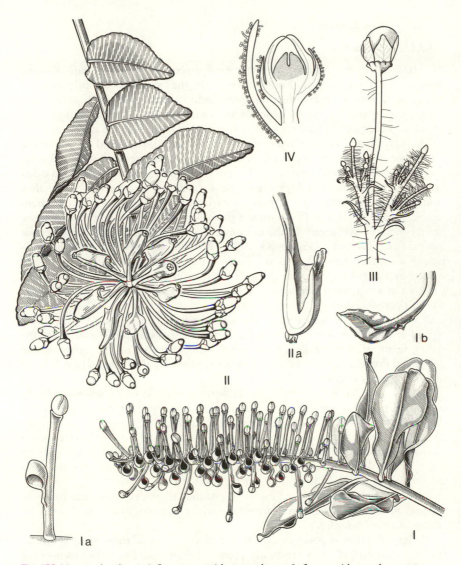

Fig. 175 Norantea brasiliensis, inflorescence with saccate bracts, Ia flower with recaulescent saccate bract, Ib bract with the blade only slightly inflexed from below. II, IIa *Marcgravia umbellata*, branch with terminal inflorescence with saccate bracts (centre), which are congenitally fused with the rudimentary flowers, IIa one such bract in sagittal cross-section, the rudimentary flower at the distal end. III *Cotinus coggygria*, cymose partial inflorescence, in which only the primary flower is fully developed (after Troll), IV *Harpagophytum procumbens*, bud of a secondary flower with subtending prophyll in longitudinal section, the flower bud modified to form a nectary. IV after Ihlenfeldt and Hartmann.

2.3.3   Pseudanthia

2.3.3.1   *Introduction*

Pseudanthia can be defined briefly as "inflorescences which imitate a single flower". This would not, however, include the majority of inflorescences where the flowers maintain their complete individuality, but which present elements of a structure of higher order by means of a more or less marked division of labour in respect of features such as the display organs or the nectaries.

The latter is the case for example for the enlarged, corolloid and coloured calyx lobes which occur in the inflorescences of various members of the Rubiaceae. In these the special structure only concerns some outwardly-directed calyx lobes of peripherally placed (primary-) flowers, but there is no doubt that they contribute to the attraction of pollinators by optical means, as already mentioned in section 1.4.2.6. These are not therefore arbitrarily distributed, but are linked with the symmetry of the inflorescence as a whole, which enhances the optical effect (Weber 1955, Leppik 1977). We can also mention here the creation of a cockade-like colour pattern spread over a complete inflorescence and caused by a regular systematic change of colour in flowers which are clustered together in a capitate inflorescence, as is found for example in forms of *Lantana camara* (Verbenaceae).

A similar effect is achieved by the formation of conspicuous bracts in the region of the inflorescence, as is found in the Dogwood *Cornus suecica*, or the enlargement of the corolla or the corolloid formation of the calyx of the peripheral flowers in the corymbs of many species of *Viburnum* (section 1.4.3.9) or of *Hydrangea* (section 1.4.2.6).

A division of labour in which certain floral primordia of an inflorescence do not develop into normal flowers, but instead develop their nectar-producing disc (together with the rudimentary carpels) as the only functional organ, has already been mentioned in section 1.7.1 concerning the partial inflorescences of *Harpagophytum*. This phenomenon can now be made clearer by reference to Fig. 175 IV. What we are concerned with are the secondary and possibly also the tertiary flowers of the cymose partial florescences[*].

Even more remarkable is the formation of the saccate, nectariferous bracts in the inflorescences of the Marcgraviaceae, which we would like to discuss by means of two examples.

Norantea brasiliensis possesses racemose inflorescences of more than 30 cm in length, at the end of plagiotropically growing branches. The inflorescences are therefore oriented more or less horizontally (Fig. 175 I). The long-tubular flowers are all directed almost vertically upwards, whereas their subtending bracts hang downwards. The latter are divided into a long narrow petiole and a

[*]In the dithyrsoid inflorescences of the Wig Tree (*Cotinus coggygria*), only the terminal flowers of the thyrsoid sections and the primary flowers of the dichasially branched partial inflorescences reach maturity (and possibly in vigorous shoots the secondary flowers too). All other flowers atrophy as buds (Fig. 175 III). As the fruit ripens, however, their pedicels are elongated just as much as the pedicels of the fertile flower. In addition numerous long hairs grow out on the pedicels of the rudimentary flowers, so that the numerous branches of the fruiting inflorescence become finely and densely fringed, which is the reason for the name "Smoke Tree". Whether it is justified to interpret this kind of structure as an adaptation to distribution by wind (see section 3.4.2) is open to doubt.

deeply bowl-shaped blade (Ia). The concavity of the blade is brought about during the development of the bract when the surface of the blade curves inwards from the lower side (Planchon and Triana 1863, Goebel 1933, p. 1590). Since the orifice always points upwards because of the pendent orientation of the bracts, these can serve as containers for the nectar which is secreted by two large, inward-facing mesophyllary nectaries (for the anatomical structure see Weber 1956).

The flowers, inflorescence rachis and the bracts are shining red in colour, so that all the conditions for attracting and feeding hummingbirds are satisfied.

In *Marcgravia umbellata* the concavity of the under side of the bract is so marked that the bracts assume the appearance of small jug-shaped containers (Fig. 175 II). The inflorescence, as the name already suggests, remains short and forms an umbel. The saccate bracts only appear, however, in the terminal portion of this umbel, where they are recaulescent and fused with the pedicels of the flowers which proceed from their axils, whilst the flowers themselves are abortive, and only occur in rudimentary form as the somewhat thickened ends of the containers (Fig. 175 IIa). On the other hand, the widely spreading fertile flowers in the lower part of the umbel do not possess any bracts. However, as in *Norantea*, two transversely inserted scale-like prophylls are present, which are closely appressed to the calyx. The petals are fused together in a calyptra, which is shed in its entirety at anthesis. Once again the brightly coloured bracts produce quantities of nectar. According to our current view, the important difference between this and *Norantea* is the radiate structure of the inflorescence as a whole, which is divided into a central and sterile but nectariferous region, and a peripheral ring of the flowers which project out above it. Thus the inflorescence as a whole shows some resemblance to an individual radiate flower.

From the full range of examples of more advanced pseudanthia – as are found for example in Saururaceae (Rohweder and Treu-Koene 1971), Dipsacaceae, Calyceraceae, Xyridaceae, Rapataceae, Cyperaceae (Schultze-Motel 1959), Centrolepidaceae (see Hamann 1962), and in the Araceae – we wish to select for further consideration only the forms of pseudanthia in the Umbelliferae, the cyathium of the Euphorbiaceae-Euphorbieae and the cephalia and syncephalia of the Compositae.

2.3.3.2   Structure of pseudanthia in the Umbelliferae

We have already mentioned (section 1.4.3.4) that enlarged corollas often occur in many members of the Umbelliferae in the peripheral flowers of the secondary umbels (umbellules) and the so-called umbels (= double umbels!). The enlargement of the corolla happens in such a way that the petals which face outwards in the umbel or umbellule are often considerably more strongly developed than the others, and indeed, according to their position with respect to the outer margin, the two lobes of the deeply bifid petals can even grow unequally, so that such petals are asymmetric (Fig. 37 IV). The double umbels of *Heracleum austriacum* and *Orlaya grandiflora* in Fig. 176 IV, V may serve as examples of the formation of such "rayed marginal flowers". The petals of the *Heracleum* often show an asymmetric form, whereas the much more marked

development of the outwardly directed petals in *Orlaya* always takes place uniformly.

Although the development of the petals in this case is clearly related to the form of the inflorescence as a whole, examples such as these might be regarded as only the first steps towards the construction of a pseudanthium. At the same time it is a fact that the incorporation of individual flowers in the complex structure of a pseudanthium by means of the formation of unequal corollas is scarcely carried any further than this preliminary stage.

An entirely different mechanism for the formation of pseudanthia is indicated by the inflorescences of *Astrantia*, *Hacquetia epipactis* and many species of *Bupleurum*. In our descriptions of this we refer principally to the thorough and comprehensive investigations of Froebe (1964, 1971a,b, 1979, 1980a,b), which include the surprising conclusion that the inflorescences of the apparently primitive Hydrocotyloideae are not in any way derived from the compression of racemose polytelic florescences, but from monotelic thyrsoids (1971b).

In the *Astrantia* species, e.g. *A. major* (Fig. 176 I) and *Hacquetia epipactis* (Fig. 177 I) the inflorescence is made conspicuous by the almost perianth-like formation and the greenish-white or yellowish-green colouring of the involucre which surrounds the inflorescence. On the other hand, in *Bupleurum ranunculoides* (Fig. 177 III) and related species it is due to the perianth-like appearance and yellowish colouring of the involucels, which bring about a remarkable likeness to the flowers of many species of *Ranunculus*.

On closer inspection, the umbel-like inflorescence segments of *Astrantia major* can be seen to be composed of long-pedicelled male flowers with only rudimentary gynoecia and shortly pedicelled hermaphrodite flowers with well developed ovaries. There is a certain regularity determining this distribution of the sexes: "the narrow peripheral ring of flowers is male, the intermediate zone hermaphrodite and the apical flower group again male only" (Froebe 1964, p. 357). This is the case for the umbel which terminates the main axis (primary umbel); in the lateral umbels (secondary and tertiary) the hermaphrodite flowers become more and more repressed in relation to the central zone. From the way in which the hermaphrodite flowers are arranged in respect to the leaves of the involucre and the peripheral male flowers, and particularly from the sequence in which the flowers are initialised during the development of the inflorescence, it turns out that this is in fact a disguised double umbel. Within a primary umbel it is possible to distinguish about 8 highly reduced umbellules, each inserted in the axil of an involucral leaf, each of which comprises one hermaphrodite flower surrounded by three males. The hermaphrodite flowers occurring further inwards, which can similarly each be assigned to the axil of an involucral leaf, are each associated with only one or two male flowers. At the apex or towards the centre of what is therefore an inflorescence representing a double umbel "the single flowered umbellules revert to being male" (p. 361). At the same time the occurrence of intermediate forms between the herma-

Fig. 176 I *Astrantia major*, flowering shoot. II, III *Sanicula europaea*; complete inflorescence, before flowering (II) and individual umbel (III) after flowering. IV *Heracleum austriacum* and V *Orlaya grandiflora*, double umbel seen from above. I, V after Troll, II, III after Froebe, IV photo Th. Stützel.

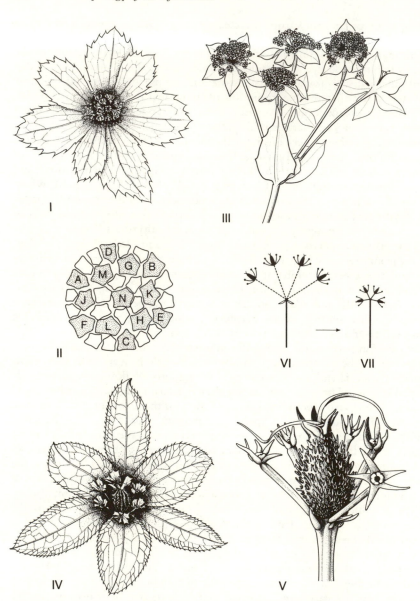

Fig. 177 I, II *Hacquetia epipactis*, umbel seen from above (I) and in a diagram (II), where the hermaphrodite flowers are designated by capital letters. III *Bupleurum ranunculoides*, umbel. IV *Actinolema eryngioides*, umbel seen from above. V *Sanicula coerulescens*, umbellule, consisting of a central hermaphrodite flower and five male flowers. VI-VII schematic representation of the reduction process leading from the umbel of *S. coerulescens* (VI) to that of *S. europaea* (VII). II, IV-VII after Froebe.

phrodite flowers with well-developed ovaries and the male flowers is not uncommon (see Troll 1957, Fig. 273).

In this case therefore the pseudanthia are formed by the perianth-like structure of the involucre and a high degree of "condensation" of a double umbel, in which the "peduncles" of the umbellules are no longer visible and the umbellules themselves are greatly reduced. In the peripheral zone of the double umbel they are still many-flowered and "from 1 to 3 male flowers still occur besides the hermaphrodite one. Further inwards towards the centre of the umbel the umbellules become weaker: firstly single-flowered and hermaphrodite and then single-flowered and male" (Froebe 1964, p. 361).

The structure of the pseudanthium in *Hacquetia epipactis* is rather more easily seen, although in this case not only the peduncles of the umbellules are reduced, but also the pedicels remain very short, so that the flowers are compressed into a head. The inflorescence corresponds "morphologically to a double umbel where the central umbellules are specially favoured" (Froebe 1964, p. 356). Here the hermaphrodite flowers represent the favoured central flowers of umbellules. Only the hermaphrodite flower N of the central umbellule in Fig. 177 II is still surrounded by a ring of male flowers, whereas the umbellules with the hermaphrodite flowers A to F which are positioned in the axils of the involucral leaves are reduced to the hermaphrodite flower and two males. The hermaphrodite flowers G to M are the central flowers of umbellules whose male flowers are lacking entirely.

This effect is found in an intensified form in *Actinolema eryngioides*. In the impressive umbel illustrated in Fig. 177 IV a male flower is inserted in the axil of each of the 6 involucral leaves. Each of them would correspond to a reduced umbellule. Correspondingly the central hermaphrodite flower may be regarded as the only remaining flower, the central one, from the terminal umbellule. At first sight the inflorescence as a whole has the appearance of a single hermaphrodite flower.

The inflorescences of the Sanicle (*Sanicula europaea*, Fig. 176 II) have a peculiar structure in which the involucral leaves are greatly reduced. Once again this is a case of a much modified and contracted double umbel. The term double umbel must by no means be referred to the complete inflorescence as illustrated in Fig. 176 II. This is rather like the complete inflorescence of *Astrantia* in Fig. 176 I, composed of a terminal and several lateral "long-peduncled" double umbels, one of which is reproduced in Fig. 176 III. Its 4 hermaphrodite flowers correspond to the favoured central flowers of 4 umbels, which additionally contain 2 to 4 male flowers. This follows from the arrangement of the flowers in relation to the involucral leaves as well as from the ontogeny of the inflorescence. An individual umbellule – from *Sanicula coerulescens* in fact, where the hermaphrodite flower is surrounded by as many as 5 male flowers – is shown in Fig. 177 V.

It is well known that in the Umbelliferae there are also capitulate or cylindrical inflorescences whose form results, similarly as takes place in the Compositae, not merely by compression of the flowers, but by a pronounced (primary) thickening of the inflorescence axis. This is the case in the capitula of the species of *Eryngium* (Fig. 178), where numerous individual flowers are packed closely together on the surface of a clavate or capitate "receptacle". From the fact that many of these

flowers exhibit a supernumerary leaflet besides their subtending bract and, from the observation that the flowers in the apical region of the head may even bear an involucel of up to 5 leaves (Troll and Heidenhain 1952, p. 155), it is concluded that each capitulum corresponds to a double umbel, where the peduncles of the umbels are rudimentary and where the umbellules are reduced to a single flower each. Since in this case the involucral leaves are strongly developed, the heads of *Eryngium* species still give the impression of a single flower, in spite of their often considerable size. This is often enhanced by the contrasting colouring of the involucral leaves, which in Sea–holly, *Eryngium maritimum*, in its alpine relative, *E. alpinum*, and in *E. amethystinum* is amethyst–blue.

Given the above considerations, the following are the principal factors relevant to the formation of pseudanthia in the Umbelliferae: 1. the "loss of individuality of the flowers or sub–units of the inflorescence", as we find in practice in *Hacquetia*, *Actinolema* or in a different way, in *Eryngium*, 2. the "delegation of the display function" to the peripheral leaf organs, e.g. the peripheral petals of the umbels or umbellules, the leaves of the involucels in *Bupleurum ranunculoides*, or of the involucre in *Astrantia*, *Hacquetia* and *Actinolema*, and 3. the "pattern formation of the 'petaloids' in the sense of a replacement for the corolla (pseudo–corolla)" (Froebe and Ulrich 1979). To these Froebe adds 4. the "dominance of the pseudo–corolla over the disk" and 5. the "contrast with the surroundings". It is still remarkable that the method of pseudanthium formation by means of the enlargement of the marginal petals or corolla of marginal flowers, which is so successful in the Compositae, has hardly proceeded beyond its early stages in the Umbelliferae.

2.3.3.3 The cyathium of the Euphorbiaceae-Euphorbieae

The cyathium (floral receptacle) of the Euphorbieae represents a high degree of integration of extremely modified flowers into a flower–like system of organs.

It is the cyathia at the apex of the main axis and at the ends of the 5 branches in the inflorescence of the Sun Spurge (*Euphorbia helioscopia*, Fig. 179 I) which give the appearance of flowers. The superficial resemblance to a flower is mainly brought about by the fact that the 5 bracts are united into a cup–shaped involucre, as is shown for the Cypress Spurge (*E. cyparissias*) in Fig. 179 IV. The calyx–like involucre of the Pereae or the petaloid and coloured involucre in *Dalechampia* may perhaps be regarded as early stages of this. On the edge of the receptacle and between the tips of the 5 bracts of the *Euphorbia* cyathium 5 or 4 oval or – in the case of Cypress Spurge – lunate glands (nectaries) are inserted, which are interpreted as being the equivalent of stipules of the bracts. Each of the bracts bears a cincinnoid male inflorescence in its axil (Fig. 179 VI), whose flowers, however, are reduced to a single stamen. The pedunculate single and likewise perianth–less female flower arises from the centre of the cyathium and hangs out and over its margin. In this location the gland which is usually positioned between the bracts is usually lacking (except in the terminal cyathium of the inflorescence). The reason why the individual stamens and also the

Fig. 178 I *Eryngium bromeliifolium*, distal portion of an inflorescence. II *E. giganteum*, axial section through a capitulum. Due to Troll.

long-peduncled ovaries may really be regarded as reduced flowers without perianths and not as the parts of a single flower is that the stamens exhibit an obvious constriction which separates the filament from its pedicel (Fig. 179 V, Str). In the genus *Anthostema* in the same family, which also forms cyathia, there is still a definite perianth present at this position (Fig. 179 II), and moreover this is also formed in the female flower of the cyathium (Fig. 179 III). In the genus *Actinostemon*, which is similarly related, larger male flowers with up to 9 stamens together still possess a rudimentary 3-merous perigon, whereas in the weaker ones this is lacking and the number of stamens may be restricted to one. In other cases the pedicel and filament of the reduced male *Euphorbia* flowers can quite often be distinguished by the fact that the pedicel is hairy, robust and brown-coloured, whereas the filament on the other hand is glabrous, delicate and pale. Furthermore the male flowers, even when they are reduced to a single stamen, often still have subtending bracts in the form of tiny subulate lobes. The female flower is pendulous during the flowering period, but then becomes erect again – a phenomenon which is well-known from the geotropic conversion of flowers, but not for an ovary within a flower. Moreover a perigon-like structure still occurs at the base of the female flower in some *Euphorbia* species, and in

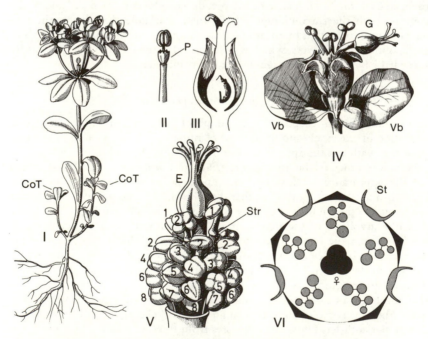

Fig. 179 I *Euphorbia helioscopia*, flowering plant, CoT cotyledonary shoots, II, III *Anthostema senegalense*, male (II) and female (III, longitudinal section) flowers, P perigon. IV *Euphorbia cyparissias*, cyathium with prophylls (Vb), V *E. lathyris*, cyathium after removal of the cup-shaped involucre, the arabic numerals denote the order of development of the male flowers, each of which consists of just one stamen, Str constriction between filament and pedicel, E (or G in IV) the female terminal flower. VI *E. peplus*, diagram of a cyathium. After Schmeil (I), Baillon (II, III) and Troll (IV-VI).

Euphorbia pauciflora this can even consist of two alternating 3-merous whorls. In yet further cases there is just a suggestion of the perigon as a slight fringe.

The interior of the cyathium is more or less divided into chambers at the base by 5 septa which are decurrent in front of the glands and which are often markedly fimbriate. Quite often a longitudinal furrow can be seen between the marginal teeth. It is assumed that the septa originate from the "back to back" fusion of prophylls which are combined from the neighbouring male partial inflorescences. In favour of this assumption is the fact that in *Euphorbia paucifolia* these partial inflorescences are surrounded by two bracts which are still widely separated (Mansfeld 1929).

The statement that the 5 "stamen groups", each of which is placed in front of the leaf organs of the cyathium, are in fact cincinnoid partial inflorescences, is confirmed both by the fact that during ontogeny, the differentiation proceeds outward and in a zigzag fashion, and by the order of dehiscence, i.e. the order in which the individual anthers open.

Thus, in a taxonomic group which is very often characterised by unisexual flowers, after reduction of the individual flowers even to the highest possible degree – to just a single stamen or a 3-locular ovary – organ systems of higher order are derived which are highly reminiscent of the hermaphrodite flowers of other plants and are pollinated by insects (mainly Diptera) or birds (as in *Poinsettia*, see below). But this is not the end of the story! The cyathium can undergo further transformation, as for example in the florist's *Euphorbia fulgens*, where the glands between the 5 bracts of the cyathium which are interpreted as equivalents of stipules develop flat, petaloid appendices which spread at right angles, and are bright red in colour (Fig. 180 I, II). The superficial likeness of such a cyathium to a sympetalous hermaphrodite flower may be said to be perfect.

In the South American genus *Pedilanthus* (Slipper Spurge) the similarly bright red coloured cyathia even undergo a further, zygomorphic modification (Fig. 180 III to VI). This takes place in such a way that two of the 5 leaf organs which form the cyathium are so greatly enlarged that they occupy not just the lower side of the zygomorphic cyathium but extend a long way up at the sides (Fig. 180 VI). The nectary is lacking from between these two leaf organs. The three other bracts are closely pushed together on the upper side of the cyathium, like the four remaining nectaries, which are moved away somewhat to the outside. They are enclosed by an "involucral tube", which extends backwards like a spur, and could represent the product of fusion of 4 of those glandular appendages which are developed like petals in *Euphorbia fulgens*. The zygomorphic structure of the cyathium is also expressed in the position of the female flower, which is somewhat shifted away from the centre, and in the yet more favoured development of the 3 posterior cincinni of male flowers (Fig. 180 III), which are not only more vigorous, but also contain more flowers (stamens) than the two anterior ones. The cyathium behaves like a protogynous flower: the style of the female flower elongates first from out of the beak-like tip of the cyathium and its tip with the receptive stigma lobes curves upward (III), then the stamens of the male flowers are exserted whilst the style of the female flower becomes deflexed (IV, V). In the native habitats of the genus *Pedilanthus* the pollination of the highly nectariferous cyathia is carried out by hummingbirds (see Porsch 1923).

The cyathia of *Euphorbia* (*Poinsettia*) *pulcherrima* are arranged in cincinni and are also zygomorphic. This is because they possess only one very large cup-shaped nectary, which is placed on the side of the cyathium which faces the subtending bract. Since at this point two out of the four nectaries are usually positioned in the cyathia of other species, Goebel (1931) suggests that "in *Euphorbia pulcherrima* these are fused to form the large involucral gland", and all the more so because two correspondingly small glands occasionally appear in the same position. The shining yellow glands make a striking colour contrast with the bright red bracts, which surround the ensemble of cyathia like a red star, and it is these which give it the German vernacular name of "Christmas Star" (Weihnachtsstern). In the more robust inflorescences these are apparently arranged in the form of a double ring, in which the leaves of the outer ring are twice the length of those of the inner. From the morphological point of view, there are two prophylls for each of the cyathia (which are themselves arranged in cincinni) and what happens is that one of these assumes a form approaching that of a foliage leaf, whilst the other remains small and sterile. According to Porsch (1923) the red pigment is in the cells of the epidermis on both sides of the leaf, and only found there. Since the entire mesophyll of the bracts is composed of a spongy parenchyma which is poor in chlorophyll and has a remarkable number of intercellular spaces, we have another example of the "cat's-eye effect" explained in section 1.4.3.1. If the pigment is lacking, then the large bracts are white, which is a permanent character in one form of this plant. The nectaries produce so much nectar that the nectaries overflow – this can readily

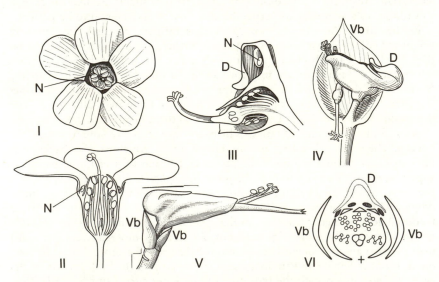

Fig. 180 I, II *Euphorbia fulgens*, cyathium seen from above (I) and in axial section (II). III, IV *Pedilanthus bracteatus*, the dorsiventral cyathium in median section (III) and after removal of one of the two bracts (prophylls). V *Pedilanthus tithymaloides*, cyathium in male condition viewed laterally. VI Diagram of the cyathium of *Pedilanthus*. N Nectary, D "involucral tube", Vb prophyll. II after Troll, III-V after Porsch, VI after Goebel; adapted.

be observed in houseplants. In its native habitats in Mexico and Central America the cyathia are pollinated by hummingbirds.

In the Crown of Thorns, *Euphorbia milii (E. splendens)*, the two broadly ovate prophylls of each cyathium are a fiery red colour and give the cyathium the appearance of a flower. We cannot give further consideration here to the large number of further modifications in form and colour of the cyathium of the Euphorbiaceae.

2.3.3.4   The cephalia and syncephalia of the Compositae

The inflorescences of the Compositae exhibit synflorescences of the polytelic type, whose florescences are derived from racemes and developed by means of compression and thickening of the inflorescence axis to form capitula. These Composite capitula must be reckoned as highly developed pseudanthia. This is particularly true of every member of the Asteroideae, whose heads contain marginal ray florets as well as central tubular florets (section 1.4.3.9), but it is also true of the Cichorioideae, which – as already mentioned (section 1.4.3.9) constantly only develop ray florets. The similarity of such a capitulum to a single flower is all the greater because it is surrounded at the base by an involucre of bracts (often called phyllaries), arranged in a rosette and often numerous, which only bear flowers in their axils in rare and exceptional cases. The capitulate inflorescence of the Composites has been called a calathide ('wicker basket') because of this.

It could be said that the involucre in these pseudanthia replaces the calyx, whilst the ray florets (or the outermost spathulate corolla lobes of the Cichorioideae) take over the role of the corolla, which is also the impression given by the arrangement of overall colour pattern. Take for example the concentric rings of different colours which extend across all components of the "false corolla" in the rosette-like heads of the Blanket Flower (*Gaillardia*). In *Chrysanthemum carinatum (C. tricolor)* these rings even have three or more colours! The ring of ray florets of the Composite capitulum therefore exhibits a kind of colour pattern which is already familiar from the corolla of a simple radial flower. The illusion of a single flower is strengthened on the quite frequent occasions when the ray florets are five in number. The capitulum does not merely imitate the external appearance of a single flower, but also exhibits a unity in its functional aspects. The ray florets (or in the Cichorioideae the outermost ligulate florets) do not wither until the last of the tubular florets (or the central ligulate florets) which mature in centripetal order, have completed anthesis.

It may be said that these capitula form a fundamental unit, which takes up the same relative position in the branching of the Composite inflorescence as does the individual flower in the inflorescences of other families. Hence inflorescences can be built out of capitula by ramification of paniculate, botryoid or thyrsoid types, the latter with cincinnoidal or helicoidal branching of their axes, which could be termed as a paniculodium, botryodium etc. (after Troll 1967, pp. 93–94). It is rather to be expected therefore that a combination of capitula to form head-like aggregates of the second order can take place, forming the so-called **cephalodium** or **syncephalium**. However, before we discuss this possibility in more detail, we would like to turn our attention to a special structural form of the Composite capitulum.

In Fig. 182 I a capitulum of *Helipterum roseum* is illustrated as seen from above. At first sight it appears to be equivalent to a capitulum of the Asteroideae, with its central tubular florets and its peripheral ray florets. On closer inspection, however, our misconception is corrected. As the axial section in Fig. 181 I shows, this capitulum is in no way composed of tubular florets and ray florets, as is the capitulum of *Achillea millefolium* illustrated beneath (Fig. 181 II), but comprises only tubular florets. What appeared to be ray florets are in fact phyllaries, but with a chaffy constituency and coloured like corollas. Apart from the oval and scale-like lowermost phyllaries, these are all characterised by being differentiated into a short, narrow base and a long-elliptic distal portion (Fig. 182 II-IV). Between these two there is a zone of articulation, by means of which the distal, petaloid section carries out hygroscopic movements: when it is dried out the under surface of the articulation shrinks more than the upper side, so that the petaloid section is turned more or less at right angles to the basal portion, which stays fixed in place. If on the other hand the underside of the articulation is moistened with water, or an increase in the humidity of the atmosphere takes place, then the petaloid section curves inwards, and the head closes up. The involucre therefore performs both the functions of a perianth in a characteristic way: it serves both for display and for protection at the same time. Since the hyaline phyllaries have already died by the time the head flowers, but do not wither, the external form of the flower head survives after the florets have withered and died. The capitula even still appear to be alive because of the hygroscopic movements of the phyllaries, which has given rise to the name "immortelle" or "everlasting" to this and other species (e.g. also *Helichrysum*). They are frequently used in bunches of dried flowers. The German vernacular name "weather thistle" (Wetterdistel) for *Carlina acaulis* has also arisen because the heads open in dry sunny weather and close in damp weather; here also the involucre has taken over the display function of the ray florets and carries out hygroscopic movements as well.

It is in the Yarrow (*Achillea millefolium*) that we first recognise a phenomenon where the capitula are reduced in size and have few flowers but still achieve a conspicuous effect in a corymbose inflorescence. This evolutionary tendency is continued so far in the Globe Thistle (*Echinops sphaerocephalus*, Fig. 183 III, IV) that a spherical inflorescence is formed from genuinely single-flowered capitula, which are all in their turn provided with a well-developed involucre (Fig. 182 VI, VII).*

In comparison to the latter, the capitulum of *Syncephalantha decipiens*, illustrated in Fig. 183 I, II as seen from the side and from above, appears to be a "normal" Composite capitulum, provided with ray florets and an involucre. The schematic axial section in Fig. 181 III shows, however, that this is in fact a **syncephalium**, in which several capitula are combined into a new morphological entity, which once again looks very much like a single capitulum. The syncephalium usually consists of a head which terminates its main axis and 5 lateral heads. Only these peripheral heads form ray florets, and moreover only

*The development of the inflorescence in *Echinops exaltatus* was investigated by Leins and Gemmeke (1979).

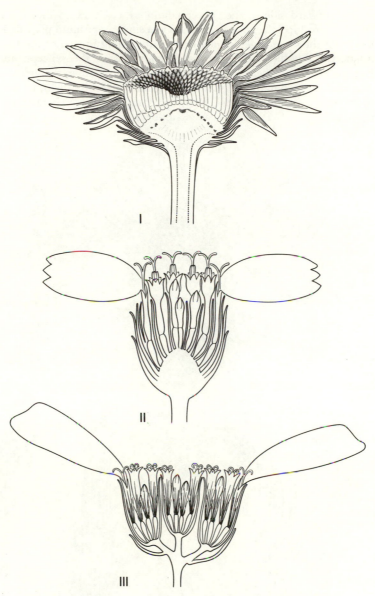

Fig. 181 I *Helipterum roseum*, section of capitulum before flowering. II *Achillea millefolium*, diagram of longitudinal section through a capitulum. III *Syncephalanthus decipiens*, pseudo-cephalium in longitudinal section, the terminal capitulum has no ray florets, the peripheral capitula have them only on the outer side. II After Warming, adapted, from Troll, I, III after Troll.

on the outward side of the syncephalium as a whole, so that the complete construct produces the same impression as a single capitulum provided with ray florets. At the same time the phyllaries (subtending bracts) of the lateral heads imitate the involucre. Of course the main axis of the syncephalium is not thickened to form a head, and the axes which terminate in the individual heads are rather short, but are still clearly recognisable. Troll (1928), to whose comments we are mainly referring here, has compared these characteristics to the juvenile stage of a many-headed inflorescence of a Butterbur (*Petasites*

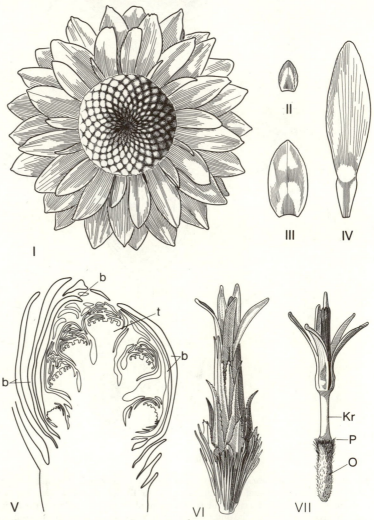

Fig. 182 I-IV *Helipterum roseum*, flowering head, seen from above (I), II-IV outer, middle and inner phyllaries. V *Petasites paradoxus*, longitudinal section through a young inflorescence, t terminal head, b bracts. VI, VII *Echinops sphaerocephalus*, individual head complete (VI) and after removal of the involucre (VII). O ovary of the individual flower, P its so-called pappus, Kr coralla tube. All due to Troll.

Fig. 183 I, II *Syncephalantha decipiens*; I flowering shoot with pseudo-cephalium, II pseudo-cephalium seen from above. III, IV *Echinops sphaerocephalus*, inflorescence at the beginning of flowering (III) and fully flowering (IV). Due to Troll.

paradoxus, Fig. 182 V). In this case also a terminal capitulum (t) terminates the inflorescence. Further capitula are placed in the axils of the uppermost bracts (b), which are all clearly in close proximity to each other because the inflorescence axis has not yet elongated. "If we assume that the inflorescence axis continues to be inhibited and therefore remains short, so that the capitula are

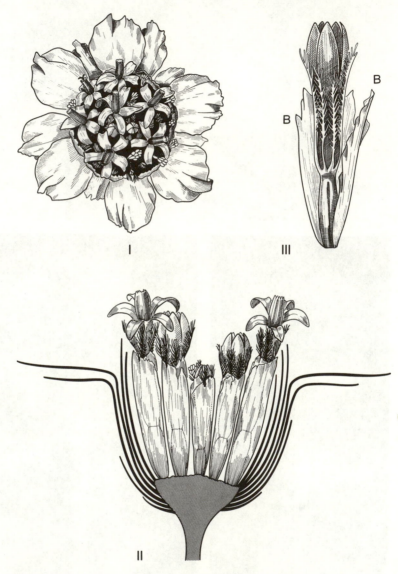

Fig. 184 Myriocephalus gracilis, pseudo-cephalium seen from above (I), in axial section (II, somewhat diagrammatic), and a single capitulum reduced to one flower (III), in which the involucre, which consists of 3 phyllaries only (B), has been pulled aside somewhat. After Troll.

densely crowded ... we then obtain a pseudo-cephalium, or false capitulum, like that which *Syncephalantha decipiens* Bartl. possesses".

The integration of the individual capitula is less far advanced in the species of Edelweiss (*Leontopodium*). By compression of the internodes numerous individual capitula are brought together into an almost head-like corymb, which is surrounded by the white-felted bracts of the capitula in the shape of a star.

In addition to the stages of integration described above, syncephalia of higher order are also formed in the Compositae, although not in such a way that they form marginal ray florets in order to simulate the habit of single flowers. Instead they normally exhibit a common involucre and a common receptacle, so that they can be referred to as cephalodia. At the same time, the individual capitula are often greatly reduced, as for example in *Flaveria repanda*, whose aggregate of heads of highly complex structure is surrounded in the manner of an involucre by the bracts which subtend the lower of the paraclades which contribute to the aggregate. According to Kunze (1969), who has thoroughly investigated this and other complex inflorescences in the Compositae, higher order aggregates of capitula occur in the tribes Vernonieae, Heliantheae, Helenieae, Inuleae, Cardueae and Mutiseae of the Asteroideae.

Among the cephaloid Composite inflorescences there are also "everlasting heads of second order" (Troll), namely in *Myriocephalus* and *Angianthus*, two genera of the Inuleae-Angianthinae. According to Troll (1928), the most perfect development of this exists in *Myriocephalus gracilis*. Once again, when seen from above (Fig. 184 I), the aggregate of capitula resembles a normal Composite capitulum. An axial section through the inflorescence (II) shows, however, that numerous capitula are inserted on a disc-shaped, broadened inflorescence axis and are surrounded by a common involucre. Here too the involucral leaves are chaffy in nature, and the central ones are also differentiated into a basal portion which is closely appressed to the head and a terminal portion which is at right angles to it and is broader, dark yellow-coloured and petaloid. The individual capitula in *Myriocephalus gracilis* are usually one-flowered (few-flowered in other species) and surrounded by an involucre of a few hyaline phyllaries. In this case the terminal capitulum is lacking, so we have to regard the complete inflorescence as a polytelic truncate synflorescence, similar to that of *Ligularia przewalskii*. In contrast to *Echinops sphaerocephalus* (Fig. 184 III) the individual capitula develop in acropetal order. The same phenomenon as in the "everlasting" heads of *Helipterum* is therefore repeated at the level of the heads of second order. Hence the capitula of the Compositae provide an impressive example of certain organisational tendencies which are repeatedly successful, regardless of the environmental conditions!

3 The flower as a formal and functional entity – aspects of the biology of pollination and dispersal

3.1 General

We have already touched on the subject of **floral biology** or **floral ecology** whilst discussing nectaries as well as the structure and form of the stigma and the conducting tissue.

It would, however, be beyond the scope of this book if we were to try to present anything approaching a complete review of the numerous mechanisms which serve to bring about the pollination of the stigma and the fertilisation of the ovules, to hinder or to directly enable self-pollination, and to finally guarantee the effective dispersal of the seeds or fruits. There already exist comprehensive accounts, not only on the subject of **biology of dispersal** just mentioned, but also for floral biology. In this context we will just cite the works of Cammerloher (1931), Faegri and van der Pijl (1966/82), Knoll (1956), Knuth (1898 to 1905), Kugler (1970), Müller (1873), Percival (1965), Vogel (1954 and other quoted references), Werth (1956a) as well as Müller (1955), van der Pijl (1972) and Ulbrich (1926).*

We can do no more now than cast a glance over the great variety of functional mechanisms and the interplay of the comparative-morphological and -anatomical floral structures already considered. It must be emphasised that all the different adaptations to particular methods of pollination, especially the frequently highly-developed specialisations for distinct pollinating animals as well as the establishment of special methods of fruit or seed dispersal are not usually confined to individual characters, but concern the totality of the floral organs. This includes their position and form, colours and scent, their mechanisms for providing nourishment, their flowering time and duration, the order of dehiscence of the anthers (see Fig. 70 II), and the morphology and anatomy of the gynoecium as well as the arrangement of the flowers themselves and their position on the plant. This is often combined with extensive co-adaptation to the form and behaviour of the animal partner. Nevertheless, the variety of ways in which the flowers and the inflorescence can be formed are not to be seen merely as an adaptation to pollinators or to the particular method of fruit

---

* Christian Konrad Sprengel (1750–1816, "Das entdeckte Geheimniß der Natur im Bau und in der Befruchtung der Blumen" ('The discovery of Nature's secret concerning the structure and fertilisation of flowers'), 1793) is regarded as the founder of floral ecology, but the discovery of the sexuality of plants is attributed to Rudolph Jacob Camerarius (1665–1721, "De sexu plantarum epistula", 1694).

dispersal. In particular, the coordinated development of the floral spur and the nectar-secreting stamen appendages in *Viola* (section 1.7.3), the corresponding formation of organs or parts of organs of quite different origins in the cyathia of *Pedilanthus* or in other pseudanthia, and the formation of two-lipped "secondary flowers" in the Zingiberaceae (section 1.5.13), immediately show that the development of the "gestalt" of the flower is governed by its own laws. What "need" does there exist for the flowers of the genus *Ornithoboea* in the Gesneriaceae to "imitate" the form of an orchid flower (Weber 1979)? It is evident again and again, "*daß die Mannigfaltigkeit der Gestaltungsverhältnisse eine größere ist als die der Lebenbedingungen* (that the intrinsic variety of morphological characters is greater than that due to environmental conditions)" (Goebel 1933, p. 1832).

An excellent example of the way in which the form of the flower is determined by inner laws is provided by Vogel (1969) in an investigation of a series of floral spurs of orchids, which are created in very different ways by the synorganisation of heterogeneous parts of the perigon. In certain species of *Rodriguezia* the nectariferous spur is formed by "bending together and fusing the free longitudinal margins of lateral sepals to form a sheath", enclosing a secretory appendage of the labellum. "In *Batemania*, *Cochleanthes* and *Chondrorhyncha* each of the lateral sepals is separately forming a false spur, separated by longitudinal folding"; the pair of spurs which are nectar-bearing by their own secretion, or which are clothed with feeding hairs are connected with the inner parts of the flower. *Comparettia* and *Neokoehleria* form their nectar-providing equipment "by co-adaptation of a genuine spur formed from a pair of sepals and nectar-producing appendages of the labellum". The spur of *Cryptocentrum* is even "considered to be a congenital and unilateral fusion of a labellum spur with the inner wall of a compound spur formed from the lateral sepals." Vogel is right to ask: "What has caused the orchids in question to depart from the established principle of the normal nectar-secreting and -bearing labellum spur and replace it with functional but very intricate alternative forms of co-adaptive spur structure?" According to Vogel the explanation might be that the appropriate subtribes possess predominantly nectarless perfume flowers (see section 3.3.4.2 (b)) and could have returned secondarily to the formation of nectar flowers. Even given the above answer to the question, it is also clear that – "the organism of the plant, and also the flower, is an autonomous system, whose features have not been 'collected' by using the random mutations of natural selection alone" as expressed by Vogel (1959, p. 501) in the discussion of similar examples in species of the Ophrydeae from the Cape.

3.1.1 Self-pollination and cross-pollination

At first we must state that in many species of flowering plants, as well as in lower plants, there has resulted a definite balance in operation between sexual and asexual reproduction. Asexual reproduction can take place in the floral region by means of the asexual development of seeds (agamospermy), and also in the vegetative region by means of stolons, bulbils and other organs of vegetative reproduction. On the one hand, this balance guarantees genetic stability, which permits a population to survive in a particular locality and, on the other hand, a certain degree of genetic plasticity, which maintains adaptability to changes in the environmental conditions through a huge number of possible recombinations of old and new characters via sexual reproduction. A similar relation exists between 1. self-pollination (autogamy), i.e. the pollination of the stigma with the pollen from anthers of the same flower which is possible in hermaphrodite flowers, or other flowers of the same plant (idiogamy, geitonogamy), and 2. cross-pollination (allogamy), i.e. pollination with the pollen of other plants.

3.2   Mechanisms of self-pollination

Whilst cross-pollination leads to an increase in the proportion of recombinations, self-pollination on the other hand leads to a reduced range of variation in populations by inbreeding. This can be an advantage for annual weeds or ruderal plants which only colonise a locality for a short time, since individual plants can then quickly build up large populations by autogamy. The repeatedly successful dispersal of ineradicable weeds such as Shepherd's-purse or Groundsel (*Senecio vulgaris*) is evidence of this, as well as for example the amphitropical distribution of certain annual autogamous species of the Polemoniaceae in North and South America. Those species which are distributed in semiarid open habitats in California – certain species of *Gilia* and *Polemonium micranthum* – have probably been favoured by the fact that their seedcoats become viscid when moistened with water and are transferred by desert birds to similar localities in the desert zones of Peru, Chile and Patagonia, where on account of their autogamy they were able to found new populations. (In populations such as this which derive from a single individual the conditions for the divergence of the gene pool from that of the source population are of course especially favourable.)

In environments which are poor in pollinators because of the unfavourable conditions, **autogamy** is often the only possible means of sexual reproduction, as in high alpine or arctic regions and deserts (Hagerup 1932, 1951).

To the extent that a species is obligately self-pollinated, it has no need for organs to attract pollinators and to provide nutrition for them. Therefore the flowers are often inconspicuous, and scent and nectar are often lacking. Not uncommonly the anthers are arranged so that the pollen grains are shaken out onto the stigma of the same flower by the vibration caused by the wind (as in the pendent campanulate flowers of the Ericaceae, see for example Amberg (1912) on *Arctostaphylos alpinus*). The anthers often come into contact with the stigma during developmental and other bending movements of the filaments. In the densely packed flowers of the Umbelliferae the pollination of a stigma from the anthers of a neighbouring flower is possible (**geitonogamy**). In *Potentilla erecta, Erica cinerea, Galium harcynicum, Euphrasia borealis, Rhinanthus minor* and other plants in arctic regions the pollination can have already taken place in the bud stage, shortly before the sepals and petals open (Hagerup 1951).

The above phenomenon introduces us to what is called **cleistogamy**, in which pollination *always* takes place within the flower while it remains closed. This phenomenon has been known since the time of Dillenius in species of

Viola, in which case opening or **chasmogamous** flowers occur alongside the cleistogamous flowers. Usually the cleistogamous flowers are formed later than the chasmogamous ones, and occur on younger parts of the stem: in *Viola mirabilis* on the elongated parts of the stem (the chasmogamous ones are "basal"), and in *Viola odorata* on stolons. They are usually only shortly pedicelled, and the corolla is much reduced or completely lacking. The number of stamens which develop fully is often diminished, and the two posterior pollen sacs, or even three pollen sacs, can be lacking. The cleistogamous flowers therefore clearly present inhibited organs, which may be determined by photoperiodic factors (Goebel 1904, Borgström 1939, Holdsworth 1966), or by other factors as well (see Uphof 1938). All the same, it is often these flowers alone which produce any fruit! In other words, they differ "from the usual type of inhibited structures which are very frequent in flowers, by the fact that although the development of the flower is inhibited at an earlier or later stage, the sexual organs do ripen" (Goebel 1904, p. 786). Pollination takes place in the unopened flowers, either because the anthers split open and empty the pollen grains onto the stigma placed immediately in front of them, or else the anthers do not open. In the latter case the pollen grains germinate inside the anthers and force their pollen tubes through the walls of the anther onto the stigma. As mentioned before (section 1.5.2) the formation of the endothecium as a "fibrous layer" can be reduced or completely inhibited in cleistogamous flowers. – Other well known examples of cleistogamy are the late-summer flowers of *Oxalis acetosella*, a series of grasses (Hackel 1906), *Asarum europaeum* (Werth 1952), *Juncus bufonius* and *Lamium amplexicaule*. In the latter, cleistogamous flowers are often formed at the beginning and at the end of the flowering period, and then often have a better fruit set than the chasmogamous ones. Several species of *Commelina* form inflorescences on their epigeal shoots with chasmogamous flowers and inflorescences with cleistogamous flowers on their underground rhizomes, or reduced cincinni with a few cleistogamous flowers (see Uphof 1934). Cleistogamy can be a mendelian character (according to the investigations of Stomps (1948) on species of *Oenothera*). In cases where no cross-pollination has taken place it is not uncommon for subsequent self-pollination to be brought about by bending movements of the filaments or the style branches, as for example in many Composites (see below); in *Hibiscus trionum* the stigma is brought into contact with the anthers within only a few hours after the flower has opened by a curving movement of the style (Butterose, Grant and Lott 1977). Apart from this, cross-pollination by an visiting insect can of course also occur. The prerequisite for the success of self-pollination is in fact that no incompatibility should exist between the pollen grains and the stigma (see below).

3.3   Mechanisms which facilitate cross-pollination

3.3.1   Self-incompatibility, hercogamy and dichogamy

The development of viable seeds often fails to occur if the stigma is pollinated with pollen grains from the same flower – or what physiologically amounts to the same thing – with pollen grains from another flower of the same plant (idiogamy). This usually depends on inhibition of germination of the pollen grain or of the growth of the pollen tube by the stigma and the pollen tube conducting tissue of the stylar canal, which can bring about a certain "filtering effect". This **self-incompatibility** is controlled by what are called the **self-sterility genes**, whose effect is often determined additively via multiple alleles.

Self-fertilisation is often prevented by this genetic incompatibility alone. However in addition to this there are numerous mechanisms by which cross-pollination is furthered or even effected systematically. The first of these, where the anthers and the stigma ripen at different times, is called **dichogamy**. Two forms of this can be distinguished; the relatively uncommon **protogyny** (e.g. in Plantain, *Plantago*) and **protandry**, in both of which style and anther movements often play an important role. In the case of protandry, the anthers ripen and dehisce before the stigma becomes receptive, which is particularly well exemplified by the Compositae. In this case as already explained (section 1.5.9), the style is enclosed by a tube composed of the five fused anthers, which are facing inwards. The style bears stylar hairs on its outer surface and as it elongates, these push the pollen grains out of the tube (sometimes also by a shrinkage of the filaments which contract on a contact stimulus; *Centaurea*). After this the lobes of the stigma spread out and expose the receptive inner surfaces to pollination. If no cross-pollination (or idiogamy) has taken place, then self-pollination can occur subsequently as the stigma lobes curl downwards (Fig. 44 V) and pick up pollen from the outer surface of the style.

A further possibility for promoting cross-pollination consists in the spatial separation of the stamens and the stigmas: **hercogamy**. The cases of dimorphic and trimorphic **heterostyly** provide textbook examples of this, as is known for example in Primroses or in Purple-loosestrife (*Lythrum salicaria*). In *Primula praenitens* (*P. sinensis*) or *P. elatior* (Fig. 185 I, II) there are long-styled plants with anthers deeply inserted in the corolla tube (pin), and short-styled plants with anthers placed well above the stigma at the mouth of the corolla tube (thrum). Since the (constant) flower visitors always penetrate to the same depth in the corolla tube, the pollen of the higher anthers is usually transferred to the higher stigmas, and the pollen of the lower anthers to the lower stigmas. It has been known since Darwin that the highest level of fruit set and viability of seeds is from these "legitimate" pollinations. In both the above species moreover the stigmatic papillae are longer (or larger), and the pollen grains smaller in the long-styled form than in the short-styled form (see Ernst 1953 about the compatibility problem in this situation).

In Purple-loosestrife (Fig. 185 III-V) there even occurs trimorphic heterostyly with long-, medium- and short-styled flowers with stamens of appropriately different lengths. The inheritance of heterostyly itself depends on a combination of mendelian factors (see von Ubisch 1925). A review of heterostylous plants is found in Knuth (1898, p. 61; see also Vogel 1955).

The phenomenon which is known as **enantiostyly (enantiomorphy)**, is said to have a similar effect. In this case the dimorphism consists in the fact that the style and all or only a proportion of the stamens are exserted from the flower on opposite sides (see Knuth 1898/1905). Examples of this occur in species of *Cassia* (Müller 1883), species of *Solanum* (*S. rostratum*, see Bowers 1975), according to Ornduff and Dulberger (1978) as well as in genera of the

Haemodoraceae (*Wachendorffia*, etc.) and in Tecophilaceae (*Cyanella*), in which all the flowers of an inflorescence are either "righthand-styled" or "lefthand-styled", and there are always only a few flowers which are open at the same time.

A spatial separation of stamens and ovaries is provided naturally when unisexual flowers are formed. In this case there is the well-known distinction between

Monoecism: male and female flowers on the same plant,
Dioecism: male and female flowers on different plants,
Polygamy: unisexual and hermaphrodite flowers in various proportions. This is subdivided into:

1. **andromonoecism**, i.e. hermaphrodite and male flowers on the same plant (*Filipendula ulmaria*, etc.); 2. **gynomonoecism**, i.e. hermaphrodite and female flowers on the same plant (many Composites); 3. **trimonoecism**, i.e. hermaphrodite, male and female flowers on the same plant (*Aesculus hippocastanum*);4. **androdioecism**, i.e. hermaphrodite and male flowers on different plants (*Polygonum bistorta*); 5. **gynodioecism**, i.e. hermaphrodite and female flowers on different plants (many species of *Valeriana*, Fig. 45 IV-VI); 6. **trioecism**, i.e. hermaphrodite, male and female flowers on different plants (ash, Fig. 50). In some species even different forms of distributions occur.

The transfer of pollen grains in cross-pollination can be effected by wind (**anemophily**), by water (**hydrophily**) as well as by animals (**zoophily**).

3.3.2   Anemophily

Anemophilous plants are usually characterised by the reduction of the normal means of attracting and rewarding animal visitors and particularly by the lack of display equipment. Large and conspicuous sepals and petals could possibly be a hindrance to the spreading of pollen grains and their free access to the stigma. On the other hand a large number of pollen grains is usually produced. The grains are often small and should be distributed as quickly and as freely as possible in the air (hence there is no viscin). For this reason they are often actually shaken out of anthers which hang out of the flowers on long filaments, and the pedicels may also be mobile (e.g. Hemp). The number of stamens is frequently increased. In order to catch the pollen grains which are brought along by the wind, the stigmas are often quite large and feathery, and as freely accessible as possible. In many of our anemophilous catkin-bearing plants (hazel, alder, birch) it is not the individual stamens but the complete pendulous inflorescence (catkin) which spreads the pollen grains; also the fact that these woody plants flower before their leaves develop is an advantage for dispersing the pollen into the air and blowing it towards the stigmas.

The distribution of pollen in the air can also be effected by what is called an **explosion mechanism**, which is known from many members of the Urticaceae and in the Annual Mercury (*Mercurialis annua*). In the Urticaceae, as for example in the Common Nettle (*Urtica dioica*) the stamens are curved inwards during the bud stage because of vigorous growth of the underside of the filament (Fig. 185 VI). As a result of the increasing turgidity the cells on the inner side of the filament become increasingly stressed, but their anthers are pressed against the rudimentary ovary in the centre of the flower. If, however, the anthers open in sunny dry weather, the filaments snap back elastically and eject the pollen grains (Fig. 185 VII). In Annual Mercury the three tepals bend back so far, by means of swelling tissue at the base on the upper surface, that the flowers snap away from their short pedicels and spring off, so that the anthers burst open and the pollen grains fly out.

3.3.3   Hydrophily

There are relatively few plants in which pollination is carried out by means of water. In the Zosteraceae, Zannichelliaceae, Cymodoceaceae and Najadaceae as

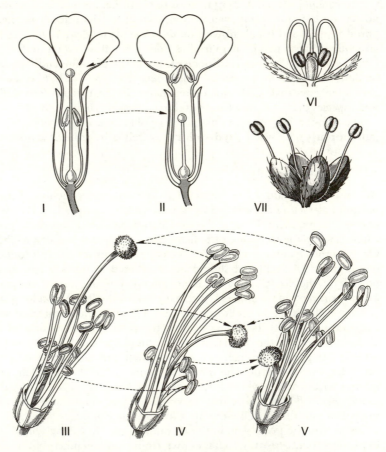

Fig. 185 I, II *Primula elatior*, heterostyly, I long-styled, II short-styled form; III-V *Lythrum salicaria*, trimorphic heterostyly, legitimate modes of pollination indicated by arrows (after H. Meierhofer, partly modified). VI, VII *Urtica dioica*, explosion mechanism in the male flower, stamens in VI still in the curved and tight-strung position, exploded in VII. After Walter as well as Wettstein and Schnarf, altered.

well as the Ceratophyllaceae, the pollen transfer takes place **underwater**, since these all live as submerged water plants. There are also examples in the Callitrichaceae (*Callitriche hamulata*, submerged-geitonogamous; *C. hermaphroditica*, pollen grains without exine!) and perhaps in Pondweeds which live completely or substantially submerged (*Potamogeton lucens* etc.?) and *Halophila* in the Hydrocharitaceae. In such cases the perianth is usually highly reduced.

The classic example of a flowering plant which is pollinated underwater is Eelgrass (*Zostera marina*). The hermaphrodite flowers are arranged in spikes, which are enclosed in sheaths of their strap-shaped leaflike pherophylls. Each flower consists of a stamen and an ovary with two long filiform stigma branches; the perigon which is still present in *Potamogeton* has been almost completely reduced. The flowers are protogynous. At flowering time the margins of the leaf sheath which encloses the spike gape apart, so that the long stigma branches can protrude. They can then readily catch the pollen grains which are floating past, since these are filiform and over 2 mm long and easily wrap themselves round the stigma branches. Similar arrangements apply in the dioecious genus *Phyllospadix* (Zosteraceae), the genus *Posidonia* in the Potamogetonaceae (hermaphrodite flowers) and the dioecious Cymodoceaceae, which all likewise form filiform pollen grains, and which in the latter family can even reach a length of over 6 mm and which possess neither sexine nor apertures (Ducker and Knox 1976). In *Zannichellia* and *Najas*, however, the pollen grains are spherical to elliptic (and similarly inaperturate), but the pollen tubes germinate when the grains are still in the open anthers*, which increases their buoyancy and facilitates their capture by the stigmas. In the Zannichelliaceae the latter are large and dish-shaped (Fig. 77 I, II). In the genus *Halophila* in the Hydrocharitaceae, which lives totally submerged, chains of pollen grains are formed from linear tetrads.

In other water plants which normally live underwater pollination takes place at the water surface. A familiar example is the dioecious Tapegrass (*Vallisneria spiralis*) in the Hydrocharitaceae, which agrees with other members of the family in this respect. The small male flowers (which like the female flowers, are still provided with a heterochlamydeous perigon) detach themselves under water and are pushed up to the surface, where they open. The female flowers have long pedicels, so that they also come up to the surface of the water. The male flowers can then come into contact with the female ones, so that pollen transfer is possible. After pollination has taken place the female flowers are once again withdrawn below the water surface as their pedicels are rolled up in spirals (hence the vernacular name in German, "Sumpfschraube = pond coil"). In *Hydrilla lithuanica* this phenomenon is linked with an explosion mechanism, by means of which the pollen grains are distributed over a radius of about 20 cm. In Canadian Waterweed (*Elodea canadensis*) the male flowers are also brought up to the water surface by elongation of their pedicels. In yet other cases the pollen, which is resistant to wetting, is released from the anthers underwater and floats up to the surface where it can now float across to the female flowers which have reached the surface on their long pedicels (*Ruppia spiralis, Callitriche hermaphroditica*).

---

* In the anthers of submerged water plants the formation of a fibrous layer is often completely inhibited.

The pollen of Hornwort (*Ceratophyllum*) is readily wetted and its specific density corresponds closely to that of the surrounding water, which quickly becomes permeated with free-floating pollen grains shortly after the anthers have emptied. This seems to ensure that the stigmas, which likewise remain below the water surface, receive their pollen (see Knoll 1956).

3.3.4   Zoophily (zooidiogamy)

3.3.4.1   *General prerequisites*

The great majority of flowering plants are pollinated by animals, and most of these by insects. This is of course only possible if at the same time certain conditions are fulfilled:

1. There must be some means which cause animals to seek out the flowers regularly, by some **attraction** – colour, scent, form of the flowers or of the inflorescence, and for some **reward** – pollen, nectar, feeding hairs etc.
2. The flowers must withstand the mechanical stresses and be of such a form that regular contact with the pollen and stigma is possible.
3. The pollen must be able to adhere to certain positions on the bodies of the flower visitors. This may be effected either on the one hand by viscin or by viscidia attached to pollinia, or on the other by surface properties of the visitor which correspond to the plant.
4. The transfer of this pollen to the stigma of another flower must be ensured by the behaviour of the visitor and the form of its body.

All this requires that the flower and the flower visitor must be adapted to one another in many different ways. Repeatedly therefore during the course of evolution reciprocal **co-adaptations** have evoked close dependencies between particular animals or groups of animals and particular groups of related plants.

From the increasing specialisation for the attraction of particular pollinators and the continual improvement of devices for the guaranteed transfer of pollen there also results an increase in the reliability of pollination with smaller quantities of pollen. Hence in wind-pollinated plants the numerical ratio of pollen grains to fertilised ovules is often about $10^6:1$, whereas in highly specialised insect-pollinated plants, such as certain orchids, a ratio of almost 1:1 can occur (see section 1.5.14).

It is quite common for particular physical types of flowers (Vogel 1954) – disc-shaped flowers, bowl-shaped flowers, funnel-shaped flowers, bell-flowers, labiate flowers, butterfly flowers, bottle-brush flowers and flowers in capitula or spadices – to be favoured by particular groups of pollinators. They are almost never specialised to one group of pollinators alone, however. The groups of pollinators in question are almost exclusively insects, birds and bats.

The various means by which the petals make themselves attractive (see section 1.4.3.1), such as colour and pattern (nectar guides!) or the secretion of various perfumes, have a different and limited effect corresponding to the physiology of perception of the different pollinators, which belong to different groups of animals. Also the offer of various rewards, such as the abundant production of pollen or the formation of nectaries or fruit bodies is naturally specified for particular pollinators. The at least temporary "flower constancy" of many pollinators, whether it be basically a "breaking in" process to continually repeating signals (colours of certain flowers) or the result of certain structural peculiarities of flowers and pollinators, is an important factor as a guarantee of regular pollination (see also Grant 1950).

In regard to the formation of nectar guides Osche (1979) has put forward the hypothesis that the primary optical signal of flowering plants for the attraction of the pollinators could have been the pollen or the androecium which produces it, in which the form and colour became decisive for the signalling effect. In support of this argument was adduced the fact that in those groups of Angiosperms which are regarded as primitive, the principal nutrient which is offered is the pollen (see section 3.3.4.2). "Also after the production of nectar as an additional attraction or even as the main reward the primary signalling function of the androecium survives". The effect of this primary signal could then have been replaced by the formation of nectar guides when the anthers and the means of providing rewards became hidden more deeply within the flower, or even within a more or less closed corolla tube. "Nectar guides are therefore secondary signals and represent *mimetic copies of signals*" (*dummies* for the androecium, stamen or pollen).

We have already mentioned (section 1.4.3.1) that perfume is often produced by special scent glands. Vogel (1962) thoroughly investigated these organs, the **osmophores**, which "represent a special kind of plant gland". They "consist of a localised, multicellular, outwardly-directed epithelium, which over a short period of time produces pollinator-attracting volatile secretions of various compositions, by means which obviously require a rapid and conspicuous use of reserve materials".

A differentiation of osmophores into an "emission layer", the epidermis, and a "production layer", several layers of cells in the underlying tissue, can be recognised not only by reference to histology but also in relation to their function. The production layer can also be combined with parenchymatous storage tissue. This storage tissue, or also the production layer as well, is usually abundantly filled with reserve starch granules before the emission of scent, but often within a few hours during the production and emission of the perfume these are used up.

Vogel succeeded in showing, by means of selective vital staining with neutral red, that quite a number of flowers "only emit scent with certain *precisely circumscribed morphological elements*". These scent-producing areas (Fig. 186) therefore play a similar role as olfactory guides to that played optically by the nectar guides.

The (seismonastic) sensitivity which many stamen filaments show to a tactile stimulus (African Hemp, *Sparmannia africana*, *Helianthemum apenninum*, etc.) may also have a role to play in the transfer of pollen. This suggestion is probably correct in the case of *Centaurea* species, e.g. *C. montana*, in which the filaments contract to a touch stimulus, so that the stigma which is surrounded by the anther tube suddenly pushes pollen grains out of the tube. In *Berberis* and *Mahonia* the sensitive upper surface of the base of the filament lies in direct proximity to the nectaries, which are positioned at the base of the petals. On the other hand, the pollen ejection device in the *Catasetum* species which are native in South America is not a case of a movement due to sensitivity, but is a mechanical process. The flowers of this orchid genus are unisexual and their arrangement is either monoecious or dioecious. In the male flowers there are two horn-like processes ("antennae") which originate from the base of the gynostemium and which block the path of the insect visitors (Euglossine bees, see section 3.3.4.2 (b)). As soon as they touch these antennae, the pollinarium, which consists of the two pollinia (section 1.5.14) of both thecae provided with a pedicel and a retinaculum, is flung out with considerable force, and usually stays attached to the head of the insect. "At the base of the *Catasetum* antennae there is tissue present which is in a state of tension, and has predisposed lines of abscission. The slightest touch of the antennae is sufficient to cause a pull on this tissue and make the lines of abscission split open. The strong forces which are

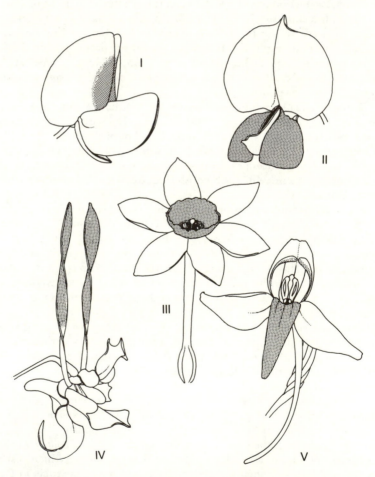

Fig. 186 The use of vital staining with neutral red to reveal scent-producing areas. I *Lupinus cruckshanksii* (standard, and also scent guide), II *Spartium junceum* (wings), III *Narcissus jonquilla* (corona), IV *Dendrobium minax* var. (lateral petals), V *Platanthera bifolia* (labellum and anterior margins of lateral petals). After St. Vogel (1962).

present equalise themselves in an instant, the pedicel of the pollinarium detaches itself and rolls up, and the pollinia and the retinaculum are projected out of the flower as a whole by the recoil" (Knoll 1956). This is therefore a case of a turgor mechanism by means of which the pollinaria can be ejected as much as 80 cm away.

Sometimes insects are not seeking out flowers for feeding purposes, but in order to lay their eggs there. This happens in such a way that the pollination of the flower is accidentally, or even "deliberately" achieved. A famous example of this is the pollination of *Yucca* species by the moth *Pronuba yuccasella*. The Yucca Moth is nocturnal, and after mating the female seeks out *Yucca* flowers which are conspicuous in the dark on account of the shining white perigon and a slight scent. The sticky pollen is then scraped out of the anthers and formed into a large ball, up to several millimetres across. The female flies over to another *Yucca* flower, carrying this heavy ball, where she lays her eggs in the ovary and inserts the pollen which she has brought into the stigma, so that the pollination of the ovules is ensured. The developing larvae of the moth feed on the growing seeds, but sufficient undamaged seeds are left over to maintain and distribute the relevant species of *Yucca*. We have already studied an even more complex example of the coupling of the reproduction and development of a pollinator with the pollination and fertilisation of a flowering plant in the case of *Ficus* species (section 2.3.2).

3.3.4.2   Pollination by insects (entomogamy)

(a) Beetle-pollinated flowers (Coleopterophilae, Cantharophilae)

Amongst the insects the beetles (Coleoptera) represent not only an ancient group, but also probably the oldest pollinators. It is therefore remarkable that beetle-pollinated flowers are found within families which are derived to a different degree, but they are especially represented as a dominant character of many families of the Magnoliales (Gottsberger 1977).

Since the mouthparts of the Coleoptera are usually suspended vertically in relation to the longitudinal axis of the body (prognathous), their length is necessarily limited. This restricts beetles to searching for food by looking in open disc-shaped or bowl-shaped flowers only. Since they use their biting mouthparts to chew at all parts of the flower, tepals or petals, stamens and carpels, they often damage these flowers considerably. Naturally the principal food is pollen, which is usually abundantly available (pollen flowers), besides which, however, there are occasionally readily accessible nectar or feeding bodies (e.g. on the stamens of *Calycanthus*, section 1.7.1). The flowers are often robust and provided with only minimal optical attractions. They are white or greenish to brownish in colour. The visitors are attracted by odours of fresh or rotten fruit. Besides primitive forms of cantharophilous flowers, which we can still find today in the genus *Drimys* in the Winteraceae (with a sweet scent nevertheless), there are also specialised forms of beetle flowers in the families of the Magnoliales. This is the case for example in the genus *Guatteria* in the Annonaceae, in which the petals turn yellow during anthesis and become inflexed. The beetles which are attracted by the fruity smell not only eat the flower parts, but also make use of the "pollination chamber" formed by the petals for mating and the laying of eggs, in such a way that the pollination of the flower occurs at the same time (Gottsberger 1970, 1977). Cantharophilous flowers are also found in many members of the Rosales, Dilleniales (*Paeonia*), Centrospermae etc., although not as the only mode of adaptation of pollination.

In some plants, such as for example in species of *Cassia*, only a proportion of the stamens in a flower produce fertile pollen, and the development of pollen in the remainder is more or less inhibited, and so the high nutrient content of the latter can be used as fodder for the pollinators. Such "feeding anthers" occur in many species, such as in the Melastomataceae, Commelinaceae and Capparidaceae. (Unequal development of the anthers – heteranthy – is in any case not rare, as in species of *Verbascum*, for instance, see section 1.5.5.)

(b) Bee-flowers (Melittophilae)

These may take the form of narrowly- or widely-tubular flowers, but in particular many dorsiventral flowers are adapted to pollination by wasps or various kinds of bees. This is especially true for many labiate, personate or papilionaceous flowers which have a landing-place for the flower visitor and which often have some special closing mechanism to protect the nectar – as for example the

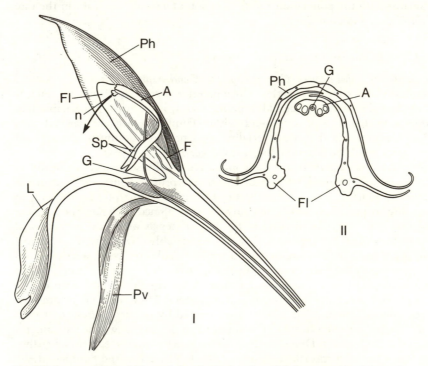

Fig. 187 Roscoea purpurea. I Flower in longitudinal section, slightly to one side of the median, so that the fertile stamen is not sectioned, II upper lip in cross-section, somewhat below the tip of the anther, Ph posterior (median) tepal of the inner whorl of perianth segments, Pv one of the two anterior (lateral) tepals of the inner whorl of perianth segments; Fl lateral staminodes of the outer whorl of stamens ("wing"), L labellum, A anther of the fertile stamen, Sp its basal spurs, F filament, G style, n stigma. After Troll.

swelling on the lower lip of Snapdragon (section 1.4.3.8). Yellow, violet or blue flower colours and nectar guides are the rule together with a sweet scent. The position of the stigma and the anthers is adjusted for nototribal pollination (see section 1.4.3.8).

Many different pollination mechanisms are also found in bee-flowers, such as the lever mechanism of *Salvia* already explained (section 1.5.3), whose counter-part in the monocotyledons is in *Roscoea purpurea* of the Zingiberaceae. The latter was thoroughly investigated by Troll (1929), to whose illustration we shall refer (Fig. 187 I). "The outer ring of the perianth is completely hidden within the subtending leaf of the flower; but the inner ring is not very conspicu-ously represented either, and the formation of the flower's appearance is almost entirely taken over by the petaloid staminodes. Of these the two anterior are fused into a spathulate and slightly doubly emarginate labellum, as in other Hedychieae, and this projects a long way out of the flower and is curved downwards. It forms the *lower lip*". As in the flower of *Curcuma australasica* already illustrated (Fig. 74 V) the upper lip of this species is composed of the staminodes of the outer whorl of stamens and the median member of inner whorl of tepals (Ph). The latter forms a concave hood holding the two lateral staminodes inside it. These are highly asymmetric in form, as shown in Fig. 187 II, i.e. the parts which are directed inwards are significantly larger than the outer halves and are locked together behind the fertile stamen, whilst the outwardly directed halves are folded over outwards like a seam.

The anther of the fertile stamen is versatile and inserted on a slender but broad articulation at the tip of the filament, so that unlike those versatile anthers with a filiform articulation which can turn in any direction, it can only move in the direction of the median. This is definitely an advantage for the pollination process. The two thecae are extended into long sterile spurs which are adjacent to the filament until shortly before anthesis. Whilst the perianth segments and the petaloid staminodes are opening and spreading out the anther becomes transversely oriented on the filament as a result of intensive growth of the upper surface of the filament in the region of the zone of articulation, but at the same time the upper fertile portion of the anther curves inwards. The result is the arrangement illustrated in Fig. 187 I. The style is very delicate and lies with its upper section in the ventral groove between the two thecae, and as the anther grows it comes to be "embraced, so to speak" by the two thecae (Fig. 187 II). "By this means it becomes fixed in place, and from now on it has to keep step with any movements which the anther may make". Its stigmatic region – which is formed into a "tubular stigma" similar to that in *Viola* (see section 1.6.16) – projects beyond the anthers.

As a consequence of the twisting of the anther on its filament the two thecal spurs now block the entry to the interior of the flower. An insect forcing its way to the nectar must, as in *Salvia* push against this obstruction. However, since the anther can be rotated downward on its broadly articulated filaments, it gives way to the pressure, so that the fertile part of the anthers together with the orifice of the tubular stigma is pushed in the direction of the arrow in Fig. 187 I, against the back of the insect, and hence pollination is effected.

Whilst on one hand the form of a labiate flower and the formation of a lever mechanism in *Salvia* and *Roscoea* are achieved in very different ways, we do on the other hand often find in the Papilionaceae very different kinds of pollination mechanism even though the actual floral diagram (Fig. 7 VI) is very similar. We have already discussed the pronouncedly zygomorphic structure of the corolla in section 1.4.3.4 (Fig. 37 I, II). The two lower petals, which are combined to form the "keel", enclose the stamens and the single carpel. Of the 10 (5 + 5) stamens usually 9 or sometimes all 10 have their filaments fused into a tube which encloses the carpel and which secretes nectar on its inner surface at the base. The keel, and the wings which are often attached to it by means of a notch or fold, serve as a landing platform for the insect visiting the flower. Under the weight of a sufficiently heavy insect (honey or humble bee) these yield and release the stigma and anthers – or also on occasions only the pollen.

In Broom (*Sarothamnus scoparius*), the upwardly curved stamens (in particular the 5 longer ones) together with the long style are held together like a coiled spring by the mutually fused margins of the keel (Fig. 188 II). If the wings and keel are pushed downwards by an insect, then the upper margins of the keel separate from one another, progressively from the base to the tip. Then first the 5 shorter stamens spring out and spread pollen onto the underside of the insect, and then the long stamens and the style break out of the keel and strike it on the back, so that the insect is normally enveloped in a cloud of pollen (**explosion mechanism**). The Spanish Broom, *Spartium junceum*, and other species also behave in a similar way.

The flowers of the Dyer's Greenweed, *Genista tinctoria*, function with what is called the **snap mechanism**. The keel and wings, which (as in Broom) are fastened together by means of corresponding folds f and f' (Fig. 188 VIII, IX), are pressed downwards by a zone of swelling tissue at their base, but are firmly attached at the tip of the keel by the style. When there is pressure from an insect the two petals of the keel become detached from one another at the the tip, so that the keel and wings snap downwards (Fig. 188 VIII).

In Lupins (*Lupinus*), Bird's-foot-trefoils (e.g. *Lotus corniculatus*, Fig. 188 V, VI), Dragon's-teeth (*Tetragonolobus maritimus*) and quite a few other papilionaceous flowers the upper margins of the keel are completely united except for a small open slit at the beak-shaped and elongated tip. Either all the filaments, or only 5 of them (Fig. 188 VII), are so greatly clavately swollen below the anthers that they completely close off the tip of the keel from below. The anthers empty out their pollen into the tip of the keel. If an insect now presses from above onto the wings and keel, which are here once again united, then the thickened stamen filaments act like a piston and push an adequate mass of pollen out of the opening in the keel (like the action of a "**grease-gun**"). The process by which pollen is applied to the ventral surface of the insect can be repeated many times; somewhat later the stigma emerges from the orifice in the keel so that **sternotribal** pollination can take place. In the Vetches (*Vicia*) the snap mechanism is linked with the formation of "style brushes". The papilionaceous flowers therefore exhibit many different adaptations for pollination whilst having a uniform basic structure.

Instead of the usual reward given by bee-pollinated flowers, i.e. nectar and pollen, there are a large number of plant groups which offer the flower visitors a fatty oil, which is produced in the flower by special oil glands (elaiophores, see section 1.7.1). These "oil flowers" were discovered and thoroughly investigated by Vogel (1971, 1973, 1974) who discovered about 1270 examples in South American and South African species of the Scrophulariaceae (especially *Calceolaria*), Malpighiaceae, Krameriaceae, Iridaceae and Orchidaceae. In the flowers of the plants the oil glands are usually arranged transversely and in pairs. In South America three taxa of solitary-living Hairy Flower-bees with about 275 species were discovered to be the pollinators. In all these "oil bees" the female possesses special oil-collecting equipment (absorbent tufts of hair, filter-combs or scraping edges) on the first or second pair of legs. The oil is gathered up with pollen on the thighs of the rear legs and provides a calorific diet for the brood. In the Holarctic region oil flowers were recognised for the first time in members of the genus *Lysimachia* (subgenera *Eulysimachia* and *Seleucia*) in the Primulaceae, where the oil is secreted by the dense glandular hairs on the outer surface of the filament tube (Vogel 1976). The oil is collected by female mining bees

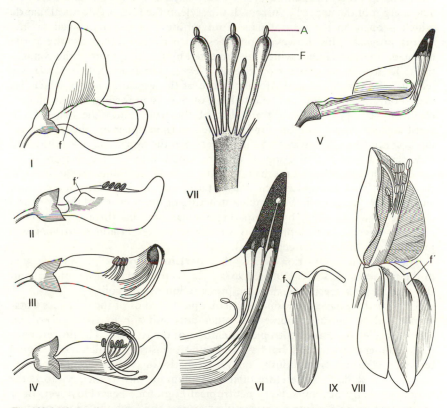

Fig. 188 I–IV. *Sarothamnus scoparius*, complete flower (I), after the standard and wings have been removed (II), III, IV position of the stamens and style in the keel before (III) and after the explosion (IV), f' notch on the keel, onto which the wing fits with the fold f. V, VI *Lotus corniculatus*, V keel with androecium and style in longitudinal section, VI tip more highly magnified, VII *Tetragonobolus maritimus*, part of stamen tube with long-clavate and short stamens, A anther, F filament. VIII, IX *Genista tinctoria*, VIII flower after the wings and keel have snapped downwards, IX right wing seen from inside, f, f' corresponding folds on the wings and keel. VII after Kugler, the rest after Knuth.

(*Macropis*), which likewise possess special collecting equipment, namely absorbent tufts on the inner flanks of the first and second pair of legs, and behave in a similar fashion to the Hairy Flower-bees referred to above.

The flowers of *Calceolaria* which function as oil flowers are in the form of "slipper flowers", which is to say that in contrast to the Snapdragon (section 1.4.3.8, Fig. 43 I to III) the inflated lower lip of the sympetalous corolla is concave and not convex. It is so strongly inflated that, it greatly exceeds the dimensions of the upper lip. In this respect the flower of *Calceolaria* resembles the flower of the Lady's-slipper (Fig. 189 I). In many species of *Calceolaria*, as for example in *C. tripartita* which was investigated in detail by Ritterbusch (1976), the central space of the lower lip is closed off by its margin, which is highly inflexed. The same is also true of the upper lip, where the margin is folded up with that of the lower lip in such a complicated fashion that a kind of "dovetailing" is created, and they are interlocked as in "tongue and groove". The margin of the upper lip surrounds the style in the form of a conical mantle which is open on the ventral side, from whose tube-like narrowed tip the stigma projects. This, however, provides space further down for the sterile thecae of the two stamens which are projected by the long arms of the connectives. These connective arms operate as a lever, similar to that in *Salvia* (see section 1.5.3). In this case the fertile thecae of the versatile anthers which are placed on short filaments can be tilted out of the casing formed by the margin of the upper lip, to the right and left of the style. Since the lower lip is constrained to move with the firmly-connected rigid upper lip, the oil-collecting bees can force their way into the slit between the upper and lower lips, and push their way to the oil glands which are found inside on the margin of the lower lip, operate the lever mechanism and carry out the pollination at the same time. Ritterbusch has undertaken a quantitative geometric analysis of the correlated growth processes in the complex structure of this flower and the structural changes related to them by comparing median sections through flowers in different stages of development. A similar study on *Pedicularis recutita* was published by Meier-Weniger (1977a).

The principle of attraction used by the "**perfume flowers**", first recognised in the full sense by Vogel (1966a, 1967a), claims to be even more unusual and still more highly specialised than the presentation of fatty oils.

Thus far these have been found in neotropical orchids of the subtribes Catasetinae and Gongorinae and several other taxa and also in *Gloxinia perennis* (Gesneriaceae) and some neotropical members of the Araceae. The flowers have only rudimentary nectaries, but have osmophores instead, "which secrete liquid terpenes with a conspicuously spicy, aromatic or terpentine-like odour" (1966a). These flowers therefore offer only fragrant oils to their visitors, the males of solitary or semi-solitary neotropical Euglossine bees. However, these creatures are attracted from considerable distances and "rub the odour-producing surfaces with their forelegs in a state of great excitement". To do this they make use of two organs which are clearly intended for collecting the fragrant oils, and which are confined to male Euglossine bees: absorbent tufts of hairs on the front foot segments (tarsal tufts) for soaking up the fragrant oil, and for storing it, spongy chitin bladders on the back legs (tibial organs), which are air-filled when empty. The oil taken up by the tarsal tufts is transferred to the back legs via the middle legs during flight. "The fragrant oil is taken up in a pan, filtered and soaked up by a fine tube into a receptacle by capillary action. From here a duct like a candle wick leads outwards to a basin with scales" (1966b). A larger bee can store up to 60 mm^3 of the fragrant oil, which is mixed with various different additives according to the individual and the species, and is probably released again by a complicated method making use of passive evaporation in a ventilating airstream created by the buzzing of the wings, and presumably serves to mark out a mating territory.

(c) Fly flowers (Myiophilae)

Fly flowers represent a very heterogeneous group (see Kugler 1955), which is a consequence of the very varied habitats occupied by the Diptera and of the different life styles of the individual Dipteran groups.

Many Diptera fly more or less regularly, although often only incidentally, to flowers and inflorescences while searching for food, in order to consume pollen or nectar there. Like most of the insects which visit flowers such Diptera favour sunny localities. They are principally looking for flowers with freely accessible nectar, which they can take up with their proboscis, which is often only short. Among plants where this is the case are many Umbellifers, species of *Valeriana* and *Valerianella*, as well as *Mentha* and *Lycopus* (i.e. Labiates with a short open corolla tube), and *Stellaria media* and other chickweeds with similarly wideopen flat flowers. Fly flowers are practically always visited by other insects besides flies. They often possess a scent of honey, but they may also give off aminoid odours.

The Hover-flies (Syrphidae) are the best-known visitors among the Diptera, and one of these, the Drone-fly (*Eristalomyia tenax*), bears a proboscis which in its extended condition is 5 mm long. Many flowers, as for example those of *Circaea* and *Veronica*, are sought after nearly exclusively by hover-flies. Kugler (1970) therefore classifies the **hover–fly flowers** as a group of its own for the purposes of floral biology. Also frequent visitors to flowers in open habitats are the Bee-flies, to which the genus *Bombylius* belongs, some of the members of which have a very long proboscis (12 mm).

One group of myiophilous plants is that which can be distinguished from the inhabitants of open, sunny localities are those "which could count rather as shade plants of the forest floor and which are specialised ... to Diptera" (Vogel 1978). Among these he counts the Golden-saxifrage (*Chrysosplenium oppositifolium*), Moschatel (*Adoxa moschatellina*) and other plants with yellow, greenish or brownish flowers and a predominantly musky or aminoid odour. Dullcoloured flowers are also found in Central European species of *Euphorbia*, species of *Saxifraga* and in *Tozzia alpina* (Kirchner 1911). A putrid odour of amines is also emitted by the green flowers of Ivy (*Hedera helix*) which appear late in the autumn, and which are regularly visited by flies. Here the nectar secreted by the disc which functions as the nectary lies freely on the surface in large drops.

Flowers with nectar which lies more or less in the open, and which emit odours which are unpleasant to humans, have been termed "nauseous flowers". The aminoid smell of Bird Cherry (*Prunus padus*) is evidently highly attractive to Diptera, likewise the fishy smell (trimethylamine) of the Rowan (*Sorbus aucuparia*). Of more interest are the **Dung-fly flowers** (Sapromyophilae), which attract dung-beetles and dung-flies by their putrid, urine-like or carrion-like odours and so simulate a suitable substrate for the laying of eggs. Their flowers are often dull in colour, or are greenish-purple flecked or are marked with a purplish-brown network of veins, which adds to the deception. The principal examples of this are numerous members of the Asclepiadaceae-Ceropegieae (*Stapelia, Hoodia, Huernia, Caralluma*), and then the pitfall flowers of Lords-and-Ladies (*Arum maculatum*, to be discussed in more detail below) and other Araceae, as well as the large dull-purple flowers of the tropical *Rafflesia* species (Rafflesiaceae) which are parasitic on flowering plants. The latter represent the largest flowers in the plant kingdom; they can be up to 1m in diameter.

Vogel (1978) has given a detailed description of the **fungus gnat flowers** which are inhabitants of cool, damp and shady biotopes in all continents, and to which similar remarks apply as for the dung-fly flowers. They attract fungus gnats (Mycetophilidae) and "mourning gnats" (Sciaridae) as well as a few other groups of Diptera, and do this largely by olfactory means, but also partly by optical signals, and in highly specialised cases even by "plastic imitations of fungal fruit bodies or parts of the same". They do not offer any nourishment to the visitors, but imitate a substrate for the laying of eggs, namely the fruiting bodies of fungi in which the larvae would normally undergo their development. The examples of this mode of floral biology investigated by Vogel include many species of *Asarum* (e.g. *A. caudatum, A. shuttleworthii*), the pitfall flowers of *Aristolochia arborea, Cypripedium debile, Arisaema* and *Arisarum* species and even a few species of the orchid genus *Masdevallia*.

The type of floral biology described above could also be assigned to the **deceptive flowers**, since they offer no reward to the flower visitor (see further Daumann 1971, Vogel 1975 as well as the literature cited by both these authors). The same is true for the **sexually deceptive flowers**, of which the best known are the species of the orchid genus *Ophrys*. The vernacular names of Bee Orchid (*O. apifera*), Fly orchid (*O. insectifera*) etc. already indicate that the form of the flowers of these orchids makes them appear very similar to certain insects. In fact they are mistaken by the males of certain insect species for the females, and so when they fly to them, pollination is effected. The attraction is often not only due to the dummy effect of the physical form but due to the scent as well, which imitates the specific sexual chemical (pheromone) of the female of that species. The same is possibly true for other, tropical orchids. Hence for example the flexibly inserted lips of *Drakea* species and the "swinging labellums" of many *Pleurothallis* species are supposed to imitate the females of Tiphiid wasps or of flies. In the case of the nectarless and insect-like flowers of *Oncidium* species it is thought that the males of the bee genus *Centris* mistake the movement of the dummies for "intruders" in their mating territory and that for this reason they attack. This shows us at the same time that deceptive flowers are not confined to fly flowers. **Pitfall** and **pinch-trap flowers** also usually do not offer any nourishment whatsoever for the pollinators, and these are also mainly pollinated by flies.

It is now time to mention the **pitfall labellums** of most *Cypripedium* species (and of course the related genera as well) which we also find in the European Lady's-slipper (*Cypripedium calceolus*). According to Vogel (1978) they can be classified as bee-deceptive flowers. The concave inflated labellum of our Lady's-slipper, resembling the shape of a clog, has an elliptic opening on the upper side, into which the gynostemium projects from above (Fig. 189 I, II). Insects (and particularly mining bees, species of *Andrena*, Daumann 1968), which slide down the smooth shining outer wall of this pit then slip and fall through the opening. The only way out they can find is by the opening to the right and the left behind the gynostemium because the walls are otherwise so steep and slippery. A line of hairs, which make it easier to climb the wall, leads to this opening. As they force their way through the narrow exit, they push against one of the two anthers and in so doing pull out a pollinium by its viscidium, which then can be brought into contact with the stigma during the "visit" to the next flower. In fact the functioning of the pitfalls is less than perfect (see Daumann 1968).

In Birthwort (*Aristolochia clematitis*, Aristolochiaceae) the entire flower is constructed as a pitfall. The flowers are placed in cincinnoid partial florescences in the axils of leafy bracts. The long perigon tube of the epigynous flower is widened above the base into the shape of a flask, which towards the tip narrows rather suddenly into a narrow tube. In the early and female stage (Fig. 189 III) this is clothed inside with downwardly directed hairs. At this time the flowers are erect and attract small Diptera with their sweetish scent. As soon as these reach the inner surface of the perigon tube, they slip down and often fall into the tube. In this the hairs act like the entrance of a lobster pot: this is because they are excentrically inserted on a thin-walled articulated cell, which when pushed upward immediately presses against the epidermis. They will give way when pushed downwards, but upwards will scarcely deflect at all, so that for the insect which has fallen into the trap, the return path is blocked. This is of course only true until the stigmas of the flower are pollinated with the pollen which has been brought in, or anyhow until the flower reaches its male stage. The anthers then open and cover the visitors with pollen, the trap hairs wither (Fig. 189 IV) and the whole flower bends over to the horizontal, so that the insects in the trap can escape. In contrast to the case of *Cypripedium*, however, the visitors are able to take up nectar, which is secreted on the inner surface of the trap (for further details see Daumann 1971a). The pitfall flowers of many other, larger-flowered tropical species of *Aristolochia* are clearly to be regarded as dung-fly flowers, on account of their colour and smell.

Pitfall flowers are found in great variety among the species of the genus *Ceropegia* in the Asclepiadaceae (Fig. 40). The pollinators – very small Diptera, but neither carrion-flies nor dung-flies – are attracted principally by the flower scents, which although not pleasantly fragrant, are not genuinely carrion-like or faecal. They are produced by osmophores, which just in the case of the *Ceropegia* species are very clearly differentiated and highly exposed at the ends of the perianth segments. It is possible that the presence of mobile ciliae on the osmo-phores plays a role of optical attraction. These are set in motion by the slightest movement of the air and create an impression of continuous shimmering. These structures were described as **shimmering bodies** by Vogel (1954, 1961) and may imitate the presence of members of the same species and so encourage the pollinators to fly down. The functioning of the pitfall trap is aided by surfaces which are suitable for landing and which are combined in different arrange-ments with smooth surfaces of different size, which lead visitors to slip into the trap. In this the specially prominent uvula (see section 1.4.3.7) which is accented by the "false nectar guides" may act as a "slippery slide" for a "guided fall". To this may be added a light-trap effect brought about by the window-like openings and the penetration of light through the diaphanous zones of the corolla. For more details see the comments made by Vogel (1961).

The best known example of a pitfall flower, Lords-and-Ladies (*Arum maculatum*), is not formed from a single flower but is a complete inflorescence which is enclosed by a large bracteose leaf organ, the spathe (Fig. 189 V). Only the upper part of this unfolds and so exposes the clavately thickened distal part of the inflorescence axis, the spadix. The rest of the spathe remains tightly rolled up, and is pinched into a waist directly below the spadix, but widens into an inflated chamber below. The inflorescence axis with its terminal spadix bears successively, in order from the bottom up, a series of zones as follows: 1) with fertile female flowers, which consist of a single monomerous ovary, then 2) sterile, long-styled female flowers, then 3) male flowers consisting of 3 or 4 stamens, and finally 4), another broader whorl of sterile, long-styled female flowers. The naked and usually violet-coloured spadix serves as an osmophore, and emits a urine-like odour (see Vogel 1962) coupled with a considerable rise of temperature (as much as $+ 17\,^{\circ}$C in *Arum italicum*!). Small Diptera are attracted to visit by the smell, slip on the epidermis of the inner surface of the spathe, which is covered with oil droplets, drop down and fall into the chamber. They cannot escape from this at that particular time, since the long and deflexed styles of the sterile female flowers form a dense barrier (the "obstacle" flowers) and in any case the wall of the trap has slippery surfaces. When the stigmas are pollinated and the inflorescence makes the transition to the male phase, the epidermis of the spadix withers and shrinks so that the insects can escape again, freshly laden with new pollen.

Pinch-trap flowers have a very complex structure and are characteristic of the Asclepiadaceae-Asclepioideae, as for example in the European species *Vincetoxicum hirundinaria = Cynanchum vincetoxicum* and the genus *Asclepias* (Fig. 190). The flowers are 5-merous with the exception of the dimerous and apocarpous gynoecium. In the flower illustrated in Fig. 190 I the 5 sepals and 5 petals are reflexed so that the 5 stamens are prominent. The anthers are placed on short filaments and each bears a hood-like nectary on its dorsal surface, and from the inner surface of this projects a horn-like curved appendage (Fig. 190 III). The complete dorsal anther appendage is interpreted as a nectary and according to Kunze (1979) it is initiated much later than the other parts of the anther, and in many members of the Asclepioideae, as for example in *Tylophora* or *Marsdenia*, it is replaced by a simple scale. The nectaries of the 5 stamens taken together form a kind of corona. Within each anther, only the two anterior pollen sacs are formed. As in the majority of orchids the pollen grains are stuck together into a pollinium. A secondary fusion between the anthers and the gynoecium takes place at a later stage of development (VII). The two carpels are in fact postgenitally united with one another in the distal zone and form a large 5-cornered stigmatic head (II). On each one of these corners, which lie between the narrowly winged anthers, a gland is formed at an early stage (VII). From each of these glands there run two divergent stripes of glandular epithelium to the two nearest pollen sacs. These glands on the ridge of the stigmatic head exude a secretion which, after it has hardened, forms what are called the "glands" or corpuscula (Kl). The secretion from the two stripes of glandular epithelium hardens to form the so-called "caudiculae" or "translators" (t). After the disappearance of the anther wall above the pollen sacs these translators become stuck to the pollinia in which the pollen grains from the respective pollen sacs are combined and so unite the pollinia of two neighbouring pollen sacs with the corpusculum. At this time the glandular secretion as a whole separates easily from the stigmatic head (for further details see Schnepf, Witzig and Schill 1979). The winged and broadened anther margins of the adjacent stamens form folds in which the limbs of the nectar-gathering insects become entangled. As they then pull their limbs from the "pinch-trap", they at the same time pull out the corpuscula with the pollinia hanging from them, and can then transfer them

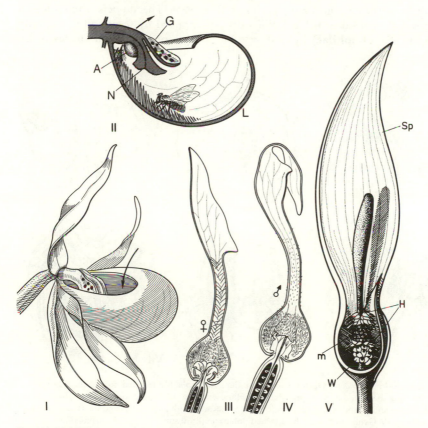

Fig. 189 Pitfall flowers. I, II *Cypripedium calceolus*; I flower, II its labellum (L) and gynostemium (G) in median section, in order to show the pitfall labellum; the arrows indicate the path taken by the insect through the openings for entry and exit, N stigma, A anthers (after Meierhofer), III, IV *Aristolochia clematitis*, pitfall flowers in female (III) and male states (IV, shown upright instead of the natural horizontal position), in median section. V *Arum maculatum*, inflorescence functioning as a pitfall trap, Sp spathe, H long-styled "obstacle" flowers, W fertile female, m male flowers. III to V after Firbas.

to the stigmas of other flowers. – In *Ceropegia* the formation of these pinch-traps can be combined with the occurrence of pitfall flowers.

(d) Lepidopterous flowers (Lepidopterophilae)

Lepidopterous flowers are often specialised for pollination by butterflies or moths by the formation of long, narrow corolla tubes. Often it is only these insects with their long probosces which are capable of sucking up the nectar which is hidden so deeply in the flower. In this is found the biological significance of such phenomena as the long hypanthia of many *Oenothera* species (see section 1.3.2), e.g. *Oenothera biennis* or *O. missouriensis* (hypanthium 14 cm long!), both of which flower at night. The flowers in question are often "stalked plate flowers", i.e. they have a long corolla tube with a broad and spreading limb, as is the case in many species of *Oenothera*. The flowers often adopt a horizontal or pendent habit. Pollinators are attracted by scent and in butterfly flowers (**Psychophilae**) by yellow, blue, purple or red colour shades or by

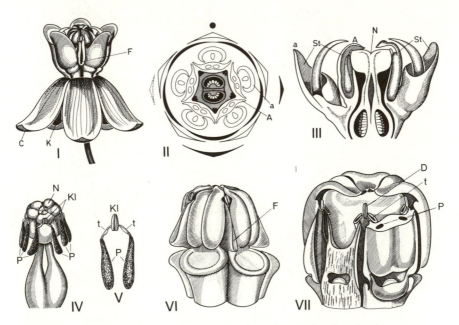

Fig. 190 Asclepias syriaca. I Open flower, calyx (K) and corolla (C) reflexed, so that the stamens surrounding the gynoecium are prominent. II Floral diagram. III Median section through androecium and gynoecium, N stigmatic head, St stamens with anthers (A) and dorsal cone-shaped nectary (a). IV Gynoecium with the attached pollinaria (P) dissected out. V A corpusculum (Kl) with two translators (t) and pollinia. VI Androecium after the cone-shaped dorsal appendages which serve as nectaries have been removed, F fold between the winged and broadened margins of two neighbouring anthers. The tarsi of the insects get caught in these folds and the corpusculum with the pollinia hanging from it is torn out as the tarsus is withdrawn. VII Androecium in young stage; the anther at the front right has been sectioned, and almost all of the forward part of the anther at the front left removed, in order to show the gland which produces the corpusculum (D) and the stripes of glandular epithelium which lead to the pollen sacs; the secretion exuded by these gives rise later on to the translators. After Engler (I, II, IV, V), Eichler (II), Müller (VI) and Payer (VII).

white. Diptera with long probosces occur as pollinators of such flowers as well as butterflies. Naturally amongst moth flowers (hawk-moth flowers, **Sphingophilae**, and moth flowers, **Phalaenophilae**) white flowers are dominant. These often first open in the evening, and then often emit a very powerful scent, as is already familiar to us in the Perfoliate Honeysuckle (*Lonicera caprifolium*) or in other species of Honeysuckle (*L. periclymenum*). The pollinators usually absorb the nectar whilst they hover in front of the flower, or at most while they cling on by their front legs. The length of proboscis which is needed to reach the nectar is often astonishing; up to 8 cm in European Sphingidae, and 25 cm in the South American hawk-moth *Cocythius cluentius*. The example of the Madagascan orchid *Angraecum sesquipedale* (*Macroplectrum sesquipedale*) whose spur has a length of 32 cm is well known. For this a pollinating hawk-moth was predicted and found: *Xanthopan morgani f. praedicta*, with a proboscis 22.5 cm long.

There are other types of flowers which are pollinated by hawk-moths, as well as long-tubed flowers of the "stalked plate" type. One such type is the bottle-brush flowers, i.e. flowers or inflorescences with a large number of long, widely-spreading and brightly coloured or white stamens, as are found in species of *Capparis* (see Werth 1956a). Sphingophilous flowers often wither by the next morning, as is the case in *Oenothera missouriensis* or *Capparis spinosa*.

3.3.4.3   Pollination by birds (ornithogamy)

Bird flowers are quite common in the tropical and subtropical zones, and can be recognised by a number of characteristics which contrast with those of the insect flowers:

1. The landing platform which is typical of many insect flowers is missing, since the flower-visiting birds either hover (hummingbirds) or make use of adjacent twigs.
2. The attraction is due to striking colours and colour contrasts, in which bright red, blue, yellow, orange and greenish-yellow colour shades play an important part (parrot colours). As examples we will only cite *Abutilon*, *Fuchsia* or the showy inflorescences of many of the Bromeliaceae, which are often provided with red or yellow-red bracts and yellow, blue or green-iridescent flowers. Since birds have but a poorly developed sense of smell, scent is of small importance, and is often absent altogether.
3. Large quantities of often rather watery nectar are offered as a reward. This may be hidden deep within the flower, or may be so abundantly secreted that flowers or inflorescences regularly overflow with it. It is not unusual for ornithogamous flowers to be used as a drinking place.
4. The flowers must have some degree of strength, according to the size of the pollinator, in order to protect themselves against damage. The ovary is often at a considerable distance from the nectar container, or else especially protected by other floral organs against damage. Anthers and stigma must be arranged so as to facilitate pollination (see for example Fig. 43 VII). This usually takes place via the head or beak of the visiting bird. An explosion mechanism can also play a part, as for example in *Loranthus dregei* (Loranthaceae). In this species the 5 stamens are enclosed before anthesis by the 5 united and tubular tepals and kept in a state of tension. At anthesis the tepals

first come apart in the middle, so that 5 slits are created, through which the bird's beak can probe into the flower. When this happens, the tepals split apart so rapidly that the enclosed stamens unroll and the distal portions, complete with the anthers, detach themselves along a ready-made abscission line and are flung out. *Protea kilimandscharica* (Proteaceae) and *Ravenala madagascariensis* in the Musaceae are also said to possess an explosion mechanism (see Kugler 1970).

The types of flower which are very often found in bird-pollinated plants are tubular flowers, bell-flowers and bottle-brush flowers (*Eucalyptus* species, *Callistemon*, see Fig. 153 II, Fig. 154 II). The principal pollinators which are possible are the New World hummingbirds (Trochilidae), and in the Old World sunbirds (Nectariniidae) and honey-eaters (Meliphagidae, Australia, New Guinea, Micronesia, Polynesia), to which may be added various other bird groups (e.g. the flowerpeckers, Dicaeidae, in South-east Asia to the Solomons and Australia; also the Scaly-breasted Lorikeet, Trichoglossidae, from Australia to New Caledonia, the Solomons and the eastern Sunda Islands; the honeycreepers, Drepanididae in Hawaii; the Bananaquits, Coerebidae, in Central and South America, as well as birds from other groups, see the data given by Kugler 1970; Neubauer 1964 reported on the pollination of *Bombax malabricum* by Jungle Crows).

3.4.4.4   Pollination by bats (chiropterogamy)

The first reliable observational data on flower-visiting bats was published at about the end of the previous century (observations by Hart in Knuth 1897, similar conjectures were expressed previously by Beccari about 1886/90). Meanwhile it has become clear, through the work of Porsch (1931, 1932, 1941), van der Pijl (1936, 1941, 1949, 1956, 1961 etc.) and Vogel (1958, 1968/69a, b), what an important role is played by bats in the pollination of numerous plants – usually woody plants or epiphytes. Flower-visiting bats include the long-tongued bats (Glossophaginae) from the group of the insectivorous bats (Microchiroptera), which are confined to the New World, and the Old World long-tongued fruit bats (Macroglossinae) from the group of the Megachiroptera. When they visit flowers by night they are guided to a great extent by odours from the flowers, which smell of fruit or fermentation, or often also smell musky or cabbagy. In addition the well-known ultra-sonic echo-location of bats is important. In this presumably lies the explanation of the fact that the flowers are often in an exposed position, either by elevation above the tree-tops or by flagelliflory, and all the more so when the flowers are exploited via hovering. Bats often cling onto the flowers with their claws. The presence of claw marks can even be taken as evidence of chiropterogamy. On the other hand, the nature of the flowers must be firm and resistant. Bat flowers are usually characterised by the abundant secretion of nectar during the night or also by a high level of pollen production. Bats are often adapted to visiting flowers by having long, narrow muzzles and long tongues.

Among the numerous chiropterophilous plants are included besides the flagelliflorous woody plants (section 2.2.5) – the Baobab (*Adansonia*) and other members of the Bombacaceae, some from the Gesneriaceae and the Lobeliaceae, the Organpipe Cactus (*Carnegiea*) and *Lemaireocereus* in Arizona, the "bell vine" (*Cobaea*, Polemoniaceae), species of *Musa* and *Agave* as well as the epiphytic genus *Hillia* in the Rubiaceae, and certain members of the Bromeliaceae (species of *Vriesia*). They are often cases of personate or campanulate flowers or bottlebrush flowers (*Syzygium jambos*; *Inga sessilis*, Mimosaceae; *Lafoensia pacari*, Lythraceae) or spadix flowers, i.e where the inflorescence is a spadix (*Freycinetia insignis*, Pandanaceae).

From among the **mammals** there are several **phalangers** (Phalangeridae: *Acrobates pygmaeus*, Pygmy Glider; *Petaurus breviceps*, Honey Glider, etc.) which function as pollinators.

3.4   The fruit (the flower at the stage when the seeds are ripe), infructescences

3.4.1   The systematics of the fruit

It is not just by chance that when speaking figuratively, we use the expression "the fruits of our efforts". The fruit represents the biological outcome reached by the flower (or also the inflorescence) as a result of the interplay of all the different mechanisms that have been discussed so far. For this reason alone it would now seem useful to finish by giving at least a summary of the different possible ways in which the fruit can be constructed. This is all the more so since other floral organs besides the carpels are often involved to very different degrees in the formation of the fruit. It has therefore proved to be appropriate to define the **fruit** as the **flower at the stage when the seeds are ripe.***

During the development of the fruit the wall of the ovary which encloses the seeds is transformed into the **pericarp**, which usually undergoes an anatomical differentiation into an outer **exocarp**, often with only one layer of cells, and an inner **endocarp** with a multicellular layer of **mesocarp** in between.

According to whether the fruit is produced from a flower with free carpels, i.e. an apocarpous (choricarpous) gynoecium, or from a flower with a syncarpous gynoecium, it is possible to distinguish between **simple fruits** and **compound fruits**. Further characters for a classification of fruit types which correspond to some extent to natural relationships are produced by the different ways in which the fruit opens to release the seeds, or in which other kinds of disseminules are formed, as well as the appropriate differences in the anatomical differentiation of the pericarp.

3.4.1.1   Simple fruits

When they are ripe the individual seeds of **dehiscent fruits** are released by the opening of the fruit casing, whereas in **indehiscent fruits** they remain enclosed by the fruit wall, or by parts of it, and are dispersed in this form.

(a) Dehiscent fruits, seeds expelled or dropped
The opening of the fruit casing often takes place by means of hygroscopic tensions, which arise as the fruit wall dries out. The essential prerequisite for the

---

* See also, however, the criticisms of this expressed by Winkler (1940).

various types of dehiscence of such fruits is the different distribution and structure of zones of sclerenchymatous tissue in the wall of the fruit and the varying
orientation of the microscopic fibrils in the walls of successive layers of cells,
which can therefore contract in different directions on drying (Steinbrinck
1873, see also in addition Hartl 1957b, Schnetter, Hilger and Richter 1979). The
tensions which result from this give rise to the dehiscence of the fruit wall in a
characteristic way (xerochasy). In quite a number of plants this is so sudden and
violent that the seeds are flung out for some distance, as for example up to 90 cm
in *Cardamine impatiens*, but not so far in *C.hirsuta* (such plants or fruits are
termed "ballistic")*. The opposite to xerochasy is **hygrochasy**. Hygrochastic
fruits open when they are moistened and close up again when they are dry.
Examples of this are found in the Cruciferae in species of *Lepidium* and *Iberis*, in
the Labiatae in species of *Prunella* and *Salvia*, and also in the genus *Plantago*. In
the fruits of some species of *Iberis*, *Thlaspi* or *Prunella* and others "falling
raindrops provide the energy to make diaspores jump out of cups" by inducing
or moving a lever mechanism. Such 'rain ballists' are found in dry regions,
"where showers thus provide dispersal and the possibility of germination" (van
der Pijl 1982, p. 72). The opening of the fruit by so-called valves is widespread
in the Aizoaceae. In all cases the dehiscence of the fruit is essentially based upon
unequal tensions in dead tissue. In contrast to these **dry ballistic fruits** there are
also those which keep their fleshy consistency until they are ripe and open by
means of differences of turgor between the inner and outer layers of the pericarp: **fleshy ballistic fruits** . Examples are the Balsams, especially *Impatiens
noli-tangere* and the cucumber *Cyclanthera explodens*. The fruit of the Balsams
consists of 5 carpels and is 5-locular, and when ripe the parenchymatous cells in
the outer region of the fruit wall are highly turgescent. The efforts of this tissue
to expand longitudinally are, however, counteracted by a buttress formed by
the stretched fibrous cells of the innermost layers of the fruit wall, until the
bonds between the 5 carpels are slackened. Then the slightest touch is enough to
make the 5 fruit valves roll up explosively, so that the seeds are thrown a long
way off (Fig. 193 I). By way of contrast to this, in the slightly twisted (zygomorphic) fruit of *Cyclanthera explodens* the more turgescent tissue lies on the
inner side of the convex wall, which is suddenly flung back (Fig. 193 III, IV).

If the fruit consists of a single carpel only, then this may open either by a
ventral suture (**follicle**, Fig. 191 *Delphinium*), or in addition to that along the
midrib or "dorsal suture" (**pod**, Fig. 191, *Laburnum*). Follicle and pod (as well as
the drupe of the cherry and the berries of the Baneberry, *Actaea*) can be
interpreted as reduced forms of compound fruits, in which the number of
carpels is limited to one i.e. "**monocarpellate fruits**".

Several carpels are always involved in the structure of the capsule. The
longitudinally dehiscent capsule which is formed in this way opens partly or
completely along the lines of fusion between the carpels (**septicidal**, Fig. 191,
Veratrum), or along the middle lines of the carpels (**loculicidal**), or often in

---

* The dehiscence of the fruit in the genus *Cuphea* in the Lythraceae is particularly unusual, in that
the placenta twists itself out of the breaking fruit, after it has gained a considerable degree of
independence via the early breakdown of the partition wall of the bicarpellate, dorsiventral
gynoecium. "The movement is effected by a zone of growth on the central face of the placenta"
(Baum and Leinfellner 1951).

combination with breakages on the septum walls (**septifragal**, Fig. 191, *Iris*). A special form of the capsule is the siliqua, which consists of two carpels whose valves become detached from a "housing" (replum) formed from the fused carpel margins and their robust placentas (Fig. 191, *Chelidonium*).

In **poricidal capsules** (Fig. 191, *Papaver*) only small openings are formed, these being very varied in size and number, and in pyxidia (Fig. 191, *Anagallis*) the complete apical section is circumscissile, i.e. it detaches itself by means of a suture which encircles all the carpels. In *Genlisea hispidula* (Lentibulariaceae) in addition to the lid there is even a ring of pericarp separated by a further circumscission, and it is possible for a spiral circumscission to occur (Stopp 1958b).

(b) Indehiscent fruits

According to the constitution of the pericarp these may occur variously as **berries** with a juicy-fleshy pericarp (Fig. 191, *Atropa*), as **nuts** with a completely sclerenchymatous pericarp (Fig. 191, *Corylus*) and as **drupes** with a sclerenchymatous endocarp and a fleshy mesocarp (Fig. 191, *Olea*). Such drupes may contain more than one stone, according to the number of carpels, each of which however always contains only one seed (*Sambucus*, Elder; *Arctostaphylos uva-ursi*, Bearberry). In the Coconut (*Cocos nucifera*) the mesocarp is fibrous. The berry with a rind, or **hesperidium**, such as the Pumpkin (*Cucurbita pepo*), cucumber (*Cucumis sativus*) or orange (*Citrus sinensis*), etc. is provided with a relatively firm exocarp and a soft mesocarp within. The endocarp is developed as a "fruit pulp" in which the seeds are embedded.

Among indehiscent fruits are the special types which are typical of the Compositae or to some extent the Valerianaceae, the **achene** (Fig. 191, *Carduus*) and the grass fruit, the **caryopsis** (Fig. 191, *Triticum*), in which the coat of the single seed fuses (or is said to fuse, see Wagenitz 1976) with the pericarp during ripening. In addition the **lomentaceous fruits** and the **schizocarps**, comprising several seeds, are also classified here. In the former, the loment (Fig. 191, *Ornithopus sativus*, Serradella) and the biloment (Radish, *Raphanus*) break up into single-seeded sections, enclosed in fragments of the carpel wall. In the latter, they fall apart into mericarps by septicidal divisions, corresponding to the individual carpels (Fig. 191, *Acer*, maples; the fruits of the Umbelliferae; Mallow "cheeses"). The fruits of the Labiatae and the Boraginaceae (see section 1.6.8), which fall apart into four single-seeded "mericarpic nutlets" can also be classified as lomentaceous fruits. Moreover there are quite a number of "ballistic" fruits in the Labiatae, in which the nutlets are ejected individually.

3.4.1.2 Compound fruits (Fig. 192)

Compound fruits can similarly be subdivided, according to whether the individual carpels enclose several or only one ripe seed. In the first case the carpels open in the manner of follicles: **compound follicle** (Fig. 192, *Trollius*). In a second case each carpel can be developed with sclerenchyma: **compound nutlet** (Fig. 192, *Geum*). The Cinquefoils (*Potentilla*) form compound nutlets, whereas in the closely related strawberry (*Fragaria*, Fig. 192), the receptacle becomes fleshy. In raspberries and blackberries (*Rubus*, Fig. 192) the individual carpels develop as drupes and together form a **compound drupe**.

The carpels of a compound drupe can also be temporarily or permanently enclosed in receptacular tissue. This can be the case not only when the carpels

SIMPLE FRUITS

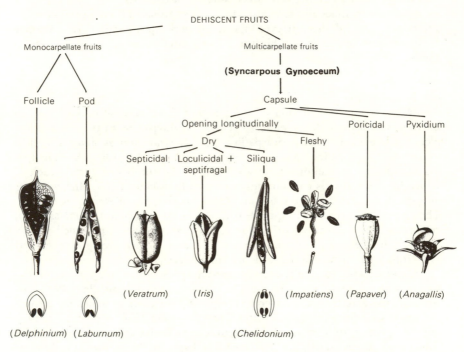

DEHISCENT FRUITS

Monocarpellate fruits   Multicarpellate fruits

(Syncarpous Gynoeceum)

Follicle   Pod   Capsule

Opening longitudinally   Poricidal   Pyxidium

Dry   Fleshy

Septicidal   Loculicidal +   Siliqua
         septifragal

(*Veratrum*)   (*Iris*)   (*Impatiens*)   (*Papaver*)   (*Anagallis*)

(*Delphinium*)   (*Laburnum*)   (*Chelidonium*)

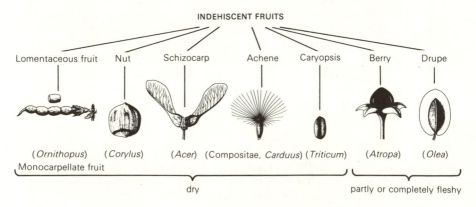

INDEHISCENT FRUITS

Lomentaceous fruit   Nut   Schizocarp   Achene   Caryopsis   Berry   Drupe

(*Ornithopus*)   (*Corylus*)   (*Acer*)   (Compositae, *Carduus*) (*Triticum*)   (*Atropa*)   (*Olea*)

Monocarpellate fruit

dry   partly or completely fleshy

Fig. 191 Types of fruit in flowers with syncarpous gynoecia (after Troll, Duchartre, Firbas, Rauh, Baillon, Schimper, Weber, Knoblauch, and some originals). The monocarpellate fruits are derived from apocarpous gynoecia.

take the form of nutlets (the rose–hip, *Rosa*, Fig. 192; see also section 1.3.2), but also when the carpel is developed like a drupe (Fig. 192, *Mespilus*, Medlar) or in follicles. The latter is represented by the fruit of *Nuphar* and also by the Apple, to which we have already given a thorough discussion in the context of pseudo-syncarpy (see section 1.6.13).

The pedicel of the fruit can undergo considerable anatomical changes as the fruit ripens. The increasing weight of the fruit often requires an increase in the sclerenchymatous parts, especially in plants with hanging fruits. An enlargement of the phloem and xylem portions frequently corresponds to the increased nutritional needs of the ripening fruit (e.g. pear, quince). In addition to this there is often also some development of fleshiness in the pedicels during the formation of fruit, which we want to say more about later.

Finally it should be mentioned that certain differences can occur in the fruits of one and the same plant. **Heterocarpy** of this kind is found between the fruits on the periphery and those more towards the centre in the capitulum of the Marigold (*Calendula*). We have already discussed the heterocarpy occurring in the fruits of the genus *Fedia* in the Valerianaceae (section 1.4.2.7, Fig. 33 VIII to X).

The **disseminules** can therefore consist of individual seeds (dehiscent fruits), of complete (indehiscent) fruits or also broken off parts of fruits or of mericarps (lomentaceous fruits, schizocarps, nutlets of the compound fruits). It is also possible, however, for the fruits of a complete inflorescence to remain combined in an **infructescence**, which is dispersed as a whole. This is already the case in the infructescences of many species of *Valerianella* and *Fedia* (Fig. 33 VIII) and in not a few of the steppe-inhabiting species of the Chenopodiaceae, in which the complete plant or parts of it take on a spherical shape when dead

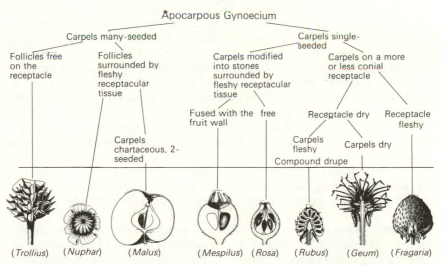

COMPOUND FRUITS

Apocarpous Gynoecium

Carpels many-seeded — Carpels single-seeded

Follicles free on the receptacle | Follicles surrounded by fleshy receptacular tissue | Carpels modified into stones surrounded by fleshy receptacular tissue | Carpels on a more or less conial receptacle

Fused with the fruit wall | free | Receptacle dry | Receptacle fleshy

Carpels chartaceous, 2-seeded | Carpels fleshy | Carpels dry

Compound drupe

(*Trollius*) (*Nuphar*) (*Malus*) (*Mespilus*) (*Rosa*) (*Rubus*) (*Geum*) (*Fragaria*)

Fig. 192 Forms of fruit from flowers with apocarpous gynoecia. After Troll, Duchartre, Firbas, Rauh.

and dried. They are then blown far and wide by the wind over the flat country, scattering their fruits, until they come to rest – perhaps in a hollow or fold in the ground, where in some cases the rest of the fruits can find favourable conditions for germination. The fruits in an infructescence can also be united with one another, in such a way that the axis of the infructescence and the bracts are included in it (*Ananas*, Pineapple; Jack-fruit tree, *Artocarpus integer*, Moraceae). We have already mentioned Figs and Mulberries elsewhere (sections 1.4.3.2, 2.3.2).

3.4.2   Aspects of the biology of dispersal

Seeds, mericarps, fruits and infructescences are all adapted in various ways to different methods of dispersal. Apart from the possibility that the plant itself may actively disperse the disseminules in its own surroundings i.e. autochory, the effective factors of dispersal are water (hydrochory), wind (anemochory) and animals (zoochory), including man himself (anthropochory).

Apart from the "ballistic" mechanisms which we have already mentioned, the **expulsion mechanisms** in species of *Oxalis* and *Viola* as well as in *Dorstenia* can be quoted as examples of **autochory**. In *Dorstenia* each of the numerous fruit stones which are united in the infructescence (see section 2.3.2) is surrounded by the fleshy pericarp as if by a pair of tweezers (see Fig. 94 III). There is a zone of swelling tissue placed underneath the wedge-shaped stone, which in its efforts to expand puts lateral pressure on the jaws of the "tweezers" adjacent to the stone. Since, however, the fleshy tissue is very thin above the stone, it bursts open under the severe pressure, so that the stone is ejected from the jaws of its "tweezers" to a distance of several metres. In the cases of *Oxalis* and *Viola* there is also a **turgor mechanism**. This is also true in another sense for the Squirting Cucumber, *Ecballium elaterium*. Here the endocarp has a pulpy consistency, displays a high degree of turgor, which causes a high tension in the surrounding mesocarp and exocarp. An abscission layer is formed at the base of the curved and hook-like pedicel, and when the fruit is ripe this ruptures on only slight provocation. As the fruit detaches itself the fleshy interior containing the seeds is squirted out by the elastic fruit walls, which have been under severe stress until now, and the fruit is impelled forwards by the recoil. On the other hand, the **catapult fruit dispersion** of *Geranium* (Fig. 193 II), like that in *Cardamine*, is brought about by **hygroscopic movements** due to the drying out of the pericarp cells. In the former, the five carpels are only fertile at the base and become detached from below from the persistent central column, coiling up towards the tip so that the seeds are catapulted away. Considerable hygroscopic forces build up as the substantially woody fruit of the Sand-box tree (*Hura crepitans*, Euphorbiaceae) dries out, until it suddenly bursts apart with a loud crack, forming numerous mericarps which correspond to the individual carpels, and which are thrown out to a distance of as much as 14 m (explosion fruit).

There are quite a number of plants which dispose of their fruits in a suitable place for germination by means of a corresponding growth of the fruit pedicel. This is the case in Ivy-leaved Toadflax (*Cymbalaria muralis*), where the pedicels become negatively phototropic after fertilisation of the flower, and push the ripening fruit into cracks of rocks or walls. The fruits of the Peanut (*Arachis hypogaea*) and the Earth Pea (*Voandzeia subterranea*) do not ripen until after the peduncles have bored deeply under the ground. In the Mediterranean *Trifolium subterraneum*, moreover, the complete flower head is forced into the ground. These **geocarpic** plants contrast to the **amphicarpic** ones, in which both aerial and subterranean fruits are developed. An example of this is *Cardamine chenopodiifolia*. Its structure is similar to that of the plant of *Lemphoria procumbens* illustrated in Fig. 131. In this case the flowers of the main florescence are already brought below the earth's surface by a corresponding curvature of the pedicels

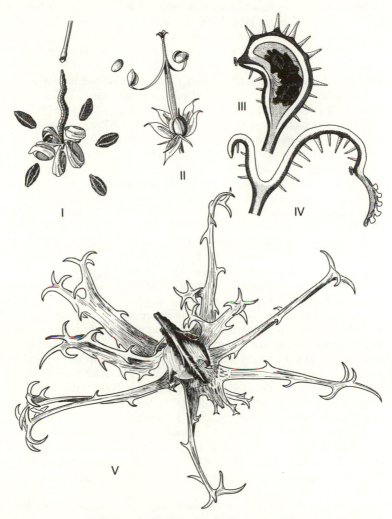

Fig. 193 I *Impatiens noli-tangere*, dehiscence of fleshy explosive fruit. II *Geranium sanguineum*, fruit ejecting the seeds. III, IV *Cyclanthera explodens*, ripe fruit in median section, before (III) and after (IV) opening. V *Harpagophytum procumbens*, fruit ("trample bur", leg. H. D. Ihlenfeldt). I, III, IV after Troll, II after Firbas.

in the bud stage. These subterranean flowers are cleistogamous and produce so-called "earth fruits", whilst the flowers of the paraclades are chasmogamous and produce "air fruits". Moreover, the seeds of the earth fruits produce larger seedlings than those of the air fruits (Troll 1951). The two forms of fruit can also be distinguished in other respects. The relationship between amphicarpy and the formation of chasmogamous and cleistogamous fruits is also clearly seen in species of *Viola* (see section 3.2). There are further examples of geocarpy in the Cruciferae (*Geococcus pusillus*, *Morisia hypogaea*) and of amphicarpy in the Papilionaceae (*Vicia angustifolia* var. *amphicarpa*, *Trifolium polymorphum*, *Amphicarpea*, etc.), Scrophulariaceae and other families. For further details and literature references see Huth 1890, Troll 1951a, b, Stopp 1954, 1958a–c.

Dispersal by water, **hydrochory**, is only possible if the fruits are buoyant and if the seeds can survive a lengthy immersion in water, or if the fruit remains sealed against the entry of water, especially sea-water. The former condition can be fulfilled by the formation of tissue with numerous intercellular spaces and air-filled or low-density parenchyma, e.g. the spongy mesocarp in species of *Potamogeton* or of *Sparganium*, see U. and D. Müller-Doblies 1977; also inflated fruiting calyces (*Hernandia ovigera*, see section 1.4.2.7) can increase the bouyancy of the fruit. A hard and impermeable exocarp, as in the Nipa palm (*Nipa fruticans*), or other impermeable tissues, can serve as a protection against damage from the effects of (sea)-water.

As adaptations to **dispersal by wind, anemochory**, we have already mentioned the **winged calyx** of the Dipterocarpaceae (Fig. 32 VI), of many of the Plumbaginaceae (Fig. 32 III), of some species of the Rubiaceae or of the genus *Triplaris* (Polygonaceae) during the discussion of calyx morphology (section 1.4.2.7). The number of these examples could be greatly multiplied. Further examples to add to this are the flattened nuts of the elms, provided with a winged margin ("for spinning through the air") or the **winged nuts** of the Rhubarb (*Rheum*) and other members of the Polygonaceae with several winged edges, and *Combretum* (Combretaceae) etc. The winged mericarps (samaras) of *Acer* (Fig. 190) show characteristic "spiralling flight." We have already commented on the ring of hairs (pappus) which acts as a "parachute" in the Compositae and the Valerianaceae (Fig. 33 I, IV, V). Also the long pinnately-divided awns of the nutlets of *Clematis*, *Pulsatilla* or *Dryas* assist dispersal by the wind ("plumed diaspores", van der Pijl 1982); similarly the inflated fruiting calyces of *Valerianella vesicaria* (Fig. 33 III) or of the Strawberry Clover (*Trifolium fragiferum*) ("flying balloons"). We have already mentioned the dispersal of the entire infructescences as a "tumble-weeds". According to Pascher (1959) the winged capsules of *Eucomis* and *Veltheimia* are "very frequently removed in their entirety, and in doing this gradually lose their seeds as they are driven before the wind". This does not apply, however, contrary to general belief, to *Anastatica hierochuntica* (Cruciferae), the "Rose of Jericho" (see Stopp 1958c).

In dispersal by animals, **zoochory**, we need to distinguish between two possible mechanisms: **epizoochory** where the seed or fruit is attached to the exterior (fur or feathers) of the animals, and **endozoochory** where the fruits or seeds are eaten and excreted again later. In the latter case the assumption is that at least a part of the seed can pass undamaged through the intestinal canal of the distributing animal, as occurs in berries and in seeds which are provided with fleshy appendages (aril, elaiosome). In this case a variety of means can serve to attract the animal dispersing agent; conspicuous colours of seeds with fleshy appendages, the colouring of berries and drupes as well as striking colour contrasts between the fruits and the bracts (*Lonicera involucrata*, fruit shining black, prophylls deep red) and what are known as "display seeds" (Mildbraed 1954 in *Paeonia*) or even certain scents.

In addition conspicuously coloured fruiting calyces, such as that of *Physalis alkekengi* (Fig. 32 IV, V) should be mentioned here. It is by no means always seed appendages or parts of the pericarp which are consumed: in *Hoslundia decumbens* (section 1.4.2.7, Fig. 33 XI) it is the fleshy sepals, in *Coriaria* (section 1.4.3.2, Fig. 37 III to X) it is the petals which become fleshy at fruiting time, in Strawberry Blite (*Blitum*) and in the Mulberry (*Morus*) it is the perigon which becomes fleshy, in the Cashew nut (*Anacardium occidentale*, Anacardiaceae) it is the pyriform and swollen pedicel, and in the fig it is the fleshy inflorescence, which is eaten by animals. In *Hovenia dulcis* in the Rhamnaceae (Japanese Raisin Tree) all the axes of the infructescence, with the exception of the epipodium of which the pedicel consists, take on a fleshy quality, and even emit scent.

The typical adaptations for epizoochory may be classified either as burs (hooks on seeds, fruit or infructescences) or adhesives (e.g. glandular hairs). In this context the formation of persistent fruiting calyces with hooked (Fig. 33 II, VI, VII, X) or glandular structures (Fig. 32 I, II) takes on a special meaning. The so-called **trample burs** represent a special form of burs, which are found in steppe and desert plants and particularly in the Martyniaceae (Proboscoidea) and the Pedaliaceae, and are especially pronounced in the latter in *Harpagophytum*. The fruits of *Harpagophytum procumbens* (Fig. 193 V) are hard, somewhat compressed woody capsules, whose margins are provided with a double row of massive processes, several centimetres long, which moreover are provided with several sharp and claw-like hooks. The plants often grow in the neighbourhood of waterholes, where mammals can easily step on these burs as they walk. Since the sharp hooks cause wounds, the animals try to get rid of these bothersome and painful burs, and in so doing tread them into the ground. Hence the plant succeeds in arranging that its disseminules reach a location which is not directly beside the mother plant, but still in the neighbourhood, i.e. in a favourable position with access to water for germination. According to Stopp (1962), however, this method of dispersal may not be the normal case. Fruits of this kind can also anchor themselves firmly in the ground with their hooks without the assistance of animals. Besides the manifold mechanisms by which fruits can anchor themselves, there are more "mechanisms for hindering dispersal", such as basicarpy, geocarpy and amphicarpy and the temporary arrest of disseminules on the mother plant − such as by part of the seed on a capsule whose opening is delayed (see Stopp 1958c, 1962 etc.). A special form of zoochory is **myrmecochory**, dispersal by ants (Sernander 1906, Uphof 1942, van der Pijl 1955). This is brought about by the fact that fleshy and nutritious appendages called **elaiosomes** appear on many seeds, mericarps or fruits and these are eaten by ants, which carry away the disseminules for this reason. A variety of morphologically quite different structures are involved in elaiosomes, e.g. outgrowths from the margin of the integument of the ovule (caruncle in Euphorbiaceae), growths on the funiculus (*Chelidonium*), the fleshy pedicel-like zone of attachment on the nutlets of *Anemone*, etc.

The multiplicity of forms of the fruit and the numerous possible ways in which the disseminules can be dispersed have nevertheless only been hinted at by this brief survey. For further information on the morphology of the fruit, in addition to the studies of Ulbrich and van der Pijl already referred to at the outset (section 3.1), the work of Brouwer and Stählin (1975) as well as the cited publications of Winkler as well as Stopp (1950a, b, 1952 etc.) are recommended.

Bibliography

References which have been added since the German edition of this book are marked with an asterisk (*).

Agthe, C. (1951) Über die physiologische Herkunft des Pflanzennektars. Ber. Schweiz. Bot. Ges. **61**, 240–273.

Amberg, K. (1912) Zur Blütenbiologie von *Arctostaphylos alpina* (L.)Sprengel. Ber. Dt. Bot. Ges. **30**, 692–703.

Arber, A. (1931a) Studies in floral morphology. I. On some structural features of the Cruciferous flower. New Phytol. **30**, 11–41.

 – (1931b) Studies in floral morphology. III. On the Fumarioideae, with special reference to the androecium. New Phytol. **30**, 317–354.

 – (1932a) Studies in flower structure. I. On a peloria of *Digitalis purpurea* L. Ann. Bot. **46**, 929–939.

 – (1932b) Studies in floral morphology. IV. On the Hypecoideae, with special reference to the androecium. New Phytol. **31**, 145–173.

 – (1933) Floral anatomy and its morphological interpretation. New Phytol.**32**, 231–242.

 – (1937) The interpretation of the flower: a study of some aspects of morphological thought. Biol. Rev. **12**, 157–184.

Artopoeus, A. (1903) Über den Bau und die Öffnungsweise der Antheren und die Entwicklung der Samen der Erikaceen. Flora **92**, 309–345.

*Ayensu, E. S. (1974) Plant and bat interactions in West Africa. Ann. Missouri Bot. Gard. **61**, 702–727.

Baillon, H. E. (1862/63) Organogénie florale des Cordiacées. Adansonia **3**. Paris.

Baker, H. G. & Baker, I. (1973) Some anthecological aspects of the evolution of nectar-producing flowers, particularly amino acid production in nectar. In Heywood V. H., ed. *Taxonomy and Ecology*, 243–264. London.

Bancroft, H. (1935) A review of researches concerning floral morphology. Bot. Rev. **1**, 77–99.

*Barabé, D. (1982) Vascularisation de la fleur de *Symplocarpus foetidus* (Araceae). Canadian J. Bot. **60**, 1536–1544.

 *– (1983) Vascularisation de la fleur de *Calla palustris* (Araceae). Canadian J. Bot. **61**, 1718–1726.

Barcianu, D. P. (1875) Untersuchungen über die Blütenentwicklung der Onagraceen. Mitt. aus d. Gesamtgebiete d. Botanik (Eds. A. Schenk and G. Luerssen) **2**, 81–129, Plate VII, Leipzig.

 – (1875) Ueber die Blüthenentwicklung der Cupheen. ibidem **2**, 179–193, Plate XI, Leipzig.

Baum, H. (1948a) Postgenitale Verwachsung in und zwischen Karpell- und Staubblattkreisen. Sitz.ber. Österr. Akad. Wiss. Mathem.-nat. Kl., Abt. I, **157**, 17–38.

 – (1948b) Über postgenitale Verwachsung in Karpellen. Österr. Bot. Z. **95**, 86–94.

 – (1948c) Die Verbreitung der postgenitalen Verwachsung im Gynözeum und ihre Bedeutung für die typologische Betrachtung des coenokarpen Gynözeums. Österr. Bot. Z. **95**, 124–128.

 – (1948d) Die Stellung der Samenanlagen am Karpell bei *Asclepias syriaca, Cynanchum vincetoxicum* und *Erythraea centaurium*. Österr. Bot. Z. **95**, 251–256.

- (1948e) Ontogenetische Beobachtungen an einkarpelligen Griffeln und Griffelenden. Österr. Bot. Z. **95**, 362–372.
- (1948f) Zur Frage des schrittweisen Überganges vom apokarpen zum coenokarpen Gynözeum. Österr. Bot. Z. **95**, 470–474.
- (1949a) Der einheitliche Bauplan der Angiospermengynözeen und die Homologie ihrer fertilen Abschnitte. Österr. Bot. Z. **96**, 64–82.
- (1949b) Das Zustandekommen "offener" Angiospermengynözeen. Österr. Bot. Z. **96**, 285–288.
- (1949c) Beiträge zur Kenntnis der Schildform bei den Staubblättern. Österr. Bot. Z. **96**, 453–466.
- (1950) Unifaziale und subunifaziale Strukturen im Bereich der Blütenhülle und ihre Verwendbarkeit für die Homologisierung der Kelch- und Kronblätter. Österr. Bot. Z. **97**, 1–43.
- (1951a) Vorläuferspitzen im Blütenbereich. Österr. Bot. Z. **98**, 280–291.
- (1951b) Die Frucht von *Ochna multiflora*, ein Fall ökologischer Apokarpie. Österr. Bot. Z. **98**, 383–394.
- (1952a) Die Bedeutung der diplophyllen Übergangsblätter für den Bau der Staubblätter. Österr. Bot. Z. **99**, 228–243.
- (1952b) Die Peltation der Außenkelchblätter von *Hibiscus costatus*. Österr. Bot. Z. **99**, 370–375.
- (1952c) Die Querzonenverhältnisse der Karpelle von *Helleborus foetidus* und ihre Bedeutung für die Beurteilung der epeltaten Karpelle. Österr. Bot. Z. **99**, 402–404.
- (1952d) Normale und inverse Unifazialität an den Laubblättern von *Codiaeum variegatum*. Österr. Bot. Z. **99**, 421–451.
- (1952e) Über die "primitivste" Karpellform. Österr. Bot. Z. **99**, 632–634.
- (1953a) Die Unabhängigkeit der diplophyllen Gestalt der Staubblattspreite von ihrer Funktion als Träger der Pollensäcke. Österr. Bot. Z. **100**, 265–269.
- (1953b) Die Karpelle von *Eranthis hiemalis* und *Cimicifuga americana* als weitere Verbindungsglieder zwischen peltaten und epeltaten Karpellen. Österr. Bot. Z. **100**, 353–357.
- (1953c) Zur Peltationsnomenklatur der Karpelle. Österr. Bot. Z. **100**, 424–426.
- (1953d) Ergänzende Mitteilungen zum Problem der unifazialen Kelchblattspitzen. Österr. Bot. Z. **100**, 593–600.

Baum-Leinfellner, H. (1953) Über unifaziale Griffel and Narben. Planta **42**, 452–460.

Baum, H. & Leinfellner, W. (1951) Die Plazenta des dorsiventralen *Cuphea*-Gynözeums. Österr. Bot. Z. **98**, 187–205.
- & Leinfellner, W. (1953a) Die ontogenetischen Abänderungen des diplophyllen Grundbaues der Staubblätter. Österr. Bot. Z. **100**, 91–135.
- & Leinfellner, W. (1953b) Bemerkungen zur Morphologie des Gynözeums der Amentiferen in Hinblick auf Phyllo- und Stachyosporie. Österr. Bot. Z. **100**, 276–291.

Baumann-Bodenheim, M. G. (1955) Ableitung und Bau bicarpellat-monospermer und pseudomonocarpellater Araliaceen- und Umbelliferen-Früchte. Ber. Schweiz. Bot. Ges. **65**, 481–510.

Beccari, O. (1886–1890) Malesia **3**.

Bechtel, A. R. (1921) Floral anatomy of the Urticales. Amer. J. Bot. **8**, 386–410, Pl. XV–XXII.

Beer, R. (1906) On the development of the pollen grain and anther in some Onagraceae. Beih. Bot. Centralbl. **19**, 286–313.

*Bell, C. R. (1971) Breeding systems and floral biology of the Umbelliferae. In Heywood, V. H.: The Biology and Chemistry of the Umbelliferae. Suppl.1 to the J. Linn. Soc., Botany, **64**, 93–108.

Bell, J. and Hicks, G. (1976) Transmitting tissue in the pistil of Tobacco: Light and electron microscopic observations. Planta **131**, 187–200.

Bennek, C. (1958) Die morphologische Beurteilung der Staub- und Blumenblätter der Rhamnaceen. Bot. Jb. Syst. **77**, 423–427.

Benner, U. & Schnepf, E. (1975) Die Morphologie der Nektarausscheidung bei Bromeliaceen: Beteiligung des Golgi-Apparates. Protoplasma **85**, 337–349.

Bernbeck, F. (1932) Vergleichende Morphologie der Urticaceen- und Moraceen-Infloreszenzen. Botanische Abhandlungen (Ed. K. Goebel) **3**.

*Bernhardt, P. (1983) Dimorphic flowers in *Amyema melaleucae*: a shift towards obligate autogamy. Bull. Torrey Bot. Club, 110, 195–202.

*Bernhardt, P. & Burns-Balogh, P. (1983) Pollination and pollinarium of *Dipodium punctatum* (Sm.) R. Br. Victorian Nat., **100**, 197–199.

*Bernhardt, P. & Calder, D. M. (1980) Comparative breeding system and adaptive floral morphologies of *Amyema linophyllum* and *Amyema preisii* (Loranthaceae). Phytomorphology, **30**, 271–289.

*Bernhardt, P. & Knox, R. B. (1983) The stigmatic papillae of *Amyema* (Loranthaceae): developmental responses to protandry and surface adaptations for bird pollination. American J. Bot. **70**, 1313–1319.

Bhattacharjya, S. S. (1954) Ein Beitrag zur Morphologie des Andröceums von *Benincasa hispida* (Thunb.). Cong. Ber. Dt. Bot. Ges. **67**, 22–25.

*Bino R. J., Dafni, A. & Meeuse, A. D. J. (1982) The pollination ecology of *Orchis galilaea* (Bornm. et Schulze) Schltr. (Orchidaceae). New Phytol. **90**, 315–319.

*Bino R. J., Dafni, A. & Meeuse, A. D. J. (1984) Entomophily in the dioecious gymnosperm *Ephedra aphylla* Forsk. (= *E.alte* C. A. Mey.) with some notes on *E.campylopoda* C. A. Mey. I. Aspects of the entomophilous syndrome. Proc. Koninkl. Nederl. Akad. Wetens., Series C **87**, 1–13.

Blaauw, A. H. (1931) Orgaanvorming en Periodiciteit van *Hippeastrum hybridum*. Verh. Kon. Akad. Wet. Amsterdam, Afd. Natuurk. Deel XXIX, Nr.1.

Bocquet, G. (1959) The structure of the placental column in the genus *Melandrium* (Caryophyllaceae). Phytomorph. **9** (3), 217–221.

– & Bersier, J. D. (1960) A propos de la vascularisation florale dans le genre *Silene* sect. *Gastrolychnis* (Fenzl) Chowdhuri. Actes Soc. Helvétique Sc. nat. d'Argovie, 113–114.

Boke, N. H. (1947) Development of the adult shoot apex and floral initiation in *Vinca rosea* L. Amer. J. Bot. **34**, 433–439.

– (1948) Development of the perianth in *Vinca rosea* L. Amer. J. Bot. **35**, 413–423.

– (1949) Development of the stamen and carpels in *Vinca rosea* L. Amer. J. Bot. **36**, 535–547.

Bolle, F. (1935) Über eine bemerkenswerte Mißbildung bei *Geum*. Notizbl. d. Bot. Gartens u. Museums Berlin, **113**, 349–354.

– (1940) Theorie der Blütenstände. Verh. Bot. Ver. Prov. Brandenbg. **80**, 53–81.

– (1954) Die Gesetze der Blütenstände. Ber. Dt. Bot. Ges. **66**, Gen. Vers. Heft., (12)- (14).

– (1967) Über blühende Kräuter. Bot. Jb. Syst. **86**, 1–4.

*Bonniers, M. G. (1879) Les nectaires. Ann. Sci. Nat. Bot. **8**, 1–112.

Borgström, G. (1939) Formation of cleistogamic and chasmogamic flowers in wild violets as a photoperiodic response. Nature (London) **2**, 514–515.

Bowers, K. A. W. (1975) The pollination ecology of *Solanum rostratum* (Solanaceae). Amer. J. Bot. **62**, 633–638.

Braun, A. (1842) Wuchsverhältnisse der Pflanzen in ihrer Anwendung auf Unterscheidung und Gruppierung der Species. Flora **25** II; 692–696.

Braun, A. (1851) Betrachtungen über die Erscheinung der Verjüngung in der Natur. Leipzig.

– (1853) Das Individuum der Pflanze. Berlin.

Bravais, L. & Bravais, A. (1837) Essai sur la disposition symétrique des inflorescences. Ann. Sci. Nat. 2. sér., Bot. **7**, 193–221, 291–348, and 8, 11–42.

– & Bravais, A. (1838) Mémoires sur la disposition géométrique des feuilles et des inflorescences. Paris.

Brenner, W. (1910) Beiträge zur Blütenbiologie. Ber.d. Realschule Basel 1909/10, 1–42.

*Briggs, B. & Johnson, L. (1979) Evolution in the Myrtaceae – evidence from inflorescence structure. Proc. Linn. Soc. N. S. W. **102**, 157–272.

Brouwer, W. & Stählin, A. (1975) Handbuch der Samenkunde für Landwirtschaft, Gartenbau und Forstwirtschaft. Frankfurt/M.

Brown, W. (1938) The bearing of nectaries on the phylogeny of flowering plants. Proc. Am. Philosoph. Soc. Philadelphia **79**, 549–595.

Buchenau, F. (1865) Der Blüthenstand der Juncaceen. Jb.wiss. Bot. **4**, 385–435, and Plates 28–30.

*Buchmann, St. L. & Buchmann, M. D. (1981) Anthecology of *Mouriri myrtilloides* (Melastomataceae: Memecyleae), an oil flower in Panama. (Reproductive Botany 7–24) Biotropica **13** (2), 7–24.

Budell, B. (1964) Untersuchungen der Antherentwicklung einiger Blütenpflanzen. Z.f. Bot. **52**, 1–28.

Bünning, E. (1953) Stammblütigkeit. -Aus der Heimat, **61**, 188–193, Plates 43 and 44.

Bugnon, P. (1929) Calicule des Rosacées et concresence congenitale. Bull. Soc. Sci. Bretag. **6**, 9–20.

Bunniger, L. (1972) Untersuchungen über die morphologische Natur des Hypanthiums bei Myrtales- und Thymelaeales-Familien II. Myrtaceae, III. Vergleich mit den Thymelaeaceae. Beitr. Biol. Pflanzen **48**, 79–156.

– & Weberling, F. (1968) Untersuchungen über die morphologische Natur des Hypanthiums bei Myrtales-Familien I. Onagraceae. Beitr. Biol. Pflanzen **44**, 447–477.

Burck, W. (1890) Ueber Kleistogamie im weiteren Sinne und das Knight-Darwin'sche Gesetz (German translation by P. Herzsohn). Ann. Jard. Bot. Buitenzorg **8**, 122–163, Plates XX–XXIII, Leiden.

Bureau, E. (1888) Sur un figuier à fruits souterrains. J.de Botanique **2**, 213–216, Pl. VII.

Buttrose, M. S., Grant, W. J. R. & Lott, J. N. A. (1977) Reversible curvature of style branches of *Hibiscus trionum* L., a pollination mechanism. Aust. J. Bot. **25**, 567–570.

Buvat, R. (1952) Structure, évolution et fonctionnement du méristème apical de quelques Dicotylédones. Ann. Sc. Nat. Bot. **11**. sér. **13**, 199–300.

Buxbaum, F. (1937) Allgemeine Morphologie der Kakteen. Die Blüte. Jb. Dt. Kakteen-Ges. **1937**, Einleitender Sonderteil, Berlin.

– (1948) Zur Klärung der phylogenetischen Stellung der Aizoaceae und Cactaceae im Pflanzenreich. Sukkulentenkunde, Jb. Schweiz. Kakt. Ges. **2**, 3–16.

– (1961) Vorläufige Untersuchungen über Umfang, systematische Stellung und Gliederung der Caryophyllales (Centrospermae). Beitr. Biol. Pflanzen **36**, 1–56.

Cammerloher, H. (1931) Blütenbiologie I. Wechselbeziehungen zwischen Blumen und Insekten. – Berlin (Sammlg. Borntraeger Bd.15).

De Candolle, A. P. (1813) Théorie élémentaire de la botanique. Paris.

– (1817) Considérations générales sur les fleurs doubles. Mém. Phys. Chim. de la Soc. d'Arcueil III, 385–404.

De Candolle, C. (1890) Recherches sur les inflorescences épiphylles. Mém. Soc. Phys. Hist.nat. Genève, Suppl.**6**, 1–37.

Carlquist, S. (1969) Toward acceptable evolutionary interrelations of floral anatomy. Phytomorphology **19**, 332–362.

Canright, J. E. (1952) The comparative morphology and relationships of the Magnoliaceae. I. Trends of specialization in the stamens. Am. J. Bot. **39**, 484–497.

Carlquist, S. (1969). Toward acceptable evolutionary interpretations of floral anatomy. Phytomorphology **19**, 332–362.

Carniel, K. (1963) Das Antherentapetum. Ein kritischer Überblick. Österr. Bot. Z. **110**, 145–176.

Carr, St G. M. & Carr, D. J. (1961) The functional significance of syncarpy. Phytomorphology **11**, 249–256.

Caspary, J. (1848) De nectariis. Diss. Elberfeld.

Cejp, K. (1926) Die Terminalblüten. Beih. Bot. Cbl. **43**, Abt. I. 101–126. Pl. I.

Čelakovský, L. J. (1874) Über die morphologische Bedeutung der Samenknospen. Flora **57**, 113–119, 129–137, 145–150, 161–173, 178–185, 201–208, 215–221, 225–238, 241–251, Pl. III.

– (1875) Über den "eingeschalteten" epipetalen Staubgefäßkreis. Flora **58**, 481–524.

– (1876) Vergleichende Darstellung der Placenten in den Fruchtknoten der Phanerogamen. Abh. Kgl. Böhm. Ges. Wiss., math.-naturw. Cl., VI. Folge, **8**, 2–74, 1 plate. Prague.

– (1878) Teratologische Beiträge zur morphologischen Deutung des Staubgefäßes. Jb. Wiss. Bot. **11**, 124–174.

– (1880) Ueber die Blüthenwickel der Boraginaceen. Flora **63**, 355–369.

– (1882) Vergrünungsgeschichte der Eichen von *Aquilegia* als neuer Beleg zur Foliolartheorie. Bot. Centralbl. **10**, 331–342, 1 plate.

– (1884a) Über ideale oder congenitale Vorgänge der Phytomorphologie. Flora **67**, 435–460.

– (1884b) Neue Beiträge zur Foliolartheorie des Ovulums. Abh. Kgl. Böhm. Ges. Wiss., math.-naturw. Cl., VI. Folge, **12**, 1–42, Plates 1 and 2.

– (1893) Gedanken über eine zeitgemäße Reform der Theorie der Blütenstände. Bot. Jb. Syst. **16**, 32–51.

– (1894) Das Reductionsgesetz der Blüthen, das Dédoublement und die Obdiplostemonie. Sitz. Ber. Königl. Böhm. Ges. Wiss. math.-Nat. Cl., 1–142, Pls. I-V.

– (1896) Über den phylogenetischen Entwicklungsgang der Blüthe und über den Ursprung der Blumenkrone. I. Theil. Sitz.-Ber. Kgl. Böhm. Ges., math.-Nat. Cl. 1896, 1–91.

– (1900) Ueber den phylogenetischen Entwicklungsgang der Blüthe und über den Ursprung der Blumenkrone, II. Theil. Sitz.-Ber. Kgl. Böhm. Ges., math.-Nat. Cl. 1900, 1–221.

Chakravarty, H. L. (1958) Morphology of the staminate flowers in the Cucurbitaceae with special reference to the evolution of the stamen. LLoydia 21, 49–87.

Chapman, M. (1936) Carpel anatomy of the Berberidaceae. Amer. J. Bot. 23, 340–348.

Chatin, A. (1870) De l'anthère. Paris.

Chrometzka, P. (1967) Über die Entwicklung der Zellennaht zwischen den Kelchblättern von *Oenothera*. Österr. Bot. Z. 114, 46–50.

*Clarke, C. B. (1881) On right-hand and left-hand contortion. J. Linn. Soc., Botany 18.

*Claßen, R. (1984) Untersuchungen zur Struktur von Pseudanthien unter besondere Berücksichtigung ihrer Analogie zu zoophilen Einzelblüten. Diss. Mathem.-naturwiss. Fak. Rhein.-Westf. Techn. Hochsch. Aachen.

Clements, F. E. & Long, F. L. (1923) Experimental pollination, an outline of the ecology of flowers and insects. Carnegie Inst., Publ.no. 336, Washington.

Cohen, L. I. (1968) Development of the staminate flower in the Dwarf Mistletoe, *Arceuthobium*. Amer. J. Bot. 55, 187–193.

Corner. E. J. H. (1946) Centrifugal stamens. J. Arn. Arboretum 27, 423–437.

Coulter, J. M. & Chamberlain, C. J. (1912) Morphology of Angiosperms. New York.

Cronquist, A. (1957) Outline of a new system of families and orders of Dicotyledons. Bull. Jard. Bot. Etat, Bruxelles 27, 13–40.

Cusick, F. (1966) On phylogenetic and ontogenetic fusions. In Cutter, E. (ed.) Trends in Plant Morphogenesis, 170–183.

Danert, S. (1958) Die Verzweigung der Solanaceen im reproduktiven Bereich. Abh. Akad. Wiss. Berlin, Kl.f. Chemie, Geolog. u. Biol., Jg.1957, Nr.6, Berlin.

Dashek, W. V., Thomas, H. R., & Rosen, W. G. (1971) Secretory cells of lily pistils. II. Electron microscope cytochemistry of canal cells. Amer. J. Bot. 58, 909–920.

Daumann, E. (1930a) Das Blütennektarium von *Nepenthes*. Beiträge zur Kenntnis der Nektarien I. Beih. Bot. Centralbl. 47, Abt. I., 1–14.

– (1930b) Das Blütennektarium von *Magnolia* und die Futterkörper in der Blüte von *Calycanthus*. Planta 11, 108–116.

– (1932) Über die "Scheinnektarien" von *Parnassia palustris* und anderer Blütenökologie. Jb.wiss. Bot. 77, 104–149.

– (1941) Die anbohrbaren Gewebe und rudimentären Nektarien in der Blütenregion. Beih. Bot. Centralbl. 61, Abt. A, 11–82.

– (1960) Über die Bestäubungsökologie der *Parnassia*-Blüte. Ein weiterer Beitrag zur experimentellen Blütenökologie. Biologia plantarum (Prague) 2, 113–125.

– (1968) Zur Bestäubungsökologie von *Cypripedium calceolus* L. Österr. Bot. Z. 115, 434–446.

– (1970) Das Blütennektarium der Monokotyledonen unter besonderer Berücksichtigung seiner systematischen und phylogenetischen Bedeutung. Fedde's Repert. 80, 463–590.

– (1971a) Zur Bestäubungsökologie von *Aristolochia clematitis* L. Preslia (Prague) 43, 105–111.

– (1971b) Zum Problem der Täuschblumen. Preslia (Prague) 43, 304–317.

– (1974) Zur Frage nach dem Vorkommen eines Septalnektariums bei Dicotyledonen. Zugleich ein Beitrag zur Blütenmorphologie und Bestäubungsökologie von *Buxus* L. und *Cneorum* L. Preslia (Prague) 46, 97–109.

*Davidson, D. W. & Morton, S. R. (1981) Myrmecochory in some plants (F. Chenopodiaceae) of the Australian arid zone. Oecologia 50, 357–366. Berlin.

Davies, P. A. (1952) Structure and function of the mature glands of the petals of *Frasera carolinensis*. Transact. Kentucky Acad. Sci. 13, 228–234.

Davis, G. L. (1966) Systematic embryology of the Angiosperms. New York-London-Sydney.

Delpino, F. (1890) Note ed osservazioni botaniche, dec.seconda. Malpighia 4, 3–34.

Demeter, K. (1922) Vergleichende Asclepiadaceenstudien. Flora 115, 130–176.

Dengler, N. G. (1972) Ontogeny of the vegetative and floral apex of *Calycanthus occidentalis*. Canad. J. Bot. 50, 1349–1356.

Dickinson, T. A. & Sattler, R. (1974) Development of the epiphyllous inflorescence of *Phyllonoma integerrima* (Turcz.) Loes.: implications for comparative morphology. Bot. J. Linn. Soc. **69**, 1–13, 7 pl.

Diels, L. (1916) Käferblumen bei den Ranales und ihre Bedeutung für die Phylogenie der Angiospermen. Ber. Dt. Bot. Ges. **34**, 758–774.

Dobat, K. (1985) Blüten und Fledermäuse. Frankfurt.

Döll, J. Ch. (1842) Über die Verwandschaft mehrerer phanerogamischer Familien. Flora **25**, 675–681.

Domin, K. (1914) Morphologische Studien über den Außenkelch und den Blütenboden der Rosaceen nach Beobachtungen an vergrünten Blüten von *Potentilla aurea*. Bull. intern. Acad. des Scienc. Prague **19**, 1.

Douglas, G. E. (1944) The inferior ovary I. Bot. Rev. **10**, 125–186.

– (1957) The inferior ovary II. Bot. Rev., **23**, 1–46.

Drude, O. (1887) Die systematische und geographische Anordnung der Phanerogamen. In Schenk, Handb.d. Botanik III, 2, 175–496, Breslau.

★Dubuc–Lebreux, M. A. & Sattler, R. (1984) Quantitative distributions of mitotic activity during early corolla development of *Solanum dulcamara* L. Bot. Gaz., **145**, 22–25. Univ. Chicago.

★Duchartre, M. P. (1845) Observations sur l'organogénie de la fleur dans les plantes de la famille des Malvacées. Ann. Sci. nat. Série 3. **4**, 123–150.

Ducker, S. C. & Knox, R. B. (1976) Submarine pollination in seagrasses. Nature (London), **263**, 705–706.

Ducker, S. C., Pettitt, J. M. & Knox, R. B. (1978) Biology of Australian seagrasses: pollen development and submarine pollination in *Amphibolis antarctica* and *Thallassodendron ciliatum* (Cymodoceaceae). Aust. J. Bot. **26**, 265–285.

Dvorak, F. (1968) A contribution to the study of the variability of the nectaries. Preslia (Prague) **40**, 13–17.

Dupuy, P. (1963) Contribution à l'étude de quelques problèmes de morphologie et de tératologie expérimentales chez les Angiospermes. Thèse Sc. Nat. Poitiers.

Eames, A. J. (1929) The role of flower anatomy in the determination of angiosperm phylogeny. Proc. Int. Congr. Plant Sci. **1**, 423–427.

– (1931) The vascular anatomy of the flower with refutation of the theory of carpel polymorphism. Amer. J. Bot. **18**, 147–188.

– (1951) Again: 'The New Morphology'. New Phytol. **50**, 17–35.

– (1961) Morphology of the Angiosperms. New York-Toronto-London.

Eber, E. (1933) Karpelbau und Plazentationverhältnisse in der Reihe der Helobiae. Mit einem Anhang über die verwandschaftlichen Beziehungen zwischen Ranales une Helobiae. Flora **127**, 277–330.

Eckardt, T. (1937a) Untersuchungen über Morphologie, Entwicklunsgeschichte und systematische Bedeutung des pseudomonomeren Gynoeceums. Nova Acta Leopold. N. F. **5**, 1–112, Pls. I-XXIII.

– (1937b) Über Blütenbau und Verwandtschaft des Strandlings, *Litorella lacustris* L. Hercynia **1**, 154–165.

– (1954) Morphologische und systematische Auswertung der Placentation von Phytolaccaceen. – Ber. Dt. Bot. Ges. **67**, 113–128, Pl. III.

– (1955) Nachweis der Blattbürtigkeit ("Phyllosporie") grundständiger Samenanlagen bei Centrospermen. Ber. Dt. Bot. Ges. **68**, 167–182.

– (1956) Das Verhältnis des Gynoeceums zur Blütenachse mit kritischen Ausblicken auf Formprobleme bei den Angiospermen. Ber. Dt. Bot. Ges. **68**, Gen.vers. H. (28).

– (1957a) Zur systematischen Stellung von *Eucommia ulmoides*. Ber. Dt. Bot. Ges. **69**, 487–498.

– (1957b) Vergleichende Studie über die morphologischen Beziehungen zwischen Fruchtblatt, Samenanlage und Blütenachse bei einige Angiospermen. Zugleich als kritische Beleuchtung der "New Morphology". N. H. Morphol. **3**, 1–91, Pls. I-XI. Weimar.

– (1963) Zum Blütenbau der Angiospermen in Zusammenhang mit ihrer Systematik. Ber. Dt. Bot. Ges. **76**, 1. Gen. Vers. H., 38–49.

– (1963) Some observations on the morphology and embryology of *Eucommia ulmoides* Oliv. Maheshvari Comm. Vol., J. Indian bot. Soc. **42A**, 27–34, P. I, II.

– (1967a) Vergleich von *Dysphania* mit *Chenopodium* und mit Illecebraceae. Bauhinia, Z. Basler Bot. Ges. **3**, 327–344.

- (1967b) Blütenbau und Blütenentwicklung von *Dysphania myriocephala* Benth. Bot. Jb. Syst. **86**, 20–37.
- (1968) Zur Blütenmorphologie von *Dysphania plantaginella* F.v. M. Phytomorph. **17**, 165–172.
- (1971) Anlegeung und Entwicklung der Blüten von *Gyrostemon ramulosus* Desf. Bot. Jb. Syst. **90**, 434–446, Pl.16.
- (1974) Vom Blütenbau der Centrospermen-Gattung *Lophiocarpus* Turcz. Phyton (Austria). **16**, 13–27.
- ★– (1976) Classical morphological features of Centrospermous families. Plant Syst. Evol. **126**, 5–25.
Eckert, G. (1966) Entwicklungsgeschichtliche und blütenanatomische Untersuchungen zum Problem der Obdiplostemonie. Bot. Jb. **85**, 523–604, Stuttgart.
★Ehrendorfer, F. (1976) Closing remarks: Systematics and evolution of Centrospermous families. Plant Syst. Evol., **126**, 99–105.
★Ehrendorfer, F. (1976) Chromosome numbers and differentiation of Centrospermous families. Plant Syst. Evol., **126**, 27–30.
Eichler, A. W. (1875/78) Blüthendiagramme I.u. II. Leipzig. Reprint Koeltz, Eppenhain 1954.
★El Hamidi, A. (1952) Vergleichend-morphologische Untersuchungen am Gynoeceum der Unterfamilien Melanthoideae und Asphodeloideae der Liliaceae. Arbeiten aus d. Inst.f.allgem. Botanik d. Univ. Zürich Serie A No.7.
Endress, P. (1969) Gesichtspunkte zur systematischen Stellung der Eupteleaceen (Magnoliales). Ber. Schweiz. Bot. Ges. **79**, 229–278, Pls. I, II.
- (1971) Bau der weiblichen Blüten von *Hedyosmum mexicanum* Cordemoy (Chloranthaceae). Bot. Jb. Syst. **91**, 39–60.
- (1972a) Aspekte der Karpellontogenese. Verh. Schweiz. Naturf. Ges. **1972**, 126–130.
- (1972b) Zur vergleichende Entwicklungsmorphologie, Embryologie und Systematik der Laurales. Bot. Jb. Syst. **92**, 331–428.
- (1973a) "Arillen" bei holzigen Ranales une ihre phylogenetische Bedeutung. Verh. Schweiz. Naturf. Ges. **1973**, 89–90.
- (1973b) Arils and aril-like structures in woody Ranales. New Phytol. **72**, 1159–1171.
- (1975a) Aspects of carpel ontogeny and phylogeny. XII Int. Bot. Congr. Leningrad 1975, Abstracts: 213.
- (1975b) Nachbarliche Formbeziehungen mit Hüllfunktion im Infloreszenz- und Blütenbereich. Bot. Jb. Syst. **96**, 1–44.
- (1976) Die Androeciumanlage bei polyandrischen Hamamelidaceen und ihre systematische Bedeutung. Bot. Jb. Syst. **97**, 436–457.
- (1977a) Blütenmorphologie – Rückblick und aktuelle Probleme. Ber. Dt. Bot. Ges. **90**, 1–13.
- (1977b) Über Blütenbau und Verwandtschaft der Eupomatiaceae und Himantandraceae (Magnoliales). Ber. Dt. Bot. Ges. **90**, 83–103.
- ★– (1982) Austrobaileyaceae – Blütenstruktur einer archaischen, isolierten Angiospermenfamilie. W.van Cotthem (ed.): Morphologie, Anatomie und Systematik der Pflanzen, 47. Ninove, Waegeman.
- ★– (1983a) Primitive, highly specialised flowers: the flowering process in the Eupomatiaceae. Acta Bot. Neerl., **32**, 343–344.
- ★– (1983b) The early floral development of *Austrobaileya*. Bot. Jahrb. Syst. **103**, 481–497.
- ★– (1984a) The role of inner staminodes in the floral display of some relic Magnoliales. Plant Syst. Evol. **146**, 269–282.
- ★– (1984b) The flowering process in the Eupomatiaceae (Magnoliales). Bot. Jahrb. Syst. **104**, 297–319.
- ★– (1985) Stamenabszission und Pollenpräsentation bei Annonaceae. Flora **176**, 95–98.
- ★– (1986) An entomophily syndrome in Juglandaceae: *Platycarya strobilacea*. Veröff. Geobot. Inst. ETH, Stiftung Rübel, Zürich **87**, 100–111.
- ★– Jenny, M. & Fallen, M. E. (1983) Convergent elaboration of apocarpous gynoecia in higher advanced dicotyledons (Sapindales, Malvales, Gentianales). Nordic J. Botany **3**, 293–300.
- ★– & Lorence, D. H. (1983) Diversity and evolutionary trends in the floral structure of *Tambourissa* (Monimiaceae). Plant Syst. Evol. **143**, 53–81.
- ★– & Sampson, F. B. (1983) Floral structure and relationships of the Trimeniaceae (Laurales). J. Arnold Arboretum **64**, 447–473.

– & Voser, P. (1975) Zur Androeciumanlage und Antherenentwicklung bei *Caloncoba echinata* (Flacourtiaceae). Plant. Syst. Evol. **123**, 241–253.

Engler, A. (1876) Beiträge zur Kenntnis der Antherenbildung der Metaspermen. Jb.wiss. Bot. **10**, 275–316, Pl. XX-XXIV.

– (1926) Angiospermae. Kurze Erläuterungen der Blüten- und Fortpflanzungsverhältnisse. In Engler, A. & Prantl, K., Die Natürlichen Pflanzenfamilien. 2. Aufl. Bd.**14a**. Leipzig.

– (1954/1964) Syllabus der Pflanzenfamilien 12 Aufl., Bd. I & II (Eds. H. Melchior & E. Werdermann). Berlin.

*Erbar, C. (1983a) Carpel development and the principle of variable proportions. Acta Bot. Neerl. **32**, 349.

*– (1983b) Zum Karpellbau einiger Magnoliiden. Bot. Jahrb. Syst. **104**, 3–31.

*– (1986) Untersuchungen zur Entwicklung der spiraligen Blüte von *Stewartia pseudocamillia* (Theaceae). Bot. Jahrb. Syst. **106**, 391–407.

*– & Leins, P. (1985) Studien zur Organsequenz in Apiaceen-Blüten. Bot. Jahrb. Syst. **105**, 379–400.

*– Ernst, A. (1933) Fortpflanzung der Gewächse. 5. Samenpflanzen in: Handwörterbuch der Naturwissenschaften 2. Aufl. **4**, 365–398.

Ernst, A. (1953) Die Relation Antherenstellung/Pollenkorngröße bei Blütendimorphismus und das Kompatibilitätsproblem. Planta **42**, 81–128.

Esau, K. (1965) Plant Anatomy, 2nd edition. London & New York.

Eyde, R. H. (1975) The Bases of Angiosperm Phylogeny: Floral Anatomy. Ann. Missouri Bot. Gard. **62**, 521–537.

*Eyde, R. H. (1975) The foliar theory of the flower. Am. Sci. **63**, 430–437.

*Eyde, R. H. & Morgan, J. T. (1973) Floral structure and evolution in Lopezieae (Onagraceae). Amer. J. Bot. **60**, 771–787.

*Eyde, R. H. & Tseng, C. C. (1969) Flower of *Tetraplasandra gymnocarpa*, Hypogyny with epigynous ancestry. Science **166**, 506–508.

Faegri, K. F. & van der Pijl, L. (1966) The principles of Pollination Ecology (3rd edn 1971). Oxford-New York-Toronto-Sydney-Braunschweig.

Fahn, A. (1952) On the structure of floral nectaries. Bot. Gaz. **113**, 464–470.

*– (1953) The topography of the nectary in the flower and its phylogenetical trend. Phytomorphology 3, 424–426. (New Delhi).

*Fallen, M. E. (1983) Morphological, functional, and evolutionary aspects of the flower in the Apocynaceae. Inaugural-Diss. philosoph. Fak. II Univ. Zürich.

Fedde, F. (1936/1960) Papaveraceae, in Engler, A. Prantl, K., & Harms, H., Die natürlichen Pflanzenfamilien, 2nd edn. Bd. **17b**, 5–145. Berlin.

Feldhofen, E. (1933) Beiträge zur physiologischen Anatomie der nuptialen Nektarien aus den Reihen der Dikotylen. Diss. Munich.

*Fey, B. S. (1981) Untersuchungen über Bau und Ontogenese der Cupula, Infloreszenzen und Blüten sowie zur Embryologie bei Vertretern der Fagaceae und ihre Bedeutung für die Systematik. Diss. Phil. Fak. II Univ. Zürich.

*– & Endress, P. K. (1983) Development and morphological interpretation of the cupule in Fagaceae. Flora, **173**, 451–468.

Fitting, H. (1942) Morphologie, in Strasburger, E., Lehrbuch der Botanik für Hochschulen, 21st edn. Jena.

Franck, D. H. (1976) The morphological interpretation of epiascidiate leaves – a historical perspective. Bot. Rev. **42**, 345–388.

Frei, E. (1955) Die Innervierung der floralen Nektarien dikotyler Pflanzenfamilien. Ber. Schweiz. Bot. Ges. **65**, 60–114.

Frey-Wyssling, A. (1935) Die Stoffausscheidung der höheren Pflanzen. Monographien aus dem Gesamtgebiet der Physiologie der Pflanzen und der Tiere. Bd.32. Berlin.

– & Agthe, C. (1950) Nektar ist ausgeschiedener Phloemsaft. Verh. Schweiz. Naturf. Ges. **130**, 175–176.

*– & Haeusermann, E. (1960) Deutung der gestaltlosen Nektarien. Ber. Schweiz. Bot. Ges., **70**, 150–162.

Friedrich, H. Chr. (1956) Studien über die natürliche Verwandtschaft der Plumbaginales, Primulales und Centrospermae. Phyton **6**, 220–263.

Froebe, H. A. (1964) Die Blümenstände der Saniculoideen (Umbelliferae). Beitr. Biol. Pflanzen **40**, 325–388.

– (1971a) Wuchsform und Infloreszenzgestaltung bei *Sanicula, Hacquetia* und *Astrantia* (Umbelliferae). Bot. Jb. **91**, 1–38.
– (1971b) Inflorescence structure and evolution in Umbelliferae. In: The Biology and Chemistry of Umbelliferae (ed. V. H. Heywood). Suppl. I, Bot. J. Linn. Soc. **64**, 157–178.
– (1979) Die Infloreszenzen der Hydrocotyloideen (Apiaceae). Reihe trop. u. subtrop. Pflanzenwelt **29**, Abh. Akad. Wiss. Lit. Mainz, math.-naturw. Kl.
– (1980) Randmusterbildung und Synorganisation bei strahlenden Apiaceendolden. Pl. Syst. Evol. **133**, 223–237.
⋆– Magin, N., Jöhlinger, H. & Netz, M. (1983) A re-evaluation of the inflorescence of *Dalechampia spathulata* (Scheidw.) Baillon (Euphorbiaceae). Bot. Jahrb. Syst., **104**, 249–260.
– & Ulrich, G. (1979) Pseudanthien bei Umbelliferen. Beitr. Biol. Pflanzen **54**, 175–206.
⋆Galle, P. (1977) Untersuchungen zur Blütenentwicklung der Polygonaceen. Bot. Jahrb. Syst. **98**, 449–489.
⋆Gammero, J. C. (1968) Observaciones sobre la biología floral y morfología de la Potamogetonácea *Ruppia cirrhosa* (Petag.) Grande (= *R. spiralis* L. ex Dum.). Darwiniana **14**, 575–607. San Isidro, Argentina.
Gavaudan, P. (1959) Sur la signification de quelques formes foliaires et la théorie des gradients de sexualisation chez les Angiospermes. Trav. Lab. Biol.végaet. Fac. Sc. Poitiers **12**, 1–14.
Gelius, L. (1967) Studien zur Entwicklungsgeschichte an Blüten der *Saxifragales* sensu lato mit besonderer Berücksichtigung des Androeciums. Bot. Jb. **87**, 253–303.
⋆Gentry, A. H. (1974) Coevolutionary patterns in Central American Bignoniaceae. Ann. Missouri Bot. Gard. **61**, 728–759.
Gerstberger, P. & Leins, P. (1978) Rasterelektronenmikroskopische Untersuchungen an Blütenknospen von *Physalis philadelphica* (Solanaceae) – Anwendung einer neuen Präparationsmethode. Ber. Dt. Bot. Ges. **91**, 381–387.
Gilg, E. (1925) Flacourtiaceae, in Engler, A. & Prantl, K., Die natürlichen Pflanzenfamilien, 2nd edn, Bd. **21**, 377–455.
Glück, H. (1919) Blatt- und blütenmorphologische Studien. Jena.
Goebel, K. von (1882) Über die Anordnung der Staubblätter in einigen Blüten. Bot. Ztg. 22–25.
– (1884) Vergleichende Entwicklungsgeschichte der Pflanzenorgane, in Schenck, A. (Ed.): Handbuch der Botanik **3**. Breslau.
– (1886) Beiträge zur Kenntnis gefüllter Blüten. Jb. wiss. Bot. **17**, 207–296.
– (1904) Die kleistogamen Blüten und die Anpassungstheorien. Biol. Centralbl. **24**, 673–697, 737–787.
– (1911) Über gepaarte Blattanlagen. Flora N. F. **3**, 248–262.
– (1923) Organographie der Pflanzen, Bd. III, Jena.
– (1924) Die Entfaltungsbewegungen der Pflanzen u. deren teleologische Bedeutung. 2. Aufl. (Ergänzungsbd. I zur Organographie der Pflanzen). Jena.
– (1928/33) Organographie der Pflanzen. 1 u.3. Teil.3. Aufl. Jena.
– (1931) Blütenbildung und Sproßgestaltung. 2. Aufl. (2. Ergänzungsband zur Organographie der Pflanzen). Jena.
Goethe, J. W. von (1790) Versuch der Metamorphose der Pflanzen zu erklären. Gotha.
Gopal, G. & Puri, V. (1962) Morphology of the flower of some Gentianaceae with special reference to placentation. Bot. Gaz. **124**, 42–57.
Gottsberger, G. (1967) Blütenbiologische Beobachtungen an brasilianischen Malvaceen. Österr. Bot. Z. **114**, 349–378.
– (1970) Beiträge zur Biologie von Annonaceenblüten. Österr. Bot. Z. **118**, 237–279.
– (1971) Colour change of petals in *Malaviscus arboreus* flowers. Acta Bot. Neerl. **20**, 381–388.
– (1974) The structure and function of the primitive Angiosperm flower – a discussion. Acta Bot. Neerl. **23**, 461–471.
– (1977) Some aspects of beetle pollination in the evolution of flowering plants. Plant. Syst. Evol. Suppl. **1**, 211–226.
⋆– (1984) Pollination strategies in Brazilian *Philodendron* species. Ber. Deutsche. Bot. Ges. **97**, 391–410.
– Schrauwen, J. & Linskens, H. F. (1973) Die Zucker-Bestandteile des Nektars einiger tropischer Blüten. Portugaliae Acta Biologica Serie A, **13**, 1–8.
Gracza, P. (1970) Comparative observations on the floral primordium of *Salvia nemorosa* L. and

Papaver somniferum L. with regard to the initial conditions of pistil organisation. Ann. Univ.scient. Budapestinensis Sect. Biol. **12**, 123–132.

Grant, V. (1950) The flower constancy of bees. Bot. Rev. **16**, 379–398.

Grant, K. A. & Grant, V. (1968) Hummingbirds and their flowers. Columbia Univ. Press, New York-London.

Grassmann, P. (1884) Die Septaldrüsen. Ihre Verbreitung, Entstehung und Verrichtung. Flora **67**, 113–144, Pls. I, II.

Grégoire, V. (1931) La valeur morphologique des carpelles dans les Angiospermes. Bull. Acad.roy. Belg. Cl. Sc. Sér.5, **17**, 1286–1302.

– (1935) Sporophylles et organes floraux, tige et axe floral. Rec. trav. bot. Néerl. **32**, 453–466,1935.

– (1938) La morphogénèse et l'autonomie morphologique de l'appareil floral. I. Le carpelle. Cellule **47** (3), 287–452.

Guédès, M. (1964) Sur l'interprétation morphologique du placenta des Solanacées. Bull. Soc. Bot. France **111**, 135–139.

– (1965) Remarques sur la notion de carpelle condupliqué. Bull. Soc. Bot. France **112**, 54–68.

– (1966a) Homologies du carpelle et de l'étamine chez *Tulipa gesneriana* L. Österr. Bot. Z. **113**, 47–83.

– (1966b) The location of the transmitting and receptive tissues in teratological carpels of *Nigella damascena* L. and its bearing on the interpretation of the so-called conduplicate carpel. Flora, Abt. B, **156**, 395–407.

– (1966c) Le carpelle du *Prunus paniculata* Thunb. (*P. serrulata* Lindl.), ses modifications morphologiques dans les fleurs doubles et sa signification. Flora, Abt. B **156**, 464–499.

– (1966d) Refléxions sur la notion de carpelle pelté. Beitr. Biol. Pflanzen **42**, 393–423.

– (1967) Sépale, carpelle et feuille végétative chez *Trifolium repens* L. Flora, Abt. B. **157**, 190–228.

– (1969) Homologies de l'étamine et du carpelle chez *Papaver orientale* L. C. R. Acad. Sci. Paris **268**, Sér. D, 926–929.

– (1972a) Stamen-carpel homologies. Flora **161**, 184–208.

– (1972b) Contribution à la Morphologie du Phyllome. Mém. Mus. Hist. Nat. Paris Nouv. Sér. B, Botanique **21**.

– (1973) Carpel morphology and axis-sharing in syncarpy in some Rutaceae, with further comments on "New Morphology". Bot. J. Linn. Soc. **66**, 55–74.

– & Dupuy, P. (1970) Further remarks on the "Leaflet Theory" of the ovule. N. Phyt. **69**, 1081–1092.

– & Schmid, R. (1978) The Peltate (Ascidiate) Carpel Theory and carpel peltation in *Actinidia chinensis* (Actinidiaceae). Flora **167**, 525–543.

Gumppenberg, O. von (1924) Beiträge zur Entwicklungsgeschichte der Blumenblätter mit besonderer Berücksichtigung der Nervatur. Bot. Arch. **7**, 448–490.

Gut, B. J. (1966) Beiträge zur Morphologie des Gynoeceums und der Blütenachse einiger Rutaceen. Bot. Jb. Syst. **85**, 151–247.

★Haas, R. (1976) Morphologische, anatomische und entwicklungsgeschichtliche Untersuchungen an Blüten und Früchten hochsukkulenter Mesembryanthemaceen-Gattungen. Dissertationes Botanicae, J. Cramer, 33, Vaduz.

Hackel, E. (1906) Über Kleistogamie bei Gräsern. Österr. Bot. Z. **16**, 81–88, 143–154, 180–186.

Hagemann, W. (1963) Weitere Untersuchungen zur Organisation des Sproßscheitelmeristems; der Vegetationspunkt traubiger Infloreszenzen. Bot. Jb. **82**, 273–315.

– (1970) Studien zur Entwicklungsgeschichte der Angiospermenblätter. Bot. Jahrb. Syst. **90**, 297–413.

– (1975) Eine mögliche Strategie der vergleichenden Morphologie zur phylogenetischen Rekonstruktion. Bot. Jb. **96**, 107–124.

Hagerup, O. (1932) On pollination in the extremely hot air at Timbuctu. Dansk Bot. Arch. **8**, 1–20.

– (1936) Zur Abstammung einiger Angiospermen durch Gnetales und Coniferae. II. Centrospermae. Kgl. Danske Vid. Selskab. Biol. Medd. **13** (6), 1–59.

– (1939) On the origin of some Angiosperms through the Gnetales and the Coniferae. IV. The Gynoecium of Personatae. Kgl. Danske Vid. Selskab. Biol. Medd. **15** (2), 1–48.

– (1951) Pollination in the Faroes – in spite of rain and poverty in insects. Dan. Biol. Medd. **18**, 1–48.

Hallier, H. (1912) L'origine et le système phylétique des Angiospermes. Arch. Néerl. Sc.s. III B **1**, 146–234.

– (1921) Zur morphologischen Deutung der Diskusgebilde in der Dikotylenblüthe. Meded. Rijks. Herb. Leiden **41**, 1–14.

Hamann, U. (1962) Beitrag zur Embryologie der Centrolepidaceae mit Bemerkungen über den Bau der Blüten und Blütenstände und die systematische Stellung der Familie. Ber. Dt. Bot. Ges. **75**, 153–171.

★– (1966) Embryologische, morphologische-anatomische und systematische Untersuchungen an Philydraceen. Willdenowia Beih. **4**. Berlin.

Hanf, M. (1935) Vergleichende und entwicklungsgeschichtliche Untersuchungen über Morphologie und Anatomie der Griffel und Griffeläste. Beih. Bot. Centralbl. **54**, A, 99–141.

Harborne, J. B. (1976) Functions of flavonoids in plants. In Goodwin, T. W., Chemistry and Biochemistry of Plant Pigments, 2nd Ed., Vol.1, Chapter 16. London-New York-San Francisco.

Harms, H. (1917) Über eine Meliacee mit blattbürtigen Blüten. Ber. Dt. Bot. Ges. **35**, 338–348.

Hartl, D. (1956a) Morphologische Studien am Pistill der Scrophulariaceen. Österr. Bot. Z. **103**, 185–242.

– (1956b) Die Beziehungen zwischen den Plazenten der Lentibulariaceen und Scrophulariaceen nebst einem Exkurs über die Spezialisationseinrichtungen der Plazentation. Beitr. Biol. Pflanzen **32**, 471–490.

– (1957a) Die Pseudosympetalie von *Correa speciosa* (Rutaceae) und *Oxalis tubiflora* (Oxalidaceae). Abh. Akad. Wiss. Lit. Mainz, math.-nat. Kl. Nr. **2**.

– (1957b) Struktur und Herkunft des Endokarps bei den Rutaceen. Beitr. Biol. Pflanzen **34**, 35–49.

– (1958) Die Übereinstimmung des Endokarps der Simaroubaceen, Rutaceen und Leguminosen. Beitr. Biol. Pflanzen **34**, 453–455.

– (1959) Die Plazenta von *Phygelius capensis* E. Meyer (Scrophulariaceae) und ihre Beziehungen zur Bignoniaceen- Plazenta. Bot. Jb. Syst. **78**, 246–252.

– (1962) Die morphologische Natur und die Verbreitung des Apicalseptums. Analyse einer bisher unbekannten Gestaltungsmöglichkeit des Gynoeceums. Beitr. Biol. Pflanzen **37**, 241–330.

– (1963) Das Placentoid der Pollensäcke, ein Merkmal der Tubifloren. Ber. Dt. Bot. Ges. **76**, 70–72.

– (1965) *Scrophulariaceae*, in Hegi, G., Flora von Mitteleuropa Bd. VI/1, 2nd edn., part 1. Munich.

★– & Severin, I. (1981) Verwachsungen im Umfeld des Griffels bei *Allium*, *Cyanastrum* und *Heliconia* und den Monokotylen allgemein. Beitr. Biol. Pflanzen **55**, 235–260.

Heel, W. A.van (1962) Miscellaneous teratological notes with some general considerations. Proc. kon. nederl. Akad. Wetensch. Amsterdam Ser. C **65**, 392–406.

– (1966) Morphology of the androecium in Malvales. Blumea **13** (2), 177–394.

– (1969) The synangial nature of pollen sacs on the strength of 'congenital fusion' and 'conservatism of the vascular bundle system', with special reference to some Malvales. Proc. kon. nederl. Akad. Wetensch. Amsterdam Ser. C **72**, 172–206.

– (1977) On the morphology of the ovules in *Salacca* (Palmae). Blumea **23**, 371–375.

– (1978) Morphology of the pistil in Malvaceae-Ureneae. Blumea **24**, 123–127.

★– (1983) The ascidiform early development of free carpels, an S. E. M. investigation. Blumea **28**, 231–270.

★– (1984a) Variation in the development of ascidiform carpels, an S. E. M. investigation. Blumea, **29**, 443–452.

★– (1984b) Flowers and fruits in Flacourtiaceae. V. The seed anatomy and pollen morphology of *Berberidopsis* and *Streptothamnus*. Blumea, **30**, 31–37.

Hegi, G. (1919) Illustrierte Flora von Mitteleuropa. I-VII (1908–1931). Bd. IV/1, Munich.

Heimlich, L. F. (1927) The development and anatomy of the staminate flower of the cucumber. Amer. J. Bot. **14**, 227–237, Pls. 24–26.

Heinsbroek, P. G. & Heel, W. A.van (1969) Note on the bearing of the pattern of vascular bundles on the morphology of the stamens of *Victoria amazonica* (Poepp.) Sowerby. Proc. kon. nederl. Akad. Wetensch. Amsterdam Ser. C **72** (4), 431–444.

Henslow, G. (1888) The origin of floral structures through insect and other agencies (2nd ed. 1893). London.

Heslop-Harrison, J. & Y. (1975) Fine structure of the stigmatic papillae of *Crocus*. Micron **6**,45–52.

Heslop-Harrison Y. (1977) The pollen-stigma interaction: pollen-tube penetration in *Crocus*. Ann. Bot. **41**, 913–922.

– & Shivanna, K. R. (1977) The receptive surface of the Angiosperm stigma. Ann. Bot. **41**, 1233–1258.

*Hess, D. (1983) Die Blüte. Eine Einführung in Struktur and Funktion, Ökologie und Evolution der Blüten, mit Anleitungen zu einfachen Versuchen. Stuttgart (Ulmer).

*Hesse, M. (1981a) Pollenkitt and viscin threads: their role in cementing pollen grains. Grana, **20**, 145–152.

*– (1981b) The fine structure of the exine in relation to the stickiness of Angiosperm pollen. Review of Paleobotany and Palynology, **35**, 81–92.

*– (1982) Zur Mechanik des Pollentransports durch blütenbesuchende Insekten. Stapfia **10**, 99–110.

Hiepko, P. (1964) Das zentrifugale Androeceum der *Paeoniaceae*. Ber. Dt. Bot. Ges. **77**, 427–435.

– (1965) Vergleichend-morphologische und entwicklungsgeschichtliche Untersuchungen über das Perianth bei den Polycarpicae (Parts I & II). Bot. Jb. **84**, 359–508.

– (1966) Zur Morphologie, Anatomie und Funktion des Diskus der *Paeoniaceae*. Ber. Dt. Bot. Ges. **79**, 233–245.

– (1975) Zur Blütenmorphologie von *Barneoudia* Gay (Ranunculaceae). Bot. Jb. Syst. **96**, 192–199.

*Hilger, H. H. (1984) Wachstum und Ausbildungsformen des Gynoeceums von *Rochelia* (Boraginaceae). Plant Syst. Evol. **146**, 123–139.

*– (1985) Ontogenie, Morphologie und systematische Bedeutung geflügelter und glochidientragender Cynoglossae- und Eritrichae-Früchte (Boraginaceae). Bot. Jahrb. Syst., **105**, 323–378.

*– Balzer, M., Frey, W. & Podlech, D. (1975) Heteromerikarpie und Fruchtpolymorphismus bei *Microparacaryum*, gen.nov. (Boraginaceae). Plant Syst. Evol. **148**, 291–312.

*– & Richter, U. (1983) Untersuchungen zur Ausbildung der Klausen und ihrer Oberflächenskulpturen bei *Paracaryum intermedium* (Boraginaceae) aus Jordanien. Beitr. Biol. Pflanzen **57**, 205–220.

Hiller, G. H. (1884) Untersuchungen über die Epidermis der Blüthenblätter. Jb. wiss. Bot. (Ed. N. Pringsheim) **15**, 411–451, Pls. XXII & XXIII.

Hillmann, A. (1910) Vergleichend-anatomische Untersuchungen über das Rosaceenhypanth. Diss. Kiel.

Hofmann, U. (1977) Die Stellung von *Stegnosperma* innerhalb der Centrospermen. Ber. Dt. Bot. Ges. **90**, 39–52.

Hofmeister, W. (1851) Vergleichende Untersuchungen der Keimung, Entfaltung und Fruchtbildung höherer Kryptogamen und der Samenbildung der Coniferen. Leipzig.

– (1868) Allgemeine Morphologie der Gewächse. Leipzig.

Holdsworth, M. (1966) The cleistogamy of *Viola cunninghamii*. Trans. Roy. Soc. New Zealand Bot. **3**, 169–174.

Holthusen, K. (1940) Untersuchungen über das Vorkommen und den Zustand der Achselknospen bei höheren Pflanzen. Planta **30**, 590–638.

Hopkins, W. G. & Hillman, W. S. (1965) Phytochrome changes in tissues of dark-grown seedlings representing various photoperiodic classes. Amer. J. Bot. **52**, 427–432.

*Hoppe, J. R. (1985) Die Morphogenese der Cyathiendrüsen und ihrer Anhänge, ihre blattypologische Deutung und Bedeutung. Bot. Jahrb. Syst., **105**, 497–581.

*– & Uhlarz, H. (1982) Morphogenese und typologische Interpretation des Cyathiums von *Euphorbia*-Arten. Beitr. Biol. Pflanzen **56**, 63–98.

Hruby, K. (1934) Zytologie und Anatomie der mittel-europäischen Salbei-Arten. Beih. Bot. Centralbl. **52**, Abt. A, 298–380.

Huber, H. (1963) Die Verwandtschaftsverhältnisse der Rosifloren. Mitt. bot. Staatssamml. München **5**, 1–48.

– (1980) Morphologische und entwicklungsgeschichtliche Untersuchungen an Blüten und Blütenständen von Solanaceen und von *Nolana paradoxa* Lindl. (Nolanaceae). Dissertationes Botanicae **55**. Cramer, Vaduz.

Hunt, K. W. (1937) A study of the style and stigma, with reference to the nature of the carpel. Amer. J. Bot. **24**, 288–295.

*Hurusawa, I. & Kakadzu, K. (1982) Vergleichende morphologische Untersuchung des Gynostemiums bei Orchidazeen der Ryukyu-Inseln, I. Teil. Bull. College of Science, Univ. of the Ryukyus **33**, 27–46.

Huth, E. (1890) Ueber geokarpe, amphikarpe und heterokarpe Pflanzen. Berlin.

Ihlenfeldt, H.-D. (1960) Entwicklungsgeschichtliche, morphologische und systematische Untersuchungen an Mesembryanthemen. Fedde's Repert. **63**, 1–104.

– & Hartmann, H. (1970) Die Gattung *Harpagophytum* (Burch.) DC. ex Meissn. Mitt. Staatsinst. Allg. Bot. Hamburg **13**, 15–69.

*Inouye, D. W. (1980) The terminology of floral larceny. Ecology **61**, 1251–1253.

Irmisch, T. (1853) Beiträge zur Biologie und Morphologie der Orchideen. Leipzig.

Jäger, I. (1961) Vergleichend-morphologische Untersuchungen des Gefässbündelsystems peltater Nektar- und Kronblätter sowie verbildeter Staubblätter. Österr. Bot. Z. **108**, 433–504.

*Jahnke, C. (1986) Der Infloreszenzbau der Cornaceen sensu lato und seine sytematischen Konsequenzen. Akad. Wiss. Lit. Mainz, math.-naturw. Kl., Trop. u.subtrop. Pflanzenwelt **57**.

*– & Froebe, H. A. (1984) Untersuchungen zur Ontologie des Lobulum Inflexum ausgewählter Apiaceen-Petalen. Beitr. Biol. Pflanzen, **59**, 75–93.

Janchen, E. (1950) Die Herkunft der Angiosperm-Blüte und die systematische Stellung der Apetalen. Österr. Bot. Z. **97**, 127–167.

*Jenny, M. (1983) Apocarpy in Sterculiaceae – structure, development, function and evolution. Acta Bot. Neerl., **32**, 344.

Johnson, L. A. S. (1972) Evolution and classification in *Eucalytpus*. Proc. Linn. Soc. New South Wales **97**, 11–29.

Johnson, L. E. B., Wilcoxon, R. D. & Frosheiser, F. I. (1975) Transfer cells in tissue of the reproductive system of Alfalfa. Canad. J. Bot. **53**, 952–956.

*Johnson, M. A. (1958) The epiphyllous flowers of *Turnera* and *Helwingia*. Bull. Torrey Bot. Club **85**, 313–323.

Jones, C. E. & Buchmann, St. L. (1974) Ultraviolet floral patterns as functional orientation cues in Hymenopterous pollination systems. Anim. Behav. **22**, 481–485.

Joshi, A. C. (1934) Morphology of the stylar canal in Angiosperms. Ann. Bot. **48**, 967–974.

– (1947) The morphology of the gynoecium. Proc., Ind. Sci. Congr. II, 1–18.

Joshi, B. M. (1976) Studies in Boraginaceae: II. Morphology of gynoecium with special reference to evolution of the parietal placentation. Geobios (Jodhpur) **3**, 76–78.

– & Rao, V. S. (1933) Floral anatomy of *Rivina humilis* and the theory of carpel polymorphism. New Phytol. **32**, 359–363.

Juhnke, G. & Winkler, H. (1938) Der Balg als Grundelement des Angiospermgynaeceums. Beitr. Biol. Pflanzen **25**, 291–324.

Junell, S. (1934) Zur Gynäceummorphologie und Systematik der Verbenaceen und Labiaten nebst Bemerkungen über ihre Samenentwicklung. Symb. Bot. Upsalienses **4**, Uppsala.

– (1938) Über den Fruchtknotenbau der Borraginazeen mit pseudomonomeren Gynäzeen. Svensk Botanisk Tidskr. **32**, 261–273.

Just, T. (1939) The typological approach to the nature of the flower, 115–131, in Wilson, C. L. & Just, T., The morphology of the flower. Bot. Rev. **5**, 97–131.

Kania, W. (1973) Entwicklungsgeschichtliche Untersuchungen an Rosaceenblüten. Bot. Jb. Syst. **93**, 175–246.

Kaplan, D. (1967) Floral morphology, organogenesis and interpretation of the inferior ovary in *Downingia bacigalupii*. Amer. J. Bot. **54**, 1274–1290.

– (1972) On the value of comparative development in phylogenetic studies – a rejoinder. Phytomorph. **21**, 134–140.

Kaul, R. B. (1967a). Development and vasculature of the flowers of *Lophotocarpus calycinus* and *Sagittaria latifolia* (Alismaceae). Amer. J. Bot. **54**, 914–920.

– (1967b) Ontogeny and anatomy of the flower of *Limnocharis flava* (Butomaceae). Amer. J. Bot. **54**, 1223–1230.

– (1968) Floral development and vasculature in *Hydrocleis nymphoides* (Butomaceae). Amer. J. Bot. **55**, 236–242.

– (1969) Morphology and development of the flowers of *Boottia cordata, Ottelia alismoides*, and their synthetic hybrid (Hydrocharitaceae). Amer. J. Bot. **56**, 951–959.

*– (1976) Conduplicate and specialized carpels in the Alismatales. Amer. J. Bot. **63**, 175–182.

Kaussmann, B. (1941) Vergleichende Untersuchungen über die Blattnatur der Kelch-, Blumen-
und Staubblätter. Bot. Archiv. **42**, 503–572.
– (1951) Morphologische und anatomische Studien an *Schizocarpa plantaginea* Hance. Planta **39**,
91–104.
– (1963) Pflanzenanatomie, unter besonderer Berücksichtigung der Kultur- und Nutzpflanzen.
Jena.
– & Neitzel, H. (1972) Ein Beitrag zur Morphologie des Gynoeceums von *Nigella damascena* L.
aus histogenetischer Sicht. Flora **161**, 30–45.
*Kircher, P. (1986) Untersuchungen zur Blüten- und Infloreszenzmorphologie, Embryologie und
Systematik der Restionaceen im Vergleich mit Gramineen und verwandten Familien.
Dissertationes Botanicae **94**. Cramer, Berlin-Stuttgart.
Kirchner, O. (1911) Blumen und Insekten. Leipzig.
Keay, R. W. J. (1954) Flacourtiaceae, in Hutchinson, J. & Dalziel, J. M., Flora of West Tropical
Africa. I/1, 185–191. London.
Kisser, J. & Hauer, A. (1955) Blütenfärbung und Ernährungszustand. Österr. Bot. Z. **102**,
572–593.
Klopfer, K. (1968a). Beiträge zur floralen Morphogenese und Histogenese der Saxifragaceae. 1.
Die Infloreszenz-Entwicklung von *Tellima grandiflora*. Flora, Abt. B., **157**, 461–476.
– (1968b) Beiträge zur floralen Morphogenese und Histogenese der Saxifragaceae. 2. Die
Blütenentwicklung von *Tellima grandiflora*. Flora, Abt. B., **158**, 1–21.
– (1969a) Beiträge zur floralen Morphogenese und Histogenese der Saxifragaceae. 3. Die
Blütenentwicklung einiger *Ribes*-Arten. Wiss. Z. Päd. Hochsch. Potsdam, **13**, 187–205.
– (1969b) Zur Ontogenese und Evolution des parakarpen Gynaeceums. Wiss. Z. Päd. Hochsch.
Potsdam, **13**, 207–243.
– (1970a) Beiträge zur floralen Morphogenese und Histogenese der Saxifragaceae. 4. Die
Blütenentwicklung einiger *Saxifraga*-Arten. Flora **159**, 347–365.
– (1970b) Beiträge zur floralen Morphogenese und Histogenese der Saxifragaceae. 5. Die
Blütenentwicklung der Gattungen *Astilbe*, *Rodgersia*, *Astilboides* und *Bergenia*. Wiss. Z. Päd.
Hochsch. Potsdam, **14**, 327–350.
– (1971) Beiträge zur floralen Morphogenese und Histogenese der Saxifragaceae. 6. Die
Hydrangeoideen. Wiss. Z. Päd. Hochsch. Potsdam, **15**, 77–95.
– (1972a) Beiträge zur floralen Morphogenese und Histogenese der Saxifragaceae. 7. *Parnassia
palustris* und *Francoa sonchifolia*. Flora **161**, 320–332.
– (1972b) Quantitative Untersuchungen zur Stellung des Gynaeceums bei verschiedenen
Saxifraga-Arten. Wiss. Z. Päd. Hochsch. Potsdam **16**, 45–49.
– (1973) Florale Morphogenese und Taxonomie der Saxifragaceae sensu lato. Fedde's Repert.
84, 475–516.
– & Ziesing, W. (1971) Entwicklungsgeschichtliche Untersuchungen an zygomorphen
Saxifragaceen-Blüten. Wiss. Z. Päd. Hochsch. Potsdam, **15**, 97–101.
Knoll, F. (1923) Über die Lückenepidermis der *Arum*-Spatha. Österr. Bot. Z. **72**, 246–254.
– (1926) Insekten und Blumen. 3. Bde. I. Zeitgemäße Ziele und Methoden für das Studium der
ökologischen Wechselbeziehungen. Abh.d. zoolog.-bot. Gesellsch. Wien **12**, 1–482 (1921).
– (1956) Die Biologie der Blüte. Berlin-Göttingen-Heidelberg.
*Knox, R. B. et al (1985) Extrafloral nectaries as adaptations for bird pollination in *Acacia
terminalis*. Amer. J. Bot. **72**, 1185–1196.
Knuth, P. (1894) Grundriß der Blüten-Biologie. Kiel-Leipzig.
– (1898–1905) Handbuch der Blütenbiologie. Leipzig.
Koelle, W. (1913) Vergleichende anatomische Untersuchungen der Liliaceen-Blumenblätter. Diss.
Kiel.
Koorders, S. H. (1897) Ueber die Blüthenknospen-Hydathoden einiger tropischer Pflanzen.
Leiden (Brill).
– (1902) Notizen mit Abbildungen einiger interessanter cauliflorer Pflanzen. Ann. Jard. Bot.
Buitenzorg **18**, 82–91.
*Kopka, S. & Weberling, F. (1984) Zur Morphologie und Morphogenese der Blüte von *Vochysia
acuminata* Bong. ssp. *laurifolia* (Warm.)Stafleu (Vochysiaceae). Beitr.z. Biol. d. Pflanzen **59**,
273–302.
*Kugler, H. (1934) Zur Blütenökologie von *Asarum europaeum* L. Ber. Deutsch. Bot. Ges. **52**,
348–354.

– (1936) Die Ausnutzung der Saftmalsumfärbung bei den Roßkastanienblüten durch Bienen und Hummeln. Ber. Dt. Bot. Ges. **54**, 394–400.

– (1952) Die spontane Bevorzugung bestimmte Farbqualitäten durch blütenbesuchende Insekten. Festschr.z. Gedenkfeier an die vor 300 Jahren erfolgte Gründung der Akademie d. Naturforscher. Veröffentl.d. Histor. Ver. Schweinfurt **2**, 58–67.

– (1955) Zum Problem der Dipterenblumen. Österr. Bot. Z. **102**, 529–541.

– (1963) UV-Musterungen auf Blüten und ihr Zustandekommen. Planta **59**, 296–329.

– (1970) Blütenökologie. 2nd edn. Stuttgart.

★– (1984) Die Bestäubung von Blüten durch den Schmalkäfer *Oedemera* (Coleoptera). Ber. Deutsch. Bot. Ges. **97**, 383–390.

Kuijt, J. & Weberling, F. (1972) The flower of *Phthirusa pyrifolia* (Loranthaceae). Ber. Dt. Bot. Ges. **85**, 467–480.

Kunze, H. (1969) Vergleichende morphologische Untersuchungen an komplexen Compositen-Blütenständen. Beitr.biol. Pflanzen **46**, 97–154.

– (1979) Typologie und Morphogenese des Angiospermen-Staubblattes. Beitr. Biol. Pflanzen **54**, 239–304.

– (1982a) Aspekte der Blütengestalt. I Die Blüte als abbildendes Organ. Elemente der Naturwissenschaft **36**, 9–26.

★– (1982b) Aspekte der Blütengestalt. II Innenraumbildung. III Die Bildung von Apparaten. Elemente der Naturwissenschaft **37**, 19–40.

★– (1984a) Koevolution von Blüten und Insekten, dargestellt am Beispiel der Ölblumen. Unterrichtsmodell für die Sekundarstufe II. Unterricht Biologie, **92**, 8. Jahrgang, 39–43.

★– (1984b) Vergleichende Studien an Cannaceen- und Marantaceenblüten. Flora **175**, 301–318.

★– (1985) Die Infloreszenzen der Marantaceen und ihr Zusammenhang mit dem Typus der Zingiberales-Synfloreszenz. Beitr. Biol. Pflanzen **60**, 93–140.

★– Lacey, E. P. & Pace, R. (1983) Effect of parental flowering and dispersal times on offspring fate in *Daucus carota*. Oecologia **60**, 274–278.

Lam, H. (1948a) Classification and the new morphology. Acta Biotheoretica Vol. **8**, Pars IV, Leiden.

– (1948b) A new system of the Cormophyta. Blumea **6**, 282–289.

– (1950) Stachyspory and Phyllospory as factors in the natural system of the Cormophyta. Svensk. Bot. Tidskr. **44**, 517–534.

Lang, B. (1977) Vergleichend-morphologische und entwicklungsgeschichtliche Untersuchungen am Gynöcium einiger *Nigella*-Arten (Ranunculaceae). Bot. Jb. Syst. **98**, 289–335.

Lange, R. (1913) Über den lippenförmigen Anhang an der Narbenöffnung von *Viola tricolor*. Ber. Dt. Bot. Ges. **31**, 268–274, Pl. XII.

★Laroca, S. (1968) Contribução para o conhecimento das relaçãoes entre abelhas e flores: Coleta de Pólen das Anteras Tubulares de certas Melastomataceae. Revista Floresta **2**, 69–73.

★Lawrence, G. H. M. (1962) Taxonomy of vascular plants. 6th printing. Macmillan, New York.

★Leereveld, H., Meeuse, A. D. J. & Stelleman, P. (1981) Anthecological relations between reputedly anemophilous flowers and syrphid flies. IV. A note of the anthecology of *Scirpus maritimus* L. Acta Bot. Neerl. **30**, 465–473.

Leinfellner, W. (1940) Das epidermale Randwachstum der Fruchtblätter. Bot. Archiv. **41**, 507–515.

– (1941) Über den unterständigen Fruchtknoten und einige Bemerkungen über den Bauplan des verwachsenblättrigen Gynoeceums an sich. Bot. Arch. **42**, 1–43.

– (1950) Der Bauplan des synkarpen Gynözeums. Österr. Bot. Z. **97**, 403–436.

– (1951a) Die U-förmige Plazenta als der Plazentationstypus der Angiospermen. Österr. Bot. Z. **98**, 338–358.

– (1951b) Die Nachahmung der durch kongenitale Verwachsung entstandenen Formen des Gynözeums durch postgenitale Verschmelzungsvorgänge. Österr. Bot. Z. **98**, 403–411.

– (1952a) Die Homologien zwischen den Kelch- und Laub-blättern der Mesembryanthemeen. Österr. Bot. Z. **99**, 295–317.

– (1952b) Zur Homologisierung der Querzonen an unifazialen Laub- und Kelchblättern. Österr. Bot. Z. **99**, 405–408.

– (1952c) Transversale Abflachungen im Spitzenbereich der Karpelle. Österr. Bot. Z. **99**, 455–468.

– (1954a) Die Kelchblätter auf unterständigen Fruchtknoten und Achsenbechern. Österr. Bot. Z. **101**, 315–327.
– (1954b) Die petaloiden Staubblätter und ihre Beziehungen zu den Kronblättern. Österr. Bot. Z. **101**, 373–406.
– (1954c) Beiträge zur Kronblattmorphologie. I. *Erythroxylon novogranatense.* Österr. Bot. Z. **101**, 428–434.
– (1954d) Beiträge zur Kronblattmorphologie. II. Die Formenmannigfaltigkeit der peltaten und diplophllen Kronblätter von *Waldsteinia geoides.* Österr. Bot. Z. **101**, 558–565.
– (1955a) Beiträge zur Kronblattmorphologie. IV. Unifaziale Vorläuferspitzen an den Kronblättern einiger Crassulaceen, Passifloraceen, Balsaminaceen, Convolvulaceen und Cucurbitaceen. Österr. Bot. Z. **102**, 73–79.
– (1955b) Beiträge zur Kronblattmorphologie. V. Über den homologen Bau der Kronblattspreite und der Staubblattanthere bei *Koelreuteria paniculata.* Österr. Bot. Z. **102**, 89–98.
– (1955c) Beiträge zur Kronblattmorphologie. VI. Die Nektarblätter von *Berberis.* Österr. Bot. Z. **102**, 186–194.
– (1955d) Beiträge zur Kronblattmorphologie. VII. Die Kronblätter einiger Linaceen. Österr. Bot. Z. **102**, 322–338.
– (1956a) Medianstipulierte Staubblätter. Österr. Bot. Z. **103**, 24–43.
– (1956b) Die blattartig flachen Staubblätter und ihre gestaltlichen Beziehungen zum Bautypus des Angiospermen-Staubblattes. Österr. Bot. Z. **103**, 247–290.
– (1956c) Die Gefäßbündelversorgung des *Lilium*-Staubblattes. Österr. Bot. Z. **103**, 346–352.
– (1956d) Inwieweit kommt der peltat-diplophylle Bau des Angiospermen-Staubblattes in dessen Leitbündelanordnung zum Ausdruck? Österr. Bot. Z. **103**, 381–399.
– (1956e) Zur Morphologie des Gynözeums von *Berberis.* Österr. Bot. Z. **103**, 600–612.
– (1957a) Die augenfällige Diplophyllie der Violaceen-Anthere. Österr. Bot. Z. **104**, 209–227.
– (1957b) Der Bündelverlauf in der dorsifixen Anthere von *Trapa natans.* Beitr. Biol. Pflanzen **34**, 83–87.
– (1958a) Zur Morphologie des Melastomataceen-Staubblattes. Österr. Bot. Z. **105**, 44–70.
– (1958b) Beiträge zur Kronblattmorphologie. VIII. Der peltate Bau der Nektarblätter von *Ranunculus,* dargelegt an Hand von *Ranunculus pallasii* Schlecht. Österr. Bot. Z. **105**, 184–192.
– (1958c) Über die peltaten Kronblätter der Sapindaceen. Österr. Bot. Z. **105**, 443–514.
– (1959a) Über die röhrenförmige Nektarschuppe an den Nektarblättern verschiedener *Ranunculus-* und *Batrachium-*Arten. Österr. Bot. Z. **106**, 88–103.
– (1959b) Besitzt die Gattung *Phylica* L. (Rhamnaceae) echt revolutive Rollblätter? Österr. Bot. Z. **106**, 577–603.
– (1960a) Petaloid verbildete Staubblätter von *Narcissus* als ein weiteres Beispiel für die Umbildung diplophyller in sekundär schlauch- oder schildförmige Spreiten. Österr. Bot. Z. **107**, 39–44.
– (1960b) Zur Entwicklungsgeschichte der Kronblätter der Sterculiaceae-Buettnerieae. Österr. Bot. Z. **107**, 153–176.
– (1960c) Zur Kenntnis des Monokotyledonen-Perigons. I. Die Perigonblätter von *Dipidax.* Österr. Bot. Z. **107**, 445–455.
– (1960d) Zur Kenntnis des Monokotyledonen-Perigons. II. Die Perigonblätter von *Ornithoglossum.* Österr. Bot. Z. **107**, 474–486.
– (1961a) Zur Kenntnis des Monokotyledonen-Perigons. III. Die Perigonblätter einiger weiterer Melanthioideen (*Melanthium, Zygadenus, Anticlea, Toxicoscordion, Veratrum* und *Kreysigia*). Österr. Bot. Z. **108**, 194–210.
– (1961b) Zur Kenntnis des Monokotyledonen-Perigons. IV. Die Perigonblätter von *Oceanorus leimanthoides* (A. Gray) Small. Österr. Bot. Z. **108**, 300–303.
– (1961c) Staubblattverwachsungen bei *Yucca filamentosa.* Österr. Bot. Z. **108**, 368–378.
– (1962a) Über die Variabilität der Blüten von *Tofieldia calyculata.* I. Zu Karpellen verbildete Staubblätter. Österr. Bot. Z. **109**, 1–17.
– (1962b) Über die Variabilität der Blüten von *Tofieldia calyculata.* II. Der Ersatz von Perigonblättern durch Staubblätter. Österr. Bot. Z. **109**, 113–124.
– (1962c) Über die Variabilität der Blüten von *Tofieldia calyculata.* III. Zusammenfassende Übersicht der vorgefundenen Abweichungen. Österr. Bot. Z. **109**, 395–430.
– (1963a) Zum Blütenbau von *Lilium tigrinum* var. *flore pleno.* Österr. Bot. Z. **110**, 177–193.

– (1963b) Zur Kenntnis des Monokotyledonen-Perigons. V. Der Bau der Perigonblätter von *Lilium*, dargelegt an Hand jener von *Lilium tigrinum* var. *flore pleno*. Österr. Bot. Z. **110**, 349–370.

– (1963c) Über die Wiederherstellung der normalen Lage der überkippten *Rhododendron*-Anthere bei petaloider Verbildung. Österr. Bot. Z. **110**, 374–379.

– (1963d) Zur Kenntnis des Monokotyledonen-Perigons. VI. Ein Vergleich der Perigonblätter von *Lloydia* und *Fritillaria* mit den Nektarblättern von *Ranunculus*. Österr. Bot. Z. **110**, 401–409.

– (1963e) Das Perigon der Liliaceen ist staminaler Herkunft. Österr. Bot. Z. **110**, 448–467.

– (1964a) Zur Formenmannigfaltigkeit der Nektarschuppe von *Ranunculus glacialis*. Österr. Bot. Z. **111**, 78–83.

– (1964b) Über die falsche Sympetalie bei *Lonchostoma* und andere Gattungen der Bruniaceen. Österr. Bot. Z. **111**, 345–353.

– (1964c) Sind die Kronblätter der Bruniaceen peltat gebaut? Österr. Bot. Z. **111**, 500–526.

– (1965a) Über die Kronblätter der Frankeniaceen. Österr. Bot. Z. **112**, 44–55.

– (1965b) Wie sind die Winteraceen-Karpelle tatsächlich gebaut? I. Die Karpelle von *Drimys*, Sektion *Tasmannia*. Österr. Bot. Z. **112**, 554–575.

– (1966a) Wie sind die Winteraceen-Karpelle tatsächlich gebaut? II. Über das Vorkommen einer ringförmigen Plazenta in den Karpellen von *Drimys*, Sektion *Wintera*. Österr. Bot. Z. **113**, 84–95.

– (1966b) Wie sind die Winteraceen-Karpelle tatsächlich gebaut? III. Die Karpelle von *Bubbia*, *Belliolum*, *Pseudowintera*, *Exospermum* und *Zygogynum*. Österr. Bot. Z. **113**, 245–264.

– (1966b) Über die Karpelle verschiedener Magnoliales. I. *Illicium* (Illiciaceae). Österr. Bot. Z. **113**, 383–389.

– (1966c) Über die Karpelle verschiedener Magnoliales. II. *Xymalos*, *Hedycarya* und *Siparuna* (Monimiaceae). Österr. Bot. Z. **113**, 448–458.

– (1966d) Über die Karpelle verschiedener Magnoliales. III. *Schisandra* (Schisandraceae). Österr. Bot. Z. **113**, 563–569.

– (1967) Über die Karpelle verschiedener Magnoliales. IV. *Magnolia* und *Michelia* (Magnoliaceae). Österr. Bot. Z. **114**, 73–83.

– (1968) Über die Karpelle verschiedener Magnoliales. VI. *Gomortega keule* (Gomortegaceae). Österr. Bot. Z. **115**, 113–119.

– (1969a) Über die Karpelle verschiedener Magnoliales. VII. *Euptelea* (Eupteleaceae). Österr. Bot. Z. **116**, 159–166.

– (1969b) Über die Karpelle verschiedener Magnoliales. VIII. Überblick über alle Familien der Ordnung. Österr. Bot. Z. **117**, 107–127.

– (1969c) Über peltate Karpelle, deren Schlauchteil außen vom Ventralspalt unvollkommen aufgeschlitzt ist. Österr. Bot. Z. **117**, 276–283.

– (1969d) Zur Kenntnis der Karpelle der Leguminosen. I. Papilionaceae. Österr. Bot. Z. **117**, 332–347.

– (1970a) Zur Kenntnis der Karpelle der Leguminosen. II. Caesalpinaceae und Mimosaceae. Österr. Bot. Z. **118**, 108–120.

– (1970b) Über die Karpelle der Connaraceen. Österr. Bot. Z. **118**, 542–559.

– (1971) Das Gynözeum von *Krameria* und sein Vergleich mit jenem der Leguminosae und der Polygalaceae. Österr. Bot. Z. **119**, 102–117.

– (1972a) Zur Morphologie des Gynözeums der Polygalaceen. Österr. Bot. Z. **120**, 51–76.

– (1972b) Das Gynöceum der Bignoniaceen. I. Über den Bau des Fruchtknotens von *Kigelia* (Crescentieae). Österr. Bot. Z. **120**, 269–277.

– (1973a) Das Gynöceum der Bignoniaceen. II. Die U-förmige Plazenta von *Schlegelia* (Crescentieae). Österr. Bot. Z. **121**, 13–22.

– (1973b) Zur Lage des wahren Karpellrandes. Österr. Bot. Z. **121**, 285–301.

– (1973c) Das Gynöceum der Bignoniaceen. III. Crescentieae (*Amphitecna*, *Colea*, *Rhodocolea*, *Ophiocolea*, *Phyllarthron*, *Phylloctenium*, *Parmentiera*, *Enallagma* und *Crescentia*). Österr. Bot. Z. **122**, 59–73.

Leins, P. (1964a) Entwicklungsgeschichtliche Studien an Ericales-Blüten. Bot. Jb. **83**, 57–88.

– (1964b) Das zentripetale und zentrifugale Androeceum. Ber. Dt. Bot. Ges. **77**, Heft, 22–26.

– (1964c) Die frühe Blütenentwicklung von *Hypericum hookerianum* Wight et. Arn. und *H. aegypticum* L. Ber. Dt. Bot. Ges. **77**, 112–123.

- (1965) Die Infloreszenz und frühe Blütenentwicklung von *Melaleuca nesophila* F. Muell. (Myrtaceae). Planta (Berlin) **65**, 195–204.
- (1967) Die frühe Blütenentwicklung von *Aegle marmelos* (Rutaceae). Ber. Dt. Bot. Ges. **80**, 320–325.
- (1971) Das Androeceum der Dikotylen. Ber. Dt. Bot. Ges. **84**, 191–193.
- (1972a) Das Karpell im ober- und unterständigen Gynoeceum. Ber. Dt. Bot. Ges. **85**, 291–294.
- (1972b) Das zentrifugale Androeceum von *Couroupita guianensis* (Lecythidaceae). Beitr. Biol. Pflanzen **48**, 313–319.
- (1975) Die Beziehungen zwischen multistaminaten und einfachen Androeceen. Bot. Jb. Syst. **96**, 231–237.
- (1979) Der Übergang vom zentrifugalen komplexen zum einfachen Androeceum. Ber. Dt. Bot. Ges. **92**, 717–719.
- ★– (1983a) Muster in Blüten. Bonner Universitätsblätter, 21–33.
- ★– (1983b) Growth in flower buds. Acta Bot. Neerl. **32**, 347.
- & Boecker, K. (1981) Entwickeln sich Staubgefäße wie Schildblätter? Beitr. Biol. Pflanzen **56**, 317–327.
- & Bonnery-Brachtendorf, R. (1977) Entwicklungsgeschichtliche Untersuchungen an Blüten von *Datisca cannabina* (Datiscaceae). Beitr. Biol. Pflanzen **53**, 143–155.
- & Erbar, C. (1980) Zur Entwicklung der Blüten von *Monodora crispata* (Annonaceae). Beitr. Biol. Pflanzen **55**, 11–22.
- ★– & Erbar, C. (1985a) Zur frühen Entwicklungsgeschichte des Apiaceen-Gynoeceums. Bot. Jahrb. Syst. **106**, 53–60.
- ★– & Erbar, C. (1985b) Ein Beitrag zur Blütenentwicklung der Aristolochiaceen, einer Vermittlergruppe zu den Monocotylen. Bot. Jahrb. Syst. **107**, 343–368.
- & Galle, P. (1971) Entwicklungsgeschichtliche Untersuchungen an Cucurbitaceen-Blüten. Österr. Bot. Z. **229**, 531–548.
- & Gemmeke, V. (1979) Infloreszenz- und Blütenentwicklung bei der Kugeldistel *Echinops exaltatus* (Asteraceae). Plat Syst. Evol. **132**, 189–204.
- Merxmüller, H. & Sattler, R. (1972) Zur Terminologie interkalarer Becherbildungen in Blüten. Ber. Dt. Bot. Ges. **85**, 294.
- & Metzenauer, G. (1979) Entwicklungsgeschichtliche Untersuchungen an *Capparis*-Blüten. Bot. Jb. Syst. **100**, 542–554.
- & Orth, C. (1979) Zur Entwicklungsgeschichte männlicher Blüten von *Humulus lupulus* (Cannabaceae). Bot. Jb. Syst. **100**, 372–378.
- ★– & Schwitalla, S. (1985) Studien an Cactaceen-Blüten I Einige Bemerkungen zur Blütenentwicklung von *Pereskia*. Beitr. Biol. Pflanzen, **60**, 313–323.
- & Sobick, U. (1977) Die Blütenentwicklung von *Reseda lutea*. Bot. Jb. Syst. **98**, 133–149.
- & Stadler, P. (1973) Entwicklungsgeschichtliche Untersuchungen am Androeceum der Alismatales. Österr. Bot. Z. **121**, 51–63.
- & Winhard, W. (1973) Entwicklungsgeschichtliche Studien an Loasaceen-Blüten. Österr. Bot. Z. **122**, 145–165.
- Leppik, E. E. (1968) Morphogenetic classification of flower types. Phytomorph. **18**, 451–466.
- (1977) Calyx-borne semaphylls in tropical Rubiaceae. Phytomorph. **27**, 161–168.
- ★Leroy, J.-F. (1983) The Origin of Angiosperms: an unrecognized ancestral Dicotyledon, *Hedyosmum* (Chloranthales), with a strobiloid flower is living today. Taxon **32**, 169–175.
- Lersten, N. (1971) A review of septate microsporangia in vascular plants. Iowa State J. Sci. **45**, 487–497.
- (1974) Morphology of discoid floral nectaries in Leguminosae, especially tribe Phaseolae (Papilionoideae). Phytomorph. **23**, 152–161.
- Letouzey, R. (1970) Manuel de Botanique Forestière Afrique Tropicale. Vol.2A. (Flacourtiacées, pp. 77–83), Nogent s/Marne.
- Linsbauer, K. (1917) C. K. Schneiders Illustriertes Handwörterbuch der Botanik. 2. Aufl. Leipzig.
- Lister, G. (1883) On the origin of the placentas in the tribe Alsineae in the order Caryophylleae. Linn. Soc. J. Bot. **20**, 423–429.
- Lüke, E. (1943) Erzeugung von Muster- und Farbänderungen an den Blüten von *Tagetes* und anderen Objekten. Diss. Göttingen.
- Lund, S. (1874) Le calice des composées. Bot. Tidskr. **2**, Sér. 2, 121–260.

*Lüttge, U. (1961) Über die Zusammensetzung des Nektars und den Mechanismus seiner Sekretion. I. Planta **56**, 189–212.

Luyten, J. & van Waveren, J. M. (1938) De Orgaanvorming van *Leucojum aestivum* L. Mededeel. Landbouwhoogeschool, Deel **42**, Verh. **1**, Wageningen.

*Macior, L. W. (1974) Behavioral aspects of coadaptations between flowers and insect pollinators. Ann. Missouri Bot. Gard. **61**, 760–769.

Magin, N. (1977) Das Gynoeceum der Apiaceae – Modell und Ontogenie. Ber. Dt. Bot. Ges. **90**, 53–66.

*– (1983) Die Nektarien der Apiaceae. Flora **173**, 233–254.

Mair, O. (1977) Zur Entwicklungsgeschichte monosymmetrischer Dikotylen-Blüten. Dissertationes Botanicae **38**. Cramer, Vaduz.

*Mansfeld, R. (1929) Beitrag zur Morphologie des *Euphorbia*-Cyathiums. Ber. Deutsch. Bot. Ges. **46**, 674–677.

Maresquelle, H. J. (1970) La thème évolutif des complexes d'inflorescences. Son aptitude à susciter des problèmes nouveaux. Bull. Soc. Bot. Fr. **117**, 1–4.

Markgraf, F. (1936) Blütenbau und Verwandtschaft bei den einfachsten Helobiae. Ber. Dt. Bot. Ges. **54**, 191–229.

– (1942) Beobachtungen über den Rosenkelch. Decheniana, Festschr., Bd. 101 AB, 100–107.

– (1970) Der Bauplan der Angiospermen-Blüte. Neujahrsblatt d. Naturforsch. Ges. Zürich **173**, Jg.115 der Vierteljahrsschrift d. Naturforsch. Ges. Zürich.

*Martin, B. F. & Tucker, S. C. (1985) Developmental studies in *Smilax* (Liliaceae). I. Organography and the shoot apex. Amer. J. Bot. **72**, 66–74.

*Mason, H. L. (1957) The concept of flower and the theory of homology. Madroño **14**, 81–95.

Masters, M. T. (1886) Prolification des Blütenstandes. In Pflanzen-Teratologie, translation by U. Dammer, 123–137, Leipzig.

Matile, P. (1977) Entwicklung einer Blüte. Neujahrsblatt d. Naturf. Gesellschaft in Zürich auf das Jahr 1978. (im Anschluß an d. Jg. 122 der Vierteljahrsschrift d. Naturf. Ges. Zürich.- 180. Stück)

Mattfeld, J. (1935) Zur Morphologie und Systematik der Cyperaceae. Proc. Intern. Bot. Congr., Amsterdam, 330–332.

– (1938a) Das morphologische Wesen und die phylogenetische Bedeutung der Blumenblätter. Ber. Dt. Bot. Ges. **56**, 86–116.

– (1938b) Über eine angebliche Drymaria Australiens nebst Bemerkungen über die Staminaldrüsen und die Petalen der Caryophyllacee. Repert. spec. nov. (Fedde), Beih. C, 147–164.

Mayr, B. (1969) Ontogenetische Studien an Myrtales-Blüten. Bot. Jb. **89**, 210–271, Stuttgart.

McCoy, R. W. (1940) Floral organogenesis in *Frasera carolinensis*. Amer. J. Bot. **27**, 600–609.

McLean, R. C. & Ivimey-Cook, W. R. (1951) Textbook of Theoretical Botany 1–4.

Meeuse, A. D. J. (1963) Stachyspory, phyllospory, and morphogenesis. Advancing Frontiers of Plant Sciences **7**, 115–156, New Delhi.

– (1966) Fundamentals of phytomorphology. New York.

– (1972) Sixty-five years of theories of the multiaxial flower. Acta Biotheoretica **21**, 167–202 (Leyden).

– (1975a) Origin of the Angiosperms – problem or inaptitude? Phytomorphology **25**, 373–379.

– (1975b) Changing floral concepts: anthocorms, flowers, and anthoids. Acta Bot. Neerl. **24**, 23–36.

*– (1980) What is polyandry? Phytomorphology **30**, 388–396.

*– (1984) Rate of dependence of *Plantago media* L. on entomophilous reproduction – preliminary report. Acta Bot. Neerl. **33**, 129–130.

Meier-Weniger, E. (1977a) Die Morphogenese der Blüte von *Pedicularis recutita* L. Ber. Dt. Bot. Ges. **90**, 67–75.

– (1977b) Untersuchungen zur Entwicklungsgeschichte der Blüten von *Pedicularis foliosa* L. und *P.recutita* L. (Scrophulariaceae). Diss. Basel.

Melville, R. (1960) A new theory of the Angiosperm flower. Nature **188**, 14–18.

– (1962) A new theory of the Angiosperm flower: I. The Gynoeceum. Kew Bull. **16**, 1–50.

– (1963) A new theory of the Angiosperm flower: II. The Androeceum. Kew Bull. **17**, 1–63.

Merxmüller, H. & Leins, P. (1971) Zur Entwicklungsgeschichte männlicher Begonienblüten. Flora **160**, 333–339.

*Meyer, E. (1819) Junci generis monographiae specimen. Diss. Göttingen.

Meyer, V. G. (1966) Flower abnormalities. Bot. Review **32**, 165–218.

Mez, C. (1889) Morphologische Studien über die Familie der Lauraceen. Verh.bot. Ver. Prov. Brandenburg **30**, (1888), 1–31.

– (1890) Morphologische und anatomische Studien über die Gruppe der Cordieae. Bot. Jb. **12**, 526–588.

Michaelis, P. (1924) Blütenmorphologische Untersuchungen an der Euphorbiaceae. Bot. Abhandl. (Ed. K. Goebel) **3**, 1–50.

Mildbraed, J. (1954) Die Schausamen von *Paeonia corallina* Retz. Ber. Dt. Bot. Ges. **67**, 73–74.

Moeliono, B. M. (1971) Caulinary or carpellary placentation among Dicotyledons. Part I Text, Part II Plates. Assen (Netherlands).

Molisch, H. (1920) Über den Wasserkelch der Blütenknospe von *Aconitum variegatum* L. Ber. Dt. Bot. Ges. **38**, 341–346.

Morf, E. (1950) Vergleichend-morphologische Untersuchungen am Gynoeceum der Saxifragaceen. Ber. Schweiz. Bot. Ges. **60**, 516–590.

Moseley, M. F. (1972a) Morphological studies of Nymphaeaceae. VI. Development of flower of *Nuphar*. Phytomorph. **21**, 253–283.

– (1972b) Some thoughts of a phylogenetic anatomist on the evolution of the flower. I. Adv. Pl. Morph. **1972**, 394–407.

Müller, F. (1883) Two kinds of stamens with different functions in the same flower. Nature **27**, 364–365.

Müller, H. (1873) Die Befruchtung der Blumen durch Insekten und die gegenseitige Anpassung beider. Leipzig.

Müller, Luise. (1893) Grundzüge einer vergleichenden Anatomie der Blumenblätter. Nova Acta Akad. Leop. -Carol., Halle, **59**, 3.

Müller, Leopoldine. (1954) Zur Biotechnik der Blüte von *Dicentra spectabilis*. Österr. Bot. Z. **101**, 221–235.

Müller, P. (1955) Verbreitungsbiologie der Blütenpflanzen. Veröff. Geobot. Inst. Rübel, Zürich,**30**.

Müller-Doblies, D. (1972) *Galanthus* ist doch sympodial gebaut! Ber. Dt. Bot. Ges. **84**, 665–682.

– (1977) Über den geometrischen Zusammenhang der monochasial Verzweigungen am Beispiel einiger Liliifloren. Ber. Dt. Bot. Ges. **90**, 351–362.

*– (1980) Notes on the inflorescence of Agapanthus. Plant Life. The Amaryllis Yearbook, 72–76. La Jolla, California.

*Müller-Doblies, D. & U. (1978) Studies on tribal systematics of Amaryllidoideae 1. The systematic position of Lapiedra Lag. Lagascalia **8**, 13–23.

*Müller-Doblies, U. (1968) Über die Blütenstände und Blüten sowie zur Embryologie von Sparganium. Bot. Jahrb. Syst. **89**, 359–450.

Müller-Doblies, U. & D. (1977) Ordnung Typhales, in Hegi, G., Illustrierte Flora von Mitteleuropa, 3. Edn. II/1 (part 4).

*Müller-Doblies, U., Albert, G. & Müller-Doblies, D. (1975) Der Blütenstand von Euphorbia fulgens Karw. ex Klotzsch und seine variablen Größen. Bot. Jahrb. Syst. **96**, 290–323.

*– & Weberling, F. (1984) Über Prolepsis und verwandte Begriffe. Beitr. Biol. Pflanzen **59**, 121–144.

Müller-Hoefs, E. (1944) Die Nervatur der Nieder- und Hochblätter. Bot. Arch. **45**, 1–92.

*Müller-Schneider, P. (1977). Verbreitungsbiologie (Diasporologie) der Blütenpflanzen. 2. Aufl. Veröff.d. Geobotan. Inst.d. Eidgenöss. Techn. Hochsch., Stiftung Rübel, in Zürich. **61**. Heft.

*Murbeck, S. (1901) Über einige amphicarpe nordwestafrikanische Pflanzen. Kongl. Vetenskaps.-Akademiens Förhandlingar **58**, 459–571.

– (1918) Über staminale Pseudapetalie und deren Bedeutung für die Frage nach der Herkunft der Blütenkrone. Lunds Univ. Arsskr. N. F. Avd. Vol.2, **14**, no. 25.

Naegeli, C. (1884) Mechanisch-physiologische Theorie der Abstammungslehre. Munich &Leipzig.

Napp-Zinn, K. (1951) Anatomische und morphologische Untersuchungen an den Involucral-und Spreublättern der Compositen. Österr. Bot. Z. **98**, 142–170.

– (1956) Beiträge zur Anatomie und Morphologie der Involucral- und Spreublätter der Compositen. Bot. Studien **6**, Jena.

Narayana, L. L & Rao, D. (1976a) Contributions to the floral anatomy of Humiriaceae 4. J. Jap. Bot. **51**, 12–15.

– & Rao, D. (1976b) Contributions to the floral anatomy of Linaceae 6. J. Jap. Bot. **51**, 92–97.

*Nelson, E. (1965) Zur organophyletischen Natur des Orchideenlabellums. Bot. Jahrb. Syst. **84**, 175–214.

*– (1967) Das Orchideenlabellum ein Homologon des einfachen medianen Petalums der Apostasicaceen oder ein zusammengesetztes Organ? Bot. Jahrb. Syst. **87**, 22–35.

Němejc, F. (1956) On the problem of the origin and phylogenetic development of the Angiosperms. Acta. Mus. Nat. Prague **12B**, 64–144.

*Netolitzky, F. (1926) Anatomie der Angiospermen-Samen. In: K. Linsbauer, Handb.d. Pflanzenanatomie II/2, 10, 1–364.

Neubauer, H. F. (1964) Die Dschungelkrähe, *Corvus macrorhynchus* Wagler, als Bestäuber der Blüten von *Bombax malabricum* DC. Ber. Dt. Bot. Ges. **77**, 219–223.

*– (1971) The development of the achene of *Polygonum pennsylvanicum*: Embryo, endosperm and pericarp. Amer. J. Bot. **58**, 655–664.

Neumayer, H. (1924) Die Geschichte der Blüte. Abh. Zool. Bot. Ges. Wien **14**, 1–110.

Nishino, E. (1976) Developmental anatomy of foliage leaves, bracts, calyx and corolla in *Pharbitis nil*. Bot. Mag. Tokyo, **89**, 191–209.

*– (1978) Corolla tube formation in four species of Solanaceae. Bot. Mag. Tokyo, **91**, 262–277.

*– (1982) Corolla tube formation in six species of Apocynaceae. Bot. Mag. Tokyo, **95**, 1–17.

*– (1983a) Corolla tube formation in the Tubiflorae and Gentianales. Bot. Mag. Tokyo **96**, 223–243.

*– (1983b) Corolla tube formation in the Primulaceae and Ericales. Bot. Mag. Tokyo **96**, 319–342.

Nordhagen, R. (1937) Studien über die monotypische Gattung *Calluna* Salisb. I. Bergens Museum Arbok 1937, Naturvid. Rekke no. 4.

Norris, T. (1941) Torus anatomy and nectary characteristics as phylogenetic criteria in the Rhoeadales. Amer. J. Bot. **28**, 101–113.

*Oostenink, W. J. (1976) Ovary wall and pericarp ontogeny in *Lychnis alba* (Caryophyllaceae). Proc. Iowa Academy of Sciences **83**, 55–62.

Ornduff, R. & Dulberger, R. (1978) Floral enantiomorphy and the reproductive system of *Wachendorfia paniculata* (Haemodoraceae). New Phytol. **80**, 427–434.

Osche, G. (1979) Zur Evolution optischer Signale bei Blütenpflanzen. Biologie in unserer Zeit **9**, 161–170.

*– (1983) Optische Signale in der Coevolution von Pflanze und Tier. Ber. Deutsch. Bot. Ges. **96**, 1–27.

Panchaksharappa, M. G. & Rudramuniyappa, C. K. (1976) Cytochemical studies of style, stigma, pollen and pollen tube in *Panicum miliaceum* Linn. J. Karnatak Univ.-Science **21**, 218–221. (Dharwar).

Pankow, H. (1959) Histogenetische Untersuchungen an der Plazenta der Primulaceen. Ber. Dt. Bot. Ges. **72**, 111–122.

– (1962) Histogenetische Studien an den Blüten einiger Phanerogamen. Bot. Studien **13**, Jena.

– (1966) Histogenetische Untersuchungen an den Blüten einiger *Oenothera*-Arten. Flora, Abt. B. **156**, 122–132.

Parkin, J. (1914) The evolution of inflorescence. J. Linn. Soc., Bot. **42**, 511–563.

– (1951) The protrusion of the connective beyond the anther and its bearing on the evolution of the stamen. Phytomorph. **1**, 1–8.

Pascher, A. (1910a) Über einen Fall weitgehender, postnuptialer Kelchvergrößerung bei einer Solanacee. Flora **101**, 268–273.

– (1910b) Über Gitterkelche, einen neuen biologischen Kelchtypus der Nachtschattengewächse. Flora **101**, 273–278, Pl. III.

– (1959, posthumous) Zur Blütenbiologie einer aasblumigen Liliacee und zur Verbreitungsbiologie abfallender geflügelter Kapseln. Flora **148**, 153–178.

– (1960, posthumous) Über die Wasserkelche von *Datura* und *Anisodus* und über das Vorkommen freien Wassers in den Bälgen von *Paeonia*. Flora **148**, 517–528.

Pauze, F. & Sattler, R. (1978) L'androcée centripète d'*Ochna atropurpurea*. Canad. J. Bot. **57**, 2500–2511.

– & Sattler, R. (1979) La placentation axillaire chez *Ochna atropurpurea*. Canad. J. Bot. **57**, 100–107.

Pax, F. (1890) Allgemeine Morphologie der Pflanzen mit besonderer Berücksichtigung der Blütenmorphologie. Stuttgart.

– & Hoffmann, K. (1934) Caryophyllaceae, in: Engler, A. & Prantl, K.: Die natürlichen Pflanzenfamilien, 2. Aufl. **16c**, 275–364. Berlin.

Payer, J. B. (1857) Traité d'Organogénie comparée de la fleur. Paris.

Payne, W. W. & Seago, J. L. (1968) The open conduplicate carpel of *Akebia quinata* (Berberidales: Lardizabalaceae). Amer. J. Bot. **55**, 575–581.

Percival, M. S. (1965) Floral biology. Oxford–London–Edinburgh–New York–Paris–Frankfurt.

Pfeffer, W. (1872) Zur Blütenentwicklung der Primulaceen und Ampelideen. Jb.wiss. Bot. (Ed. N. Pringsheim) **8**, 194–215, Pls. XIX–XXII.

Philipson, W. R. (1947) Studies in the development of the inflorescence. III. The thyrse of *Valeriana officinalis* L. Ann. Bot. **11**, 409–416, Pl. III.

van der Pijl, L. (1930) Uit het leven van enkele gevoelige tropische bloemen, speciaal van de "horlogebloemen". De Tropische Natuur, 190–196.

– (1936) Fledermäuse und Blumen. Flora **131**, 1–40.

– (1941) Flagelliflory and cauliflory as adaptations to bats in *Mucuna* and other plants. Ann. Bot. Gard. Buitenzorg **51**, 83–93.

– (1949) Flagelliflory and cauliflory as adaptations to bats in Mucuna and other plants. Ann. Bot. Gard. Buitenzorg **51**, 83–93. Leiden.

– (1952) The stamens of *Ricinus*. Phytomorph. **2**, 130–132.

– (1955) Some remarks on myrmecophytes. Phytomorph. **5**, 190–200.

– (1956) Remarks on pollination by bats in the genera *Freycinetia, Duabanga* and *Haplophragma*, and on chiropterophily in general. Acta Bot. Neerlandica **5**, 135–144.

– (1957) The dispersal of plants by bats (Chiropterochory). Acta Bot. Neerlandica **6**, 291–315.

– (1960) Ecological aspects of flower evolution. I. Phyletic evolution. Evolution **14**, 403–416.

– (1961) Ecological aspects of flower evolution. II. Zoophilous flower classes. Evolution **15**, 44–59.

– (1969) Evolutionary action of tropical animals on the reproduction of plants. Biol. J. Linn. Soc. London **1**, 85–96.

– (1972) Principles of Dispersal in Higher Plants. 2nd ed. (3rd ed. 1982). Berlin–Heidelberg–New York.

Pilger, R. (1921) Bemerkungen zur phylogenetischen Entwicklung der Blütenstände. Bericht d. Freien Vereinigung f. Pflanzengeographie und systemat. Botanik für das Jahr 1919, 69–77, Berlin.

– (1922) Ueber Verzweigung und Blütenstandsbildung bei den Holzgewächsen. Bibl. Bot. Heft, Stuttgart, **90**.

– (1933) Bemerkungen über Anthokladien und Infloreszenzen. Bot. Jb. **65**, 75–96.

Planchon, J. E. & Triana, J. (1863) Sur les bractées des Marcgraviées. Mém. Soc. imp. d. sci. natur. Cherbourg **9**, 69–88.

Plantefol, L. (1948) L'ontogénie de la fleur. Ann. Sci. Nat., Bot. Sér. **11** (9), 33–186.

– (1949) L'ontogénie de la fleur. Fondements d'une théorie florale nouvelle. Paris.

Porsch, O. (1905) Beiträge zur "histologischen Blütenbiologie". I. Über zwei neue Insektenanlockungsmittel der Orchideenblüte. Österr. Bot. Z. **55**, 165–173, 227–235, 253–260, Pl. III & IV.

– (1906) Beiträge zur "histologischen Blütenbiologie". II. Weitere Untersuchungen über Futterhaare. Österr. Bot. Z. **56**, 41–47, 88–95, 135–143, 176–180, Pl. III.

– (1923) Blütenstände als Vogelblumen. Österr. Bot. Z. **73**, 125–149.

– (1931) *Crescentia* – eine Fledermausblume. Österr. Bot. Z. **80**, 31–44.

– (1932) Das Problem Fledermausblume. Anz. Akad. Wiss. Wien, math.naturw. Kl. **3**, 27–28.

– (1934–36) Säugetiere als Blumenausbeuter und die Frage der Säugetierblume I–III. Biol.gen. **10**, 657–685, **11**, 171–188, **12**, 1–8.

– (1938–39) Das Bestäubungsleben der Kakteenblüte. I und II. Cactaceae. Jahrbücher d. Dt. Kakteenges. **1938**, 1–94, **1939**, 95–142.

– (1941) Ein neuer Typus Fledermausblume. Biol. Gen. **15**, 283–294.

– (1957) Alte Insektentypen als Blumenausbeuter. Österr. Bot. Z. **104**, 115–164.

*Porter, C. L. (1967) Taxonomy of flowering plants. 2nd ed. San Francisco & London.

Posluszny, U. & Sattler, R. (1973) Floral development of *Potamogeton densus*. Can. J. Bot. **51**, 647–656.

– & Sattler, R. (1974) Floral development of *Potamogeton richardsonii*. Amer. J. Bot. **61**, 209–216.

Prantl, K. (1887) Beiträge zur Morphologie und Systematik der Ranunculaceen. Bot. Jb. Syst. **9**, 225–273.

Proctor, M. & Yeo, P. (1973) The pollination of flowers. Collins, London.

Pryor, L. D. & Knox, R. B. (1971) Operculum development and evolution in eucalypts. Aust. J. Bot. **19**, 143–172.

Puri, V. (1941) Studies in floral anatomy. Part I. Gynaeceum Constitution in the Cruciferae. Proc. Ind. Acad. Sci., Sect. B, **14**, 166–187.

★– (1942) Studies in floral anatomy. Proc. Ind. Acad. Sci., Sect. B, **8**, 71–88.

– (1947) Studies in floral anatomy. IV. Vascular anatomy of the flower of certain species of the Passifloraceae. Amer. J. Bot. **34**, 562–573.

– (1948) Studies in floral anatomy. V. On the structure and nature of the corona in certain species of the Passifloraceae. J. Ind. Bot. Soc. **27**, 130–149.

– (1950) Studies in floral anatomy. VI. Vascular anatomy of the flower of *Crataeva religiosa* Forst., with special reference to the nature of the carpels in the Capparidaceae. Amer. J. Bot. **37**, 363–370, 1950.

– (1951) The role of floral anatomy in the solution of morphological problems. Bot. Rev. **17**, 471–553.

– (1952a) Placentation in Angiosperms. Bot. Rev. **18**, 603–651.

– (1952b) Floral anatomy and inferior ovary. Phytomorph. **2**, 122–129.

– (1954) Studies in floral anatomy. VII. On placentation in the Cucurbitaceae. Phytomorph. **4**, 278–299.

– (1961) The classical concept of angiosperm carpel: a reassessment. J. Ind. Bot. Soc. **40**, 511–524.

– (1962) Floral anatomy in relation to taxonomy. Bull. Bot. Surv. India **4**, 161–165.

– (1963) On the concept of carpellary margins. Proc. Summer School Bot., Darjeeling, 326–333. (Ed. Maheshwari, P. et al.) New Delhi.

– (1964) On the relation between ovule and carpel. J. Ind. Bot. Soc. **42A**, 189–198.

– (1966) Studies in floral anatomy. VIII. Vascular anatomy of the flower of certain species of the Asclepiadaceae with special reference to corona. Agra Univ. J. Res. Sci. **15**, II, 189–216.

– (1967) The origin and evolution of Angiosperms. J. Ind. Bot. Soc. **46**, 1–14.

– (1970) Anther sacs and pollen grains: some aspects of their structure and function. J. Palynol. **6**, 1–17.

– (1971) The Angiosperm ovule. Proc.57th Session of the Indian Science Congress, Kharagphur 1970. Part II: Presidential Address, Sect. VI: Botany, 1–36.

– (1978) On some peculiarities of the Angiosperm carpel. Acta Bot. Ind. **6** (Suppl.), I–XIV.

– & Agarwal, R. M. (1976) On accessory floral organs. J. Ind. Bot. Soc. **55**, 95–114.

Raciborski, M. (1895) Die Schutzvorrichtung der Blütenknospen. Flora **81** (Ergänzungsband), 151–194.

Radlkofer, L. (1891) Über die Gliederung der Familie der Sapindaceen. Sitzber. K. Akad. Wiss. München, math.–physikal. Cl., **20**, 105.

★Radtke, F. (1926) Anatomisch-physiologische Untersuchungen an Blütennectarien. Planta **1**, 379–418.

★Rao, P. R. M. (1983) Seed and fruit anatomy in *Eupomatia laurina* with a discussion on the affinities of Eupomatiaceae. Flora **173**, 311–319.

★Rao, V. S. (1971) The disk and its vasculature in the flowers of some dicotyledons. Bot. Notis. **124**, 442–450.

Rauh, W. (1970 & 1973) Bromelien, Vol.2. Stuttgart.

Rauh, W. & Reznik, H. (1951) Histogenetische Untersuchungen an Blüten- und Infloreszenzachsen I. Teil. Die Histogenese becherförmiger Blüten- und Infloreszenzachsen, sowie der Blütenachse einiger Rosoideen. Sitzungsber. Heidelberg. Akad. Wiss., Math.-nat. Kl. No. 3.

Reiche, E. (1891) Über nachträgliche Verbindungen frei angelegter Pflanzenorgane. Flora **74**, 435–444.

Reinsch, J. (1927) Über die Entstehung der Ästivationsformen von Kelch und Blumenkrone dikotyler Pflanzen und über die Beziehungen der Deckungsweisen zur Gesamtsymmetrie der Blüte. Flora **21**, 78–124.

Reznik, H. (1956) Untersuchungen über die physiologische Bedeutung der chymochromen Farbstoffe. Sitzber. Akad. Wiss. Heidelb., math.-naturw. Kl., Jg. 1956, **2**.

Richter, S. (1929) Über den Öffnungsmechanismus der Antheren bei einigen Vertretern der Angiospermen. Planta **8**, 154–184.

Rickett, H. W. (1944) The classification of inflorescences. Bot. Rev. **10**, 187–231.

– (1955) Materials for a dictionary of botanical terms. III. Inflorescences. Bull. Torr. Bot. Club **82**, 419–445.

Ridley, H. N. (1930) The dispersal of plants throughout the world. London.

Ritterbusch, A. (1971) Das Phänomen der Morphogenese im Bereich der Blüte, dargestellt an der Entwicklung der Krone von *Calceolaria scabiosifolia.* Sims. Diss. Basel.

– (1976) Die Organopoïëse der Blüte von *Calceolaria tripartita* R. et P. (Scrophulariaceae). Bot. Jb. Syst. **95**, 267–320.

*Robertson, R. E. & Tucker, S. C. (1979) Floral ontogeny of *Illicium floridanum*, with emphasis on stamen and carpel development. Amer. J. Bot. **66**, 605–617.

Roeper, J. A. C. (1826) Observationes aliquot in florum inflorescentiarumque naturam. Linnaea **1**, 433–466.

– (1849) Über den Blütenstand einiger Ranunculaceen. Bot. Ztg. **7**, 401–448.

– (1872) Botanische Thesen. Rostock.

Rohweder, O. (1959) Über verlaubte Blüten von *Barbarea vulgaris* R. Br. und ihre morphologische Bedeutung. Flora **148**, 255–282.

– (1963) Anatomische und histogenetische Untersuchungen an Laubsprossen und Blüten der Commelinaceen. Bot. Jb. **88**, 1–99, Pls. I-XII.

– (1965a) Centrospermen-Studien. 1. Der Blütenbau bei *Uebelinia kiwuensis* T. C. E. Fries (Caryophyllaceae). Bot. Jb. **83**, 406–418.

– (1965b) Centrospermen-Studien. 2, Entwicklung und morphologische Deutung des Gynöciums bei *Phytolacca.* Bot. Jb. **84**, 509–526.

– (1967a) Centrospermen-Studien. 3. Blütenentwicklung und Blütenbau bei Silenoideen. (Caryophyllaceae). Bot. Jb. **86**, 130–185.

– (1967b) Karpellbau und Synkarpie bei Ranunculaceen. Ber. Schweiz. Bot. Ges. **77**, 376–432, Pl. I-IV.

– (1969) Beiträge zur Blütenmorphologie und -anatomie der Commelinaceen mit Anmerkungen zur Begrenzung und Gliederung der Familie. Ber. Schweiz. Bot. Ges. **79**, 199–220, Pls. I & II.

– (1970) Centrospermen-Studien. 4. Morphologie und Anatomie der Blüten, Früchte und Samen bei Alsinoideen und Paronychioideen s. lat. (Caryophyllaceae). Bot. Jb. **90**, 201–271.

– (1972) Das Andröcium der Malvales und der "Konservatismus" des Leitgewebes. Bot. Jb. Syst. **92**, 155–167.

– & Huber, K. (1974) Centrospermen-Studien. 7. Beobachtungen und Anmerkungen zur Morphologie und Entwicklungsgeschichte einiger Nyctaginaceen. Bot. Jb. **94**, 327–359.

– & König, K. (1971) Centrospermen-Studien. 5. Bau der Blüten, Früchte und Samen von *Pteranthus dichotomus* Forsk. (Caryophyllaceae). Bot. Jb. **90**, 447–468.

– & Treu-Koene, E. (1971) Bau und morphologische Bedeutung der Infloreszenz von *Houttuynia cordata* Thunb. (Sauraceae). Vierteljahrschrift d. Naturf. Ges. Zürich Jg. **116**, 195–212.

– & Urmi, E. (1978) Centrospermen-Studien. 10. Untersuchungen über den Bau der Blüten und Früchte von *Cucubalus baccifer* L. und *Drypis spinosa* L. (Caryophyllaceae-Silenoideae). Bot. Jb. **100**, 1–25.

– & Urmi-König, K. (1975) Centrospermen-Studien. 8. Beiträge zur Morphologie, Anatomie und systematischen Stellung von *Gymnocarpos* Forsk. und *Paronychia argentea* Lam. (Caryophyllaceae). Bot. Jb. **96**, 375–409.

Rosen, W. G. & Thomas, H. R. (1970) Secretory cells of lily pistils. I. Fine structure and function. Amer. J. Bot. **57**, 1108–1114.

Roth, I. (1957a) Zur Histogenese der dorsalen "Ligula" von *Thalictrum.* Österr. Bot. Z. **104**, 165–172.

– (1957b) Das Dorsalmeristem der Deckblätter von *Beta trigyna.* Flora **144**, 635–646.

– (1959) Histogenese und morphologische Deutung der Kronblätter von *Primula.* Bot. Jb. **79**, 1–16.

– (1962a) Histogenese und morphologische Deutung der basilären Plazenta von *Armeria.* Österr. Bot. Z. **109**, 18–40.

- (1962b) Histogenese und morphologische Deutung der basalen Plazenta von *Herniaria*. Flora **152**, 179–195.
- (1977) Fruits of Angiosperms. Stuttgart.
*Rübsamen, T. (1986) Morphologische, embryologische und systematische Untersuchungen an Burmanniaceae und Corsiaceae (mit Ausblick auf die Orchidaceae-Apostasioideae). Dissertationes Botanicae, **92**. Cramer, Berlin-Stuttgart.
Sahni, B. (1921) On the structure and affinities of *Acmopyle Pancheri* Pilger. Philos. Trans. B **CCX**, 253–310.
de Saint-Hilaire, A. (1840) Lexicon de Botanique. Paris.
*Sampson, F. B. & Tucker, S. C. (1978) Placentation in *Exospermum stipitatum* (Winteraceae). Bot. Gaz., **139**, 215–222.
Samuelsson, G. (1914) Über die Pollenentwicklung von *Anona* und *Aristolochia* und ihre systematische Bedeutung. Svensk Bot. Tidskr. **8**, 181–189.
*Sands, M. J. S. (1973) New aspects of the floral vascular anatomy in some members of the Order Rhoeadales sensu Hutch. Kew Bulletin **28**, 211–256.
Sandt, W. (1925) Zur Kenntnis der Beiknospen. Bot. Abh. (Ed. K. Goebel) **7**.
Sarkany, S. & Kovacs, A. (1971) Einzelheiten in der Organisierung des Gynoeceums bei einigen Umbelliferae. Acta Botan. Acad. Sci. Hungar., **17**, 419–437.
Sassen, M. M. A. (1973) Submikroskopische Morphologie des Griffelleitgewebes. Acta Bot. Neerl. **22**, 254.
Sattler, R. (1962) Zur frühen Infloreszenz- und Blüten-entwicklung der Primulales sensu lato mit besonderer Berücksichtigung der Stamen-Petalum-Entwicklung. Bot. Jb. **81**, 358–396.
- (1965) Perianth development of *Potamogeton richardsonii*. Amer. J. Bot. **52**, 35–41.
- (1972) Centrifugal primordial inception in floral development. In Adv. Pl. Morph. 1972, Professor Puri Commem. Vol., 170–178, Surita Prakashan. Meerut, India.
- (1973) Organogenesis of flowers. Univ. of Toronto Press, Toronto.
- (1974) A new approach to gynoecial morphology. Phytomorph. **24**, 22–34.
- (1977) Kronröhrenentstehung bei *Solanum dulcamara* L. und "kongenitale Verwachsung". Ber. Dt. Bot. Ges. **90**, 29–38.
- (1978) 'Fusion' and 'continuity' in floral morphology. Notes from the Royal Botanic Garden, Edinburgh **36**, 397–405.
*- & Perlin, L. (1982) Floral development of *Bougainvillea spectabilis* Wolld., *Boerhavia diffusa* L. and *Mirabilis jalapa* L. (Nyctaginaceae). J. Linn. Soc. (Botany) **84**, 161–182.
- & Singh, V. (1973) Floral development of *Hydrocleis nymphoides*. Canad. J. Bot. **51**, 2455–2458.
- & Singh, V. (1977) Floral organogenesis of *Limnocharis flava*. Canad. J. Bot. **55**, 1076–1086.
- & Singh, V. (1978) Floral organogenesis of *Echinodorus amazonicus* Rataj and floral construction of the Alismatales. Bot. J. Linn. Soc. **77**, 141–156.
Saunders, E. R. (1925, -27, -29) On carpel polymorphism. I.-III. Ann. Bot. **39**, 123–167, 1925; **41**, 569–627, 1927; **43**, 459–481, 1929.
- (1934) Comments on "floral anatomy and its morphological interpretation". New Phytol. **33**, 127–170.
Schaefer, H. (1942) Die Hohlschuppen der Boraginaceen. Bot. Jb. **72**, 304–346.
Schaeppi, H. (1937a) Vergleichend-morphologische Untersuchungen am Gynoeceum der Primulaceen. Z.ges. Naturw. **7**, 239–259.
- (1937b) Zur Morphologie des Gynoeceums der Phytolacaceen. Flora **131**, 41–59.
- (1937c) Vergleichend-morphologische Untersuchungen am Gynoeceum der Resedaceen. Planta **26**, 470–490.
- (1939) Vergleichend-morphologische Untersuchungen an den Staubblättern der Monocotyledonen. Nova Acta Leop. N. F. **6**, 389–447.
- (1942) Morphologische und entwicklungsgeschichtliche Untersuchungen an den Blüten von *Thesium*. Mitteil. Naturw. Ges. Winterthur **23**, 41–61.
- (1951) Morphologische Untersuchungen am Gynoeceum der Steinobstgewächse. Mitt. Naturwiss. Ges. Winterthur **26**, 27–53.
- (1953a) Morphologische Untersuchungen an den Karpellen der Calycanthaceae. Phytomorph. **3**, 112–118.
- (1953b) Kelch und Außenkelch von *Rhodotypus kerrioides*. Kleiner Beitrag zur

morphologischen Wertigkeit der Blütenblätter. Arbeiten aus d. Institut f.allgem. Botanik an d. Univ. Zürich Ser. A. Nr. **6**.

- (1954) Untersuchungen über die Anzahl der Früchtblätter bei den Spierstrauchgewächsen. Mitteil. Naturw. Ges. Winterthur **27**, 1–8.
- (1957) Blütenmorphologische Untersuchungen an *Calianthemum rutifolium* (L.) C. A. Meyer. Phyton (Horn, N.-Ö.) **7**, 228–240.
- (1958) Untersuchungen über die Anzahl der Blümenblätter bei einigen Liliaceen. Bot. Jb. **78**, 119–128.
- (1967) Vergleichend-morphologische Untersuchungen an der Blütenachse und am Gynoeceum einiger Spiraeoideen (Spierstrauchgewächse). Mitteil. Naturw. Ges. Winterthur **32**, 73–104.
- (1970) Untersuchungen über den Habitus von *Aruncus, Astilbe* und einiger ähnlicher Pflanzen **46**, 371–387.
- (1971) Zur Gestaltung des Gynoeceums von *Pittosporum tobira*. Ber. Schweiz. Bot. Ges. **81**, 40–51.
- (1972) Über die Gestaltung der Karpelle von *Caltha palustris* und *Trollius europaeus*. Vierteljahrsschrift Naturforsch. Ges. Zürich Jg. **117**, 101–113.
- (1974) Vergleichend-morphologische Untersuchungen am Gynoeceum einiger Juncaceen. Vierteljahrsschrift Naturforsch. Ges. Zürich Jg. **119**, 225–238.
- (1975) Über einfache Karpelle. Bot. Jb. Syst. **96**, 410–422.
- (1976) Über die männlichen Blüten einiger Menispermaceen. Beitr. Biol. Pflanzen **52**, 207–215.
- (1977) Über den "doppelten Fruchtknoten" von *Rhodotypos*. Beitr. Biol. Pflanzen **53**, 165–175.
- & Frank, K. (1962) Vergleichend-morphologische Untersuchungen über die Karpellgestaltung, insbesondere die Plazentation bei Anemoneen. Bot. Jb. **81**, 337–357.
- & Frank, K. (1967) Vergleichend-morphologische Untersuchungen an der Blütenachse und am Gynoeceum einiger Spiraeoideen (Spierstrauchgewächse). Mitt. Naturw. Ges. Winterthur **32**, 73–104.

Schick, B. (1980) Untersuchungen über die Biotechnik der Apocynaceenblüte. I. Morphologie und Funktion des Narbenknopfes. Flora **170**, 394–432.

Schill, R. & Jäkel, U. (1978) Beitrag zur Kenntnis der Asclepiadaceen-Pollinarien. Reihe trop. u.subtrop. Pflanzenwelt **22**, Abh. Akad. Wiss. Lit. Mainz, math.-naturw. Kl.

Schlagorsky, M. (1949) Das Bauprinzip des Primulaceengynözeums bei der Gattung *Cyclamen*. Österr. Bot. Z. **96**, 361–368.

Schlitter, J. (1965) Sind die Luzuriagoideen wirkliche Liliaceen oder haben die Ericales und Ternstroemiales organphyletisch und stammesgeschichtlich Beziehungen zu primitiven Liliifloren? Ber. Schweiz. Bot. Ges. **75**, 96–109.

Schmid, R. (1972a) Floral bundle fusion and vascular conservatism. Taxon **21**, 429–446.
 * – (1972b) Floral anatomy of Myrtaceae. Bot. Jahrb. Syst. **92**, 433–489.
- (1975) Two hundred years of pollination biology: an overview. The Biologist **57**, 26–35.
- (1976) Filament histology and anther dehiscence. J. Linn. Soc. (Bot.), **73**, 303–315.
- (1977) Functional and ecological interpretation of floral and fruit anatomy. Bot. Rev. **43** (in press).

Schnarf, K. (1929) Embryologie der Angiospermen. Handbuch d. Pflanzenanatomie (Ed. K. Linsbauer), Bd. X/2. Berlin.

*Schnepf, E. (1969) Sekretion und Exkretion bei Pflanzen. Protoplasmatologia VIII, 8.
- Witzig, F. & Schill, R. (1979) Über Bildung und Feinstruktur des Translators der Pollinarien von *Asclepias curassavica* und *Gomphocarpus fruticosus*. Trop. u. subtrop. Pflanzenwelt **25**, Mainz.

Schnetter, R., Hilger, H. H. & Richter, U. (1979) Über den Bau des Perikarps der hygrochastischen Hülse von *Haematoxylon brasiletto* (Caesalpinaceae, Fabales). Bot. Jb. Syst. **101**, 135–142.

Schniewind-Thies, I. (1897) Beiträge zur Kenntnis der Septalnektarien. Jena.

*Schoch-Bodmer, H. (1930) Zur Heterostylie von *Fagopyrum esculentum*: Untersuchungen über das Pollenschlauchwachstum und über die Saugkräfte der Griffel und Pollenkörner. Berichte d. Schweizer. Bot. Ges. **39**, 4–15.

Schoute, J. C. (1935) On corolla aestivation and phyllotaxis of floral phyllomes. Verhandl. Kon. Akad. Wetensch. Amsterdam, Afd. Naturkunde 2. Sect. Deel **34**, No.4.

Schröder, H. (1934) Untersuchungen über die Beeinflussung des Blütenfarbmusters von *Petunia hybrida grandiflora* hort. Diss. Göttingen.

*Schultz-Schultzenstein (1863) Die morphologische Gesetze der Blumen-Bildung und das natürliche System der Morphologie der Blumen. Flora **21**, 13–16, 25–29, 59–63, 105–112, 118–125.

Schultze-Motel, W. (1959) Entwicklungsgeschichtliche und vergleichend-morphologische Untersuchungen im Blütenbereich der Cyperaceae. Bot. Jb. Syst. **78**, 129–170.

Schumann, K. (1889) Beiträge zur Kenntnis der Monochasien. Sitz. Ber. Akad. Wiss. Berlin **30**, 555–584.

– (1890) Neue Untersuchungen über den Blütenanschluß. Leipzig.

*Seibert, J. (1978) Fruchtanatomische Untersuchungen an Lithospermae (Boraginaceae). Diss. Botanicae **44**, J. Cramer, Vaduz.

Sell, Y. (1964) Les complexes inflorescentiels de quelques Acanthacées. Étude particulière des phénomènes de condensation, de racémisation, d'homogénéisation et de troncature. Ann. Sci. Nat., Bot. 12. Sér. **10**, 225–300.

– (1969) Les complexes inflorescentiels de quelques Acanthacées. Étude particuliére des phénoménes de condensation, de racémisation, d'homogénéisation et de troncature. Ann. Sci. Nat., Bot. **12**. Sér **10**, 225–300.

– (1976) Tendances évolutives parmi les complexes inflorescentiels. Rev.gén. Bot. **83**, 247–267.

– (1980) Physiological and phylogenetic significance of the direction of flowering in inflorescence complexes. Flora **169**, 282–294.

Sernander, R. (1906) Entwurf einer Monographie der europäischen Myrmecochoren. Uppsala.

Seybold, A. (1954) Untersuchungen über den Farbwechsel von Blumenblättern, Früchten und Samenschalen. Sitzber. Akad. Wiss. Heidelberg, math.-naturw. Kl. Jg. 1953/54, **2**.

Sigmond, H. (1930) Vergleichende Untersuchungen über die Anatomie und Morphologie von Blütenknospenverschlüssen. Beih. Bot. Centralbl. **46** 1. Abt., 1–67.

Silberbauer-Gottsberger, I. (1973) Blüten- und Fruchtbiologie von *Butia leiospatha* (Arecaceae). Österr. Bot. Z. **121**, 171–185.

– & Gottsberger, G. (1975) Über sphingophile Angiospermen Brasiliens. Plant Syst. Evol. **123**, 157–184.

Singh, V. (1975) Placentation in Euphorbiaceae. Ann. Bot. **39**, 1137–1140.

– & Jain, D. K. (1979) Floral Organogenesis in *Adenocalymna alliaceum* (Bignoniaceae). Beitr. Biol. Pflanzen **54**, 207–213.

– & Sattler, R. (1972) Floral development of *Alisma triviale*. Canad. J. Bot. **50**, 619–627.

– & Sattler, R. (1973) Nonspiral androecium and gynoecium of *Sagittaria latifolia*. Canad. J. Bot. **51**, 1093–1095.

– & Sattler, R. (1974) Floral development of *Butomus umbellatus*. Canad. J. Bot. **52**, 223–230.

Slavíková, Z. (1968a) Zur Morphologie der Blütenhülle von Ranunculaceen I. *Anemone nemorosa* L. Preslia (Prague) **40**, 1–12.

– (1968b) Zur Morphologie der Blütenhülle von Ranunculaceen II. *Myosurus minimus* L. Preslia (Prague) **40**, 113–121.

– (1974) Zur Morphologie der Blütenhülle von *Nigella arvensis* L., *N.damascena* L. und *Ranunculus illyricus* L. Preslia (Prague) **46**, 110–117.

*Skipworth, J. P. (1983) Is the achene a reduced follicle? New Zealand J. Bot., **21**, 201–204.

*– & Philipson, W. R. (1966) The cortical vascular system and the interpretation of the *Magnolia* flower. Phytomorphology **16**, 463–469.

*Slingsby, P. & Bond, W. (1981) Ants – friends of the fynbos. Veld and Flora **67**, 39–45.

*Sobick, U. (1983) Blütenentwicklungsgeschichtliche Untersuchungen an Resedaceen unter besonderer Berücksichtigung von Androeceum und Gynoeceum. Bot. Jahrb. Syst. **104**, 203–248.

Soetiarto, S. R. & Ball, E. (1969) Ontogenetical and experimental studies of the floral apex of *Portulaca grandiflora*. 1. Histology of transformation of shoot apex into the floral apex. pp. 133–140, Pl. I-VIII; 2. Bisection of the meristem in successive stages. pp. 1067–1076, Pl. I-V. Canad. J. Bot. **47**.

Sperlich, A. (1939) Excretionsgewebe, in Handbuch d. Pflanzenanatomie (Ed. K. Linksbauer) **4**, Liefg.38, Berlin.

Sprotte, K. (1940) Untersuchungen über Wachstum und Nervatur der Fruchtblätter. Bot. Archiv. **40**, 463–506.

Stadler, S. (1886) Beiträge zur Kenntnis der Nektarien und Biologie der Blüthen. Diss. Berlin.

Staedler, G. (1923) Über Reduktionserscheinungen im Bau der Antherenwand von Angiospermen-Blüten. Flora 116, 85–108, Pls. II & III.

★Stebbins, G. L. (1970) Adaptive radiation of reproductive characteristics in angiosperms. I: pollination mechanisms. Ann. Rev. Ecol. Syst. 1, 307–326.

Steinbrinck, C. (1873) Untersuchungen über die anatomischen Ursachen des Aufspringens der Früchte. Diss. Bonn.

★Stelleman, F. (1982) De betekenis van de biotische bestuiving bij Plantago lanceolata. Diss. Faculteit der Wiskunde en Natuurwetenschappen Univ. Amsterdam.

★– (1984) The significance of biotic pollination in a nominally anemophilous plant: Plantago lanceolata. Proc. Koninkl. Nederl. Akad. Wetenschappen, Series C, 87, 95–119.

Sterling, C. (1966a) Comparative morphology of the carpel in Rosaceae VII. Pomioideae: Chaenomeles, Cydonia, Docynia. Amer. J. Bot. 53, 225–231.

– (1966b) Comparative morphology of the carpel in Rosaceae IX. Spiraeolideae: Quillajeae, Sorbarieae. Amer. J. Bot. 53, 951–960.

– (1972) Comparative morphology of the carpel in the Liliaceae: Neodregeae. Bot. J. Linn. Soc. 65, 163–171.

– (1973). Comparative morphology of the carpel in the Liliaceae: Colchiceae (Colchicum). Bot. J. Linn. Soc. 66, 213–221.

– (1977a) Comparative morphology in the carpel of the Liliaceae: tepallary and staminal vascularization in the Wurmbaeoideae. Bot. J. Linn. Soc. 74, 63–69, 2 pl.

– (1977b) Comparative morphology of the carpel in the Liliaceae: Uvularieae. Bot. J. Linn. Soc. 74, 345–354, 4 pl.

★– (1979) Comparative morphology of the carpel in the Liliaceae: Tofieldineae. Bot. J. Linn. Soc. 79, 321–332.

Stomps, T. J. (1948) Kleistogamie als mendelndes Merkmal. Rec.trav.bot.néerl. 41, 118–130.

Stopp, K. (1950a) Karpologische Studien I u. II. Abh. Akad. Wiss. Lit. Mainz, math.-nat. Kl.nr. 7.

– (1950b) Karpologische Studien III. Abh. Akad. Wiss. Lit. Mainz, math.-nat. Kl.nr. 17.

– (1952) Morphologische und verbreitungsbiologische Untersuchungen über persistierende Blütenkelche. Abh. Akad. Wiss. Lit. Mainz, math.-nat. Kl.nr. 12.

– (1954) Über die Wuchsform amphicarper Phaseolus- und Amphicarpaea Arten. Österr. Bot. Z. 101, 592–600.

– (1958a) Die kongolesischen Arten der Labiaten-Gattung Aeolanthus mit zygomorph dehiszierenden Fruchtkelchen. Beitr. Biol. Pflanzen 34, 395–399.

– (1958b) Notiz über die Dehiszenzweise der Kapselfrüchte von Genlisea hispidula Stapf. Beitr. Biol. Pflanzen 34, 401–403.

– (1958c) Die verbreitungshemmenden Einrichtungen in der südafrikanischen Flora. Bot. Studien 8.

– (1962) Antitelechore Einrichtungen bei den Gattungen Sesamum, Rogeria und Psilocaulon. Beitr. Biol. Pflanzen 37, 63–76.

Stork, H. E. (1956) Epiphyllous flowers. Bull. Torr. Bot. Club 83, 338–341.

Stroebl, F. (1925) Die Obdiplostemonie in den Blüten. Bot. Archiv 9, 210–224.

★Stützel, T. (1984) Blüten- und infloreszenzmorphologische Untersuchungen zur Systematik der Eriocaulaceen. Diss. Botanicae, 71. Cramer, Vaduz.

Suessenguth, K. (1936) Über den Farbwechsel von Blüten. Ber. Dt. Bot. Ges. 64, 409–417.

– (1938) Neue Ziele der Botanik. (Über das Vorkommen getrennter Kronblätter bei den Sympetalen, pp. 32–36; Über den Farbwechsel von Blüten, pp. 74–81). Munich-Berlin.

Swamy, B. G. L. (1949) Further contributions to the morphology of the Degeneriaceae. J. Arn. Arbor. 30, 10–38, Pl. I-IV.

Takhtajan, A. (1959) Die Evolution der Angiospermen. (German translation by W. Höppner). Jena.

Tepfer, S. S. (1953) Floral anatomy and ontogeny in Aquilegia formosa var. truncata and Ranunculus repens. Univ. Calif. Publ. Bot. 25, 513–647.

★Terabayashi, S. (1977) Studies in Morphology and Systematics of Berberidaceae I. Floral Anatomy of Ranzania japonica. Acta Phytotax. Geobot. 28, 45–57.

Thomas, H. H. (1931) The early evolution of the Angiosperms. Ann. Bot. 45, 647–672.

Thompson, J. M. (1929) Studies in advancing sterility. IV. The legume. Publ. Hartley Bot. Lab. 6.

− (1934) The state of flowering known as angiospermy. Publ. Hartley Bot. Lab. **12** (VIII).

van Tieghem, P. (1868) Recherches sur la structure du pistill. Ann. Sci. Nat. Bot. Sér. V, **9**, 127–226.

− (1875) Recherches sur la structure du pistil. Mém. Acad. Sci., Paris, **21**.

− (1918) Eléments de Botanique. I. Botanique générale. 5. Ed. Paris.

★Tiemann, A. (1985) Untersuchungen zur Embryologie, Blütenmorphologie und Systematik der Rapataceen und der Xyridaceen-Gattung *Abolboda* (Monocotyledonae). Diss. Botanicae, **82**, J. Cramer, Vaduz.

★Tiffney, B. H. (1984) Seed size, dispersal syndromes, and the rise of the Angiosperms: evidence and hypothesis. Ann. Missouri Bot. Gard. **71**, 551–576.

★Tilton, V. R. & Horner, H. T. Jr. (1980) Stigma, style and obturator of *Ornithogalum caudatum* (Liliaceae) and their function in the reproductive process. Amer. J. Bot. **67**, 1113–1131.

Trapp, A. (1954) Staubblattbildung und Bestäubungsmechanismus von *Incairvillea mirabilis* Batalin. Österr. Bot. Z. **101**, 208–219.

− (1956a) Entwicklungsgeschichtliche Untersuchungen über die Antherengestaltung sympetaler Blüten. Beitr. Biol. Pflanzen **32**, 279–312.

− (1956b) Zur Morphologie und Entwicklungsgeschichte der Staubblätter sympetaler Blüten. Bot. Studien **5**, Jena.

Treub, M. (1889) Les bourgeons floraux du *Spathodea campanulata* Beauv. Ann. Jard. Bot. Buitenzorg **8**, 38–46, pl. XIII–XV.

Troll, W. (1922) Über Staubblatt- und Griffelbewegungen und ihre teleologische Deutung. Flora **115**, 191–250.

− (Ed.) (1926a) Goethes Morphologische Schriften. E. Diederichs, Jena.

− (1926b) Über die Staubblattkrümmungen der Umbelliferen. Flora **20**, 227–242.

− (1927) Zur Frage nach der Herkunft der Blumenblätter. Flora **122**, 57–75.

− (1928a) Organisation und Gestalt im Bereich der Blüte. Monographien aus dem Gesamtgebiet der Botanik Bd. **1**, Berlin.

− (1928b) Zwei merkwürdige Fälle von Saftmalbildung. Ber. Dt. Bot. Ges. **46**, 491–498.

− (1928c) Über *Spathicarpa sagittifolia* Schott. Mit einem Anhang über die Stellung der Spathicarpeen im System der Araceen. Flora **123**, 286–316.

− (1928d) Über Antherenbau, Pollen und Pollination von *Galanthus* L. Flora **123**, 321–343.

− (1928e) Zur Auffassung des parakarpen Gynaeceums und des coenokarpen Gynaeceums überhaupt. Planta **6**, 255–276.

− (1929) *Roscoea purpurea* SM., eine Zingiberacee mit Hebelmechanismus in den Blüten. Mit Bemerkungen über die Entfaltungsbewegungen der fertilen Staubblätter von *Salvia*. Planta **7**, 1–28.

− (1931) Beiträge zur Morphologie des Gynaeceums I. Über das Gynaeceum der Hydrocharitaceen. Planta **14**, 1–18.

− (1932a) Über Diplophylle und verwandte Erscheinungen in der Blattbildung. Planta **15**, 355–406.

− (1932b) Morphologie der schildförmigen Blätter. Part I & II. Planta **17**, 153–314.

− (1932c) Beiträge zur Morphologie des Gynaeceums II. Über das Gynaeceum von *Limnocharis* Humb. et Bonpl. Planta **17**, 453–460.

− (1932d) Noch einmal über *Spathicarpa sagittifolia* Schott. Kurze Mitteilung. Planta **17**, 666–668.

− (1934a) Über Bau und Nervatur der Karpelle von *Ranunculus*. Ber. Dt. Bot. Ges. **52**, 214–220.

− (1934b) Beiträge zur Morphologie des Gynaeceums. III. Über das Gynaeceum von *Nigella* und einiger anderer Helleboreen. Planta **21**, 266–291.

− (1934c) Beiträge zur Morphologie des Gynaeceums. IV. Über das Gynaeceum der Nymphaeaceen. Planta **21**, 447–485.

− (1937–43) Vergleichende Morphologie der höheren Pflanzen (1st vol. in 3 parts). Berlin. Reprint with index 1967/71 Königstein/Ts.).

− (1949) II. Ästivationsstudien an Campanulaceenblüten, in Troll, W. & Weber, H. Morphologische und anatomische Studien an höheren Pflanzen. Sitzber. Akad. Wiss. Heidelberg **6**, 31–51.

− (1950a) Über den Infloreszenzbegriff und seine Anwendung auf die krautige Region blühender Pflanzen. Abh. Akad. Wiss. Lit. Mainz, math.-nat. Kl. Nr. **15**.

− (1950b) Botanische Notizen I. Abh. Akad. Wiss. Lit. Mainz, math.-nat. Kl. Nr. **13**.

– (1951a) Botanische Notizen II. Abh. Akad. Wiss. Lit. Mainz, math.-nat. Kl. Nr. **2**.
– (1951b) Botanische Notizen III. Abh. Akad. Wiss. Lit. Mainz, math.-nat. Kl. Nr. **3**.
– (1957) Praktische Einführung in die Pflanzenmorphologie. Zweiter Teil: Die blühende Pflanze. VEB Fischer, Jena.
– (1959) Neue Beiträge zur Kenntnis der Blütenstände und Blüten von *Ceropegia*-Arten. Abh. Akad. Wiss. Lit. Mainz, math.-nat. Kl. Nr. **5**.
– (1961) *Cochliostema odoratissimum* Lem. Organisation und Lebensweise, nebst vergleichenden Ausblicken auf andere Commelinaceen. Beitr. Biol. Pflanzen **36**, 325–389.
– (1962) Über die "Prolificität" von *Chlorophytum comosum*. Beitrag zur Kenntnis einer Goethe-Pflanze. Neue Hefte z. Morphologie, Weimar, **4**, 9–68.
– (1964/69) Die Infloreszenzen, Typologie und Stellung im Aufbau des Vegetationskörpers, Vol.**I, II**. 1. Jena.
– (1950–1975) Kommission für biologische Forschung. Bericht. Akad. Wiss. Lit. Mainz, Jb. **1950**, 31–36; **1951**, 36–38; **1952**, 39–43; **1953**, 39–45; **1954**, 38–44; **1955**, 39–46; **1956**, 36–42, (1957); **1957**, 39–50, (1958); **1958**, 127–144, (1959); **1959**, 112–131, (1960); **1960**, 81–96, (1961); **1961**, 113–126, (1962); **1962**, 86–104, (1963); **1963**, 113–137, (1964); **1964**, 93–111, (1965); **1965**, 109–133, (1966); **1966**, 115–132, (1967); **1967**, 89–103, (1968); **1968**, 88–109, (1969); **1969**, 82–101, (1970); **1970**, 109–123, (1971); **1971**, 107–127, (1972); **1972**, 98–109, (1973); **1973**, 135–151, (1974); **1974**, 128–142, (1975).
– & Heidenhain, B. (1952) Studien über die Infloreszenzen von *Euphorbia Cyparissias* L. Ber. Dt. Bot. Ges. **65**, 377–382.
– & Weberling, F. (1966) Die Infloreszenzen der Caprifoliaceen und ihre systematische Bedeutung. Abh. Akad. Wiss. Lit. Mainz, math.-nat. Kl. Nr. **5**.
★– (1981) Infloreszenzstudien an Aizoaceen, Mesembryanthemaceen und Tetragoniaceen. Akad. Wiss. Lit. Mainz, Reihe Trop. u.subtrop. Pflanzenwelt **35**.
Tschirch, A. (1904) Sind die Antheren der Kompositen verwachsen oder verklebt? Flora **93**, 51–55, Pl. II.
Tucker, S. (1961). Phyllotaxis and vascular organization of the carpels in Michelia fuscata. Amer. J. Bot. **48**, 60–71.
Tucker, S. C. (1966) The gynoecial vascular supply in *Caltha*. Phytomorphology **16**, 339–342.
– (1974a) The role of ontogenetic evidence in floral morphology. Adv. Pl. Morph. **1972**, 359–369.
★– (1974b) Whorled initiation of stamens and carpels in *Saururus cernuus*. Amer. J. Bot. **61**, 66 (Abstr.).
★– (1975) Floral development in *Saururus cernuus*. I. Floral initiation and stamen development. Amer. J. Bot. **62**, 993–1007.
– (1976) Floral development in *Saururus cernuus* (Saururaceae). 2. Amer. J. Bot. **63**, 289–301.
★– (1979) Ontogeny of the inflorescence of *Saururus cernuus* (Saururaceae). Amer. J. Bot. **66**, 227–236.
★– (1982a) Inflorescence and flower development in the Piperaceae. II. Inflorescence development of *Piper*. Amer. J. Bot. **69**, 743–752.
★– (1982b) Inflorescence and flower development in the Piperaceae. III. Floral ontogeny of *Piper*. Amer. J. Bot. **69**, 1389–1401.
★– (1984a) Origin of symmetry in flowers. In: R. A. White & W. C. Dickison (eds.) Contemporary problems in plant anatomy: 351–394. Academic Press, New York.
★– (1984b) Unidirectional organ initiation in leguminous flowers. Amer. J. Bot. **71**, 1139–1148.
★– (1985) Initiation and development of inflorescence and flower in *Anemopsis californica* (Saururaceae). Amer. J. Bot. **72**, 20–31.
★– & Gifford, E. M. Jnr. (1964) Carpel vascularization of *Drimys lanceolata*. Phytomorphology **14**, 197–203.
★– (1966a) Organogenesis in the carpellate flower of *Drimys lanceolata*. Amer. J. Bot. **53**, 433–442.
★– (1966b) Carpel development in *Drimys lanceolata*. Amer. J. Bot. **53**, 671–678.
★– Rugenstein, S. R. & Derstine, K. (1984) Inflated trichomes in flowers of *Bauhinia* (Leguminosae: Caesalpinioideae). J. Linn. Soc. (Botany) **88**, 291–301.
★– & Sampson, F. B. (1979) The gynoecium of winteraceous plants. Science **203**, 920–921.
von Ubisch, G. (1925) Genetische-physiologische Analyse der Heterostylie. Bibl. Genet. **2**, 287–342.

Uhl, N. W. (1947) Studies in the floral morphology and anatomy of certain members of the Helobiae. Diss. Cornell Univ.

– (1969) Floral anatomy of *Juania, Ravenea* and *Ceroxylon* (Palmae-Arecoideae). Gentes Herbarum **10**, 394–411.

– & Moore, H. E., Jr. (1977) Centrifugal stamen initiation in Phytelephantoid Palms. Amer. J. Bot. **64**, 1152–1161.

– (1980) Androecial development in six polyandrous genera representing five major groups of Palms. Ann. Bot. **45**, 57–75.

*Uhlarz, H. (1983) Typologische und ontogenetische Untersuchungen an *Spathicarpa sagittifolia* Schott (Araceae): Wuchsform und Infloreszenz. Beitr. Biol. Pflanzen **57**, 389–429.

– & Weberling, F. (1977) Ontogenetische Untersuchungen an *Cordia verbenacea* (Boraginaceae), ein Beitrag zur Kenntnis der Syndesmien. Ber. Dt. Bot. Ges. **90**, 127–134.

Ulbrich, E. (1928) Biologie der Früchte und Samen (Karpobiologie). Berlin.

Unruh, M. (1939) Die morphologische Bedeutung des Karpells. Beitr. Biol. Pflanzen **26**, 90–124.

– (1941) Blattnervatur und Karpellnervatur. Kleiner Beitrag zur morphologischen Deutung des Karpells. Beitr. Biol. Pflanzen **27**, 232–241.

Uphof, J. C. T. (1934) Vergleichende blütenmorphologische und blütenbiologische Studien an *Commelina virginica* L. Ber. Dt. Bot. Ges. **52**, 173–180.

– (1938) Cleistogamic flowers. Bot. Rev. **4**, 21–49.

– (1942) Ecological relations of plants with ants and termites. Bot. Rev. **8**, 563–598.

*Urban, I. (1884) Zur Biologie der einseitswendigen Blütenstände. Ber. Deutsch. Bot. Ges. **3**, 406–432.

*Urmi-König, K. (1981) Blütentragende Spross-Systeme einiger Chenopodiaceae. Diss. Botanicae **63**, J. Cramer, Vaduz.

*Vasil, I. K. & Johri, M. M. (1964) The style, stigma and pollen tube – I. Phytomorphology, **14**, 352–369.

Vaughan, J. G. (1955) The morphology and growth of the vegetative and reproductive apices of *Arabidopsis thaliana* (L.) Heynh., *Capsella bursa-pastoris* (L.) Medic. and *Anagallis arvensis* L. J. Linn. Soc. (Bot.) London **55**, 279–301.

Velenovský, J. (1910) Vergleichende Morphologie der Pflanzen, 4 Vols. Part 3: Die Morphologie der Blüte der Phanerogamen, 733–1211.

Verhoog, H. (1968) A contribution towards the developmental gynoecium morphology of *Engelhardia spicata* Lechen. ex Blume (Juglandaceae). Acta Bot. Neerl. **17**, 137–150.

Vijayaraghavan, M. R. & Usha Dhar (1977) Structure of the carpel with special reference to lysigenous cavities in *Kadsura Roxburghiana*. Bot. Jb. Syst. **98**, 273–277.

Vogel, S. (1950) Farbwechsel und Zeichnungsmuster bei Blüten. Österr. Bot. Z. **97**, 44–100.

– (1954) Blütenbiologische Typen als Elemente der Sippengliederung. In Bot. Studien, Heft **1**, Jena.

– (1955) Über den Blütendimorphismus einiger südamerikanischer Pflanzen. Österr. Bot. Z. **102**, 486–500.

– (1958) Fledermausblumen in Südamerika. Österr. Bot. Z. **104**, 491–530.

– (1959) Organographie der Blüten kapländischer Ophrydeen mit Bemerkungen zum Koaptationsproblem. Abh. Akad. Wiss. Lit. Mainz, math.-nat. Kl. Nr. **6 & 7**, 267–532.

– (1960) Über die "Uvula" von *Ceropegia Sandersonii* Hook. f., zugleich über einen merkwürdigen Fall postgenitaler Verwachsung. Beitr. Biol. Pflanzen **35**, 395–412.

– (1961) Die Bestäubung der Kesselfallen-Blüten von *Ceropegia*. Beitr. Biol. Pflanzen **36**, 165–237.

– (1962) Duftdrüsen im Dienste der Bestäubung. Über Bau und Funktion der Osmophoren. Abh. Akad. Wiss. Lit. Mainz, math.-nat. Kl. Nr. **10**.

– (1963) Das sexuelle Anlockungsprinzip der Catasetinen- und Stanhopeen-Blüten und die wahre Funktion ihres sogenannten Futtergewebes. Österr. Bot. Z. **110**, 308–337.

– (1966a) Parfümsammelnde Bienen als Bestäuber von Orchidaceen und *Gloxinia*. Österr. Bot. Z. **113**, 302–361.

– (1966b) Pollination neotropischer Orchideen durch duftstoff-höselnde Prachtbienen-Männchen. Naturwiss. **53**, 181–182.

– (1967a) "Parfümblumen" und parfümsammelnde Bienen. Umschau in Wissenschaft und Technik Nr. **10**, 327.

– (1967b) *Iris fulva* Ker-Gawl., eine Kolibriblume. Dt. Iris- und Lililiges. **1967**, 1–11.

– (1968) Chiropterophilie in der neotropischen Flora. Neue Mitteilungen I. Flora, Abt. B, **157**, 562–602.

– (1969a) Chiropterophilie in der neotropischen Flora. Neue Mitteilungen II. Flora, Abt. B, **158**, 158–222.

– (1969b) Chiropterophilie in der neotropischen Flora. Neue Mitteilungen III. Flora, Abt. B, **158**, 289–323.

– (1969c) Über synorganisierte Blütensporne bei einigen Orchideen. Österr. Bot. Z. **116**, 244–262.

– (1971) Ölproduzierende Blumen, die durch ölsammelnde Bienen bestäubt werden. Naturwiss. **58**, 58.

– (1973) Öl statt Nektar: die "Ölblume". Umschau 73 (22), 701–702.

– (1974) Ölblumen und ölsammelnde Bienen. Reihe trop. u. subtrop. Pflanzenwelt. Abh. Akad. Wiss. Lit. Mainz, math.-nat. Kl. Nr. **7**, 284–547.

– (1975) Mutualismus und Parasitismus in der Nutzung von Pollenträgern. Verh. Dt. Zool. Ges. **1975**, 102–110.

– (1976) *Lysimachia:* Ölblumen der Holarktis. Naturwiss. **63**, 44.

– (1977) Nektarien und ihre ökologische Bedeutung. Apidologie **8**, 321–335.

– (1978) Pilzmückenblumen als Pilzmimeten. Part I, Flora **167**, 329–366; Part II, Flora **167**, 367–398.

⋆– (1981a) Abdominal oil-mopping – a new type of foraging in bees. Naturwissenschaften **67**, 627.

⋆– (1981b) Trichomatische Blütennektarien bei Cucurbitaceae. Beitr. Biol. Pflanzen **55**, 325–353.

⋆– (1981c) Die Klebstoffhaare an den Antheren von *Cyclanthera pedata* (Cucurbitaceae). Plant Syst. Evol. **137**, 291–316.

⋆– (1981d) Bestäubungskonzepte der Monokotylen und ihr Ausdruck im System. Ber. Deutsch. Bot. Ges. **94**, 663–675.

⋆– (1984) The *Diascia* flower and its bee – an oil-based symbiosis in Southern Africa. Acta Bot. Neerl. **33**, 509–518.

⋆– & Michener, C. D. (1985) Long bee legs and oil-producing floral spurs, and a new *Rediviva* (Hymenoptera, Melittidae; Scrophulariaceae). J. Kansas Entomological Soc. **58**, 359–364.

⋆– Westerkamp, C., Thiel, B. & Gessner, K. (1984) Ornithophilie auf den Canarischen Inseln. Plant Syst. Evol. **146**, 225–248.

⋆Vöth, W. (1982a) Die "ausgeborgten" Bestäuber von *Orchis pallens* L. Die Orchidee, **33**, 196–203.

⋆– (1982b) Blütenökologische Untersuchungen an *Epipactis atrorubens, E. helleborine* und *E.purpurata* in Niederösterreich. Mitt. Bl. Arbeitskr. Heim. Orch. Baden-Württ. **14**, 393–437.

⋆– (1983) Blütenbockkäfer (Cerambycidae) als Bestäuber von *Dactylorhiza maculata* L. ssp. *meyeri* (Rchb.f.) Tournay. Mitt. Bl. Arbeitskr. Heim. Orch. Baden-Württ. **15**, 305–330.

Waddle, R. M. & Lersten, N. R. (1974) Morphology of discoid floral nectaries in Leguminosae, especially Tribe Phaseolae (Papilionoideae). Phytomorph. **23**, 152–161.

Wagenitz, G. (1975) Blütenreduktion als ein zentrales Problem der Angiospermensystematik. Bot. Jb. Syst. **96**, 448–470.

⋆– (1976) Was ist ein Achäne? Zur Geschichte eines karpologischen Begriffs. Candollea **31**, 79–85.

⋆– & Laing, B. (1984) Die Nektarien der Dipsacales und ihre systematische Bedeutung. Bot. Jahrb. Syst. **104**, 483–507.

Wagner, R. (1901) Über den Bau und die Aufblühfolge der Rispen von *Phlox paniculata* L. Sitz.-Ber. Akad. Wiss. Wien, math.-nat. Kl., 1. Abt., **110**, 507–591.

Walker, D. B. (1975a) Postgenital carpel fusion in *Catharanthus roseus* (Apocynaceae). I. Light and scanning electron microscopic study of gynoeceal ontogeny. Amer. J. Bot. **62**, 475–467.

– (1975b) II. Fine structure of the epidermis before fusion. Protoplasma **86**, 29–41.

– (1975c) III. Fine structure of the epidermis during and after fusion. Protoplasma **86**, 43–63.

Warburg, O. (1913–22) Die Pflanzenwelt Vol.2 (1916), Leipzig & Vienna.

⋆Warming, E. (1876) Die Blüte der Compositen. Bot. Abhandlungen aus dem Gebiet der Morphologie und Physiologie **3** (2). Bonn.

– & Möbius, M. (1911) Handbuch der systematischen Botanik. 3rd edn. Berlin.

*Waser, N. H. (1983) The adaptive nature of floral traits: ideas and evidence. In: L. A. Real (ed.) Pollination biology, Orlando. Academic Press, 241–283.

Wassmer, A. (1955) Vergleichend-morphologische Untersuchungen an den Blüten der Crassulaceen. Arb. Inst. Allg. Bot. Zürich, Ser. A, **7**, 1–112.

Weber, A. (1971) Zur Morphologie des Gynoeceums der Gesneriaceen. Österr. Bot. Z. **119**, 234–305.

– (1976) Beiträge zur Morphologie und Systematik der *Klugieae* und *Loxonieae* (Gesneriaceae). II. Morphologie, Anatomie und Ontogenese der Blüte von *Monophyllaea* R. Br. Bot. Jb. Syst. **95**, 435–454.

– (1979) *Ornithoboea arachnoidea* – an "Orchid-flowered" Gesneriad. The Gloxinian **29**, 9–12.

– (1980) Die Homologie des Perigons der Zingiberaceen. Ein Beitrag zur Morphologie und Phylogenie des Monokotylen-Perigons. Pl. Syst. Evol. **133**, 149–179.

Weber, C. O. (1860) Beiträge zur Kenntnis der pflanzlichen Mißbildung. Verhandl. Naturhist. Verein f. d. Prov. Rheinld. u. Westf. **17**, 333–388, Pls. VI & VII.

Weber, E. (1929) Entwicklungsgeschichtliche Untersuchungen über die Gattung *Allium*. Bot. Archiv. **25**, 1–44.

Weber, H. (1955) Über die Blütenkelche tropischer Rubiaceen. Abh. Akad. Wiss. Lit. Mainz, math.-nat. Kl. Nr. **11**.

– (1956) Über die Blütenstände und die Hochblätter von *Norantea* Aubl. (Marcgraviaceae). Beitr. Biol. Pflanzen **32**, 313–329.

Weberling, F. (1957) Die Infloreszenzen von *Bonplandia* Cav. und *Polemonium micranthum* Benth. und ihre vermeintliche Sonderstellung unter den Blütenständen der Polemoniaceae. Beitr. Biol. Pflanzen **34**, 195–211.

– (1961) Die Infloreszenzen der Valerianaceen und ihre systematische Bedeutung. Abh. Akad. Wiss. Lit. Mainz, math.-nat. Kl. Nr. **5**.

– (1963) Homologien im Infloreszenzbereich und ihr systematischer Wert. Ber. Dt. Bot. Ges. **76**, 1. Gen. vers. H., 102–112.

– (1965) Typology of inflorescences. J. Linn. Soc. (Bot.), London **59**, 215–221, 1 pl.

– (1970) Valerianaceae, in: Hegi, G., Illustrierte Flora von Mitteleuropa. 2nd ed. VI/2. Munich.

– (1971) Die Bedeutung der Infloreszenzmorphologie für die Systematik. Ber. Dt. Bot. Ges. **84**, 179–181.

– (1977a) Beiträge zur Morphologie der Rubiaceen-Infloreszenzen. Ber. Dt. Bot. Ges. **90**, 191–209.

– (1977b) Vergleichende und entwicklungsgeschichtliche Untersuchungen über die Haarformen der Dipsacales. Beitr. Biol. Pflanzen **53**, 61–89.

*– (1983a) Fundamental features of modern inflorescence morphology. Bothalia **14**, 917–922.

*– (1983b) Evolutionstendenzen bei Blütenständen. Abh. Akad. Wiss. Lit. Mainz, math.-naturw. Kl., **1**.

*– (1985a) Zur Infloreszenzmorphologie der Lauraceae. Bot. Jahrb. Syst. **107**, 395–414.

*– (1985b) Aspectos modernos de la morfología de las inflorescencias. Boletin Soc. Argent. de Botanica **24**, 1–28.

*– & Müller, L. (1980) Persistierende Blütensporne bei *Tropaeolum*. Flora **169**, 295–298.

*– & Uhlarz, H. (1983) Zur Morphologie und Morphogenese der Blüten von *Cadaba juncea* (Sparm.) Harv. (Capparidaceae) und *Cadaba natalensis* Sond. Beitr. Biol. Pflanzen **58**, 267–281.

*Werth, E. (1943) *Crocus* und *Colchicum*, zwei blütenbiologische Paradoxa. Ber. Deutsch. Bot. Ges. **61**, 82–86.

– (1952) *Asarum europaeum*, ein permanenter Selbstbefruchter. Ber. Dt. Bot. Ges. **64**, 287–294.

– (1956a) Bau und Leben der Blumen. Stuttgart.

– (1956b) Zur Kenntnis des Androeceums der Gattung *Salvia* und seiner stammesgeschichtlichen Wandlung.

von Wettstein, R. (1890) Zur Morphologie der Staminoidien von *Parnassia palustris*. Ber. Dt. Bot. Ges. **8**, 304–309, Pl. XVIII.

– (1901/1935) Handbuch der systematischen Botanik. 1st ed. Vienna, 1901; 4th ed. Leipzig & Vienna, 1935.

*Wiens, D., Renfree, M. & Wooller, R. O. (1979) Pollen loads of honey possums (*Tarsipes spenserae*) and nonflying mammal pollination in southwestern Australia. Ann. Missouri Bot. Garden **66**, 830–838.

*Wiens, D. & Rourke, J. P. (1978) Rodent pollinators in southern Africa *Protea* species. Nature **276**, 71–73.

Wilson, C. L. (1937) The phylogeny of the stamen. Amer. J. Bot. **24**, 686–699.

– & Just, T. (1939) The morphology of the flower. Bot. Rev. **5**, 97–131.

Winkler, H. (1936) Septicide Kapsel und Spaltfrucht. Beitr. Biol. Pflanzen **24**, 191–200.

– (1939) Versuch eines "natürlichen" Systems der Früchte. Beitr. Biol. Pflanzen **26**, 201–219.

– (1940) Zur Einigung und Weiterführung in der Frage des Fruchtsystems. Beitr. Biol. Pflanzen **27**, 92–130.

– (1944) Verstehen wir das Gynöceum der Angiospermen schon? Beitr. Biol. Pflanzen **27**, 242–267.

*Wisniewski, M. & Bogle, A. L. (1982) The ontogeny of the inflorescence and flower of *Liquidambar styraciflua* L. (Hamamelidaceae). Amer. J. Bot. **69**, 1612–1624.

Wolff, C. F. (1768) De formatione intestinorum. German translation by I. F. Meckel, Halle, 1812.

Wolter, H. (1933) (In Winkler, H., Ed.) Bausteine zu einer Monographie von *Ficaria*. 8. Über Bestäubung, Fruchtbildung und Keimung bei *Ficaria verna*. Beitr. Biol. Pflanzen **21**, 219–255.

*Wolter, M. & Schill, R. (1986) Ontogenie von Pollen, Massulae und Pollinien bei den Orchideen. Akad. Wiss. Lit. Mainz, Trop. u. subtrop. Pflanzenwelt **56**.

*Wyatt, R. (1982) Inflorescence architecture: how flower number, arrangement and phenology affect fruit set. Amer. J. Bot. **69**, 585–594.

Wydler, H. (1843) Über dichotome Verzweigung der Blüthenachsen (cymöse Infloreszenz) dicotyledonischer Gewächse. Linnaea **17**, 153–192, 408–409.

– (1844) Morphologische Mittheilungen. Bot. Ztg. **2**, 609–611ff.

– (1860) Kleinere Beiträge zur Kenntniss einheimischer Gewächse. Flora **43**, 17–32, cont.

Zandonella, P. (1967) Stomates des nectaires floraux chez les Centrospermales. Bull. Soc. Bot. France **114**, 11–20.

– (1972) Le nectaire florale des Centrospermales, localisation, morphologie, anatomie, histologie, cytologie. Thèse, Lyon.

– (1977) Apports de l'étude comparée des nectaires floraux à la conception phylogénétique de l'ordre des Centrospermales. Ber. Dt. Bot. Ges. **90**, 105–125.

*Zeller, O. (1960) Entwicklungsgeschichte und Morphogenese der Blütenknospen von der Quitte (*Cydonia oblonga* Mill.). Angew. Botanik **34**, 110–120.

– (1975) Beitrag zur Morphogenese und Histogenese der Blüten von *Rosa canina* L., *Rosa rubiginosa* L., *Rosa virginiana* Mill. und cultivar Super Star (Tantau 1960). Gartenbauwissenschaft **40**, 276–284.

Ziegler, A. (1925) Beiträge zur Kenntnis des Androeceums und der Samenentwicklung einiger Melastomataceen. Bot. Archiv **9**, 398–467.

Zimmermann, J. (1932) Über die extrafloralen Nektarien der Angiospermen. Beih. Bot. Centralbl. Abt. I, **49**, 99–196.

Zimmermann, W. (1965) Die Blütenstände, ihr System und ihre Phylogenie. Ber. Dt. Bot. Ges. **78**, 3–12.

Index of scientific names

Asterisks refer to the figures

General Index